SAILING THE OCEAN OF COMPLEXITY

Sailing the Ocean of Complexity

Lessons from the Physics-Biology Frontier

Sauro Succi

OXFORD
UNIVERSITY PRESS

Great Clarendon Street, Oxford, OX2 6DP,
United Kingdom

Oxford University Press is a department of the University of Oxford.
It furthers the University's objective of excellence in research, scholarship,
and education by publishing worldwide. Oxford is a registered trade mark of
Oxford University Press in the UK and in certain other countries

Impression: 1

Published in the United States of America by Oxford University Press
198 Madison Avenue, New York, NY 10016, United States of America

British Library Cataloguing in Publication Data
Data available

Library of Congress Control Number: 2021944568

ISBN 978–0–19–289789–3(Hbk)

DOI: 10.1093/oso/9780192897893.001.0001

Printed and bound by
CPI Group (UK) Ltd, Croydon, CR0 4YY

To my family and my friends

Cominciate col fare ciò che è necessario, poi fate ciò che è possibile e all'improvviso vi sorprenderete a fare l'impossibile. (San Francesco da Assisi) *(Start by doing what's necessary, then what's possible; and suddenly you are doing the impossible.)*

Preface

"We are awashed in complexity at every scale"

(J. Wettlaufer, Presentation of the 2021 Nobel Prize in Physics)

We live in a world of utmost Complexity, outside and within us. There are thousands of billions of billions stars out there in the Universe, a hundred times more molecules in a glass of water, and another hundred times more in our body working in sync to keep us alive and well. At face value, such numbers spell certain doom for our ability to make any sense at all of the world around and within us. And yet, they don't: *Why and How*?

These, among many others, are the main questions tackled by the Science of Complexity, hence, if only through a small window, by this book as well. The aim is not to provide a full answer to such big questions, a task which is left to the professionals of the Science of Complexity [98], but simply to convey the stupor and gratitude for the gift of being able to get a grip on (some of) the awe-commanding mechanisms which keep our body going and sailing across Oceans of Biological Complexity. Without sinking.

This book develops on a trail composed of basically four parts. Part I consists of seven chapters on general properties of complex systems. We begin with some general considerations on the Science of Complexity, followed by a survey of some of the main organizing principles common to complex systems, in particular those sitting at the frontier between physics and biology. Next, we move on to a more detailed discussion of the main engine of Complexity, nonlinearity, both in its 'dark' (unpredictability) and 'bright' (structure formation) versions. Following, a chapter on networks, the fabric of Complexity and arguably the most general mathematical framework for the exploration of complex systems.

Part II consists of five chapters devoted to the 'science of change', thermodynamics. First, in its historical and rather bleak version, in which we learn that closed systems are doomed to thermal death. Then, in its more modern and shiny guise, i.e. structures which form thanks to the nonlinearities exposed by driven systems far from equilibrium, thereby offering a temporary suspension from thermal death. An intermezzo is devoted to a giant figure of all times, Ludwig Boltzmann, the man who laid down the first quantitative bridge between the microscopic world of atoms and the macroscopic world of thermodynamics. The trail goes on to visit the central character of this book, free energy, i.e. what is left of energy to produce useful structures (Order) once the heat toll (Disorder) is defrayed. In brief, the currency of the natural world.

Part III deals with the wondrous scenarios opened up by free-energy as it meets with the modern tools of statistical physics, ultradimensional landscapes, Order parameters, and funnels, to name but a few central characters. In such modern guise, free-energy offers marvellous vistas on the world of biological Complexity, and enchanting information-sieving tools to deal with fundamental biological processes, such as deoxyribonucleic acid (DNA) translocation across cell membranes and protein folding. Biology offers many more, of course, but these are the two which I had the good fortune to meet in some detail, and they are responsible for my enchantment with the modern incarnations of free-energy. For the sake of clarity, a full chapter is devoted to soft matter, the handshaking ground between physics and chemistry most relevant to biology, which also provides much of the material substrate supporting the aforementioned free-energy vistas on biological Complexity. Without forgetting the biofluid par excellence, our sister water.

Finally, Part IV consists of just two chapters discussing the connections between time and Complexity, in particular the idea that time can possibly exist, i.e. be experienced only by systems beyond a given threshold of Complexity. A threshold that humans are guaranteed to pass. The final chapter, in fact an epilogue, is a series of very personal considerations on the emotions this author experienced in walking along his own little trail for the last three decades. No need to agree, but hopefully you can share, nonetheless.

This book is addressed to a general audience with no pre-requisite other than a genuine curiosity for the Complexity of the marvellous and sometimes scary world we happen to inhabit. Some willingness to endure occasional elementary math comes by handy, but never a *sine qua non*. Indeed, I deliberately challenged the famous adage according to which every equation splits the audience in two. I did it consistently throughout the book, and if the adage is right, I will definitely end up with no audience at all. But I am confident in a better fate, namely that my reader won't be scared away by an occasional square root or a logarithm. In fact, believe it or not, logarithms hold the key to our survival in Ocean Complexity, and if we agree that everyone should have a minimal acquaintance with Shakespeare, I think the same holds true for logarithms!

A book must convey a story, possibly a personal one; if you wish, it has somehow to be its author in paper form. This is why, whenever I felt there was a personal emotion to convey, I did not hesitate to put it to words. And since I experienced emotions galore in my personal scientific trajectory, the main hope and purpose here is that the reader may be able to share them with me.

Physics and Complexity

Before we get started, I feel urged to clear the ground on what this book and this author are *not*. This book is not meant to expose the frontiers of the Science of Complexity at large, much less to credit this author as frontman of such a frontier, which is too wide and broad for any single book and possibly any single author as well. Most certainly for the present one. No, this book is much more modest in scope, as it simply aims to convey the little but crisp joys and emotions this author has experienced first-hand upon taking a look at the world of natural Complexity from a small but rich window running across the frontier between physics and biology.

To some Complexity professionals this may sound like a very narrow and outdated frontier of the Science of Complexity.[1] I am not covering any of the hot topics of modern mainstream Science of Complexity, no complex adaptive systems, no social nets, no immunology, epidemiology, criminology, let alone risk and conflict management, financial bubbles, black swans, and definitely no formula of success [5, 14, 8, 9, 10, 61, 72, 88]. No, I am talking mostly GOP 'Good-Old-Physics', to borrow from a critical referee, notably the good-old science of change, aka thermodynamics. In slightly modern guise, though ....

Indeed, I stand firmly by two points. *First*, even though my window is mostly concerned with GOP, this does not make it a single iota less relevant to *modern* Science of Complexity. *Second*, this book is not meant to surprise professionals. Let me elaborate a bit further.

Good old physics (that still works ...)

The first point is that, like all truly good stuff, evergreen GOP has spawned plenty of *new* ideas and tools. Besides being pretty elegant, these tools have a little precious extra merit: generally they work and often disclose new vistas on Complexity, see for instance N. Goldenfeld and Leo Kadanoff, Simple Lessons from Complexity, Science 284, 87 (1999). Not something to be sneezed at in any branch of science, including the Science of Complexity, no matter how modern, sexy and upbeat. Seems to me that this point is largely underappreciated in the Complexity literature. Let me give some examples.

In his highly enjoyable and informative *Simply Complexity* [50], Neil Johnson writes that physics cannot describe Complexity because physics works mostly on two main assumptions which are largely defied by complex systems: infinite time (equilibrium) and infinite number of objects (thermodynamic limit). A look at modern statistical physics shows that both 'allegations' do not really stand in front of the truth. To be sure, Johnson concedes that an entire branch of modern statistical physics, in fact the one that kept this author professionally busy for over three

[1] In a preliminary version of this book, I used the term 'complexologist', knowing that, although it does not seem to exist in the English dictionary, it actually does in the English literature, having been used in John Horgan's influential article 'From Complexity to Perplexity', *Science America.* June 1995. Quite naively, I found that the term did well in reflecting the aspiration to grand unification which characterized the most inspired and ambitious fringe of Scientists of Complexity, as they gathered around the Santa Fe Institute in the late 80s.

However, upon reading Mitchell's beautiful *A Guided Tour to Complexity* [73], I became aware of the tense reactions spurred by this article, a polemic to which I have less than any intention to partake. Except for noting that, based on a rather aggressive review received by an early draft of this book, I must have inadvertently touched some nerve, nonetheless. Unintended: I fully respect Andrè Gide statement 'one doesn't discover new lands without consenting to lose sight of the shore', to quote from Mitchell again. But I also feel that, by the very moment you lose sight of the shore, you should also be prepared to embrace a great deal of humility. Of which I could find little trace in the dismissive statement 'good old physics tools that clearly fall short of advancing our understanding', provided in the aforementioned review.

decades, deals with finite-time (non-equilibrium) behaviour. Unfortunately, he refrains from taking the next step and further acknowledging that one of the most vibrant sectors of non-equilibrium statphys deals precisely with small systems, very far from the blamed infinity [17]. In fact, it is fair to say that both finiteness-es, in time and number (size), form a central core of the kind of Complexity one meets at the physics-biology interface.

It is actually no coincidence that many modern advances of non-equilibrium statphys revolve around to the central character of this book, free-energy, a name seldom encountered in the Complexity literature. A genuine GOP subject indeed, yet one that has spawned a remarkable number of powerful ideas and *modern* methods, which serve the purpose of exploring the frontier between physics and biology eminently well. The story repeats in hard-core books on Complexity, where we read that physics can't describe Complexity because 'physics is analytic and complexity is algorithmic'. Or, that physics works with fixed boundary conditions while complex systems are all one with their changing environment, hence boundary conditions are part of those very systems [114].

Yet again, modern theoretical and computational physics is littered with situations where analyticity and smoothness are no requirement at all, as witnessed by the wealth of results from discrete lattice models across virtually all walks of statistical physics [106, 107], again part of my own professional experience. And when it comes to self-consistent boundary conditions, fluid-structure problems, which one finds galore in soft matter and biology, are a walking advertisement for self-consistent boundary conditions, all one with the systems they keep under confinement [111]. Biological bodies, say red blood cells, interacting with the carrier fluid and the deformable cardiovascular vessels, offer a sterling example in point.

The world beyond physics

A more convincing viewpoint is formulated in the very interesting paper 'Complex systems: Physics beyond physics', by Holovatch, Kenna, and Thurner [62], in which the authors maintain that (verbatim)

> Complex systems are characterized by specific time-dependent interactions among their many constituents .... such systems can be conceptualized by extending notions from statistical physics and ... they can often be captured in a framework of co-evolving multiplex network structures.

By reading further, we learn that what they mean is that physics deals with systems which evolve under the effect of time-constant laws, whereas complex systems *co-evolve* with their own governing laws. More precisely, they refer to systems living on the nodes of a network, whose connectivity changes in time, based on the change of the state of the system in the nodes. As a concrete example, among others, they discuss human societies, composed by co-evolving individuals and institutions. Individuals form institutions which formulate the laws dictating the social behaviour of the individuals who have formed them in the first place. But the distinction is still arguable. As noted by the authors themselves, the theory of general gravitation works pretty much in the same way: spacetime is curved by the energy of the particles, and the particle motion is dictated precisely by the curvature of spacetime: an adamant example of co-evolving physical system described by Einstein's theory of general gravitation.

Further on in his deep and inspiring book *A World Beyond Physics* [53], Stuart Kauffman argues that biology cannot be reduced to physics because 'the world is not a machine'. In particular, he appeals to the ability of biological systems to evolve in *unpre-statable* ways, because they play context-dependent games which change 'on-the-fly', hence cannot be pigeonholed into any pre-stateble equations, rules, or algorithms. This is a deep, fascinating, and very radical point: if I understand it correctly, physics and biology are separated by a conceptual discontinuity, in that the Complexity of biological systems escapes mathematical formalization, no matter how sophisticated. So, biological Complexity would unravel a world not only beyond physics but beyond mathematics as well!

As Stuart knows from our most enjoyable and enriching private communications, I don't agree, at least not in full. Not because I think that biology can be reduced to physics, but simply because physics is not restricted to machines, as (implicitly) surmised by his *The World is Not a Machine*. Modern statistical physics provides many examples of systems which evolve under the action of 'bootstrap' forces, i.e. forces which change on the fly, based on the actual configuration of the system they act upon. A most relevant example in point are the force fields dictating the motion of large biological molecules, such as proteins, which depend crucially on the shape they concur to dictate in the first place. Granted, these forces may not display the degree of sophistication of the games played by biological organisms, because the rules do not change in the course of the game. So, yes, there is definitely a point here.

Yet, the world beyond physics is arguably significantly smaller than Kauffman's arguments would have it. This is already true for the basic equations of physics, from Newton's formulation of classical mechanics, all the way to the sophisticated equations of modern quantum physics, such as those described in Steven Weinberg's *Dream of a Final Theory* [119]. However, it becomes even more apparent for the case of *effective* equations (rules), where effective means that these equations (rules) are not 'fundamental' in the sense of the most glorious physics tradition, namely those that describe the magnificent four (three) interactions: gravity, electroweak, and strong. No, they are *emergent* instead, i.e. they result from clever and ingenuous approximations and information-filtering procedures. As we shall see, such effective forces make the central conceptual core of this book, as they are the ones that breath fire into the modern formulations of free energy and disclose new vistas at the frontier between physics and biology.

What physical systems don't do

With all this being said and done, it is certainly true that whenever the individual units that form a complex system prove capable of *qualitatively* changing the laws they abide to, then this is something generally not covered by physics. It would be as if, at some point, material bodies would switch from gravitation, be it Newtonian or Einsteinian, to a completely different law, which cannot be reconduced to either of the two. To the best of our knowledge, this is something physical systems don't do.

With point one off our chests, next comes our short point two: this book is not meant to surprise professionals. The Complexity literature offers a large number of excellent texts which fit this bill, from Murray Gell-Mann's classic *The Quark and the Jaguar* [46], to the most inspiring Kauffman's *At Home in the Universe* [52], up to the more recent texts previously cited. The only professionals I might hope to impress are those capable of enjoying things they already know, if only seen perhaps from a slightly different angle, the angle of modern formulations of free energy.

Finally, one of the main themes of this book is that Complexity thrives around the subtle coexistence between Universality and Individuality, general principles and crucial details, (for an inspiring discussion on this point, see N. Goldenfeld and C. Woese, Biology's next revolution, nature, 445, 25, January 2007). Physics has a natural penchant for the former and Biology for the latter. Even though I have full appreciation of the crucial role of specific details in most biological processes, I realise that my physics background might have occasionally obscured such an appreciation. For that, I rely upon the generous forgiveness of the Biologists in the audience.

The trip begins, enjoy!

Contents

List of Abbreviations

3D	three-dimensional
BH	black holes
CA	cellular automata
CERN	*Conseil Européen pour la Recherche Nucléaire* (European Council for Nuclear Research)
CGMD	coarse-grained molecular dynamics
DNA	deoxyribonucleic acid
EEG	energy-entropy-gravity
EHT	Event Horizon Telescope
EXT	extinction
LB	Lattice–Boltzmann
LGCA	Lattice Gas Cellular Automata
LHC	Large Hadron Collider
LJ	Lennard-Jones
LRS	latency, ramp-up and saturation
MIT	Massachusetts Institute of Technology
MS	monotonic saturation
NAN	Not a Number
OGT	Edward Ott, Celso Grebogi, and James Yorke
OP	Order parameter
OS	oscillatory saturation
PD	period doubling
ToEE	The Theory of Everything Else
ToS	The Theory of Something
VdW	Van der Waals

Part I

Complexity

1

Introducing Complexity

I often use the same harmonies as pop music because the complexity of what I do is elsewhere.

(Ennio Morricone)

1.1 Big science and the inner frontier

On 4 July 2012, *Conseil Européen pour la Recherche Nucléaire* (European Council for Nuclear Research) (CERN) made the public announcement that the half-century-long hunt for the famous Higgs boson was finally over: at last, the ephemeral particle to which all others owe their mass had been 'captured' in the 27-km-long accelerator ring Large Hadron Collider (LHC) of the most iconic lab in the world, stretching between France and Switzerland (see Fig. 1.1). The news justly made the headlines around the world, Stockholm responding just a year later with the 2013 Nobel Prize in physics for theorists Peter Higgs and Francois Englert, who first posited the existence of such particle in the early 60s (along with Robert Brout, who sadly passed away just two years before he could share the prize).[2] On 11 February 2016, the story was repeated, but at the other extreme of the Universe: the Laser Interferometer Gravitational-Wave Observatory (LIGO) team announced that gravitational waves had finally been detected as a result of a merger between two massive black holes of twenty-nine and thirty-six solar masses, respectively. Besides providing further confirmation of Einstein's general theory of gravitation, this discovery also validated its predictions of space-time distortion in the context of large-scale cosmic events. Even more importantly, it opened up new prospects for the direct observation of the very earliest history of the baby Universe, down to a few microseconds from its birth!

And, more recently, on 10 April, 2019, the Event Horizon Telescope (EHT) captured the first image ever of a supermassive black hole, of 6.5 billion solar masses(!), located in a far distant galaxy, five hundred millions of trillions of kilometres away from us. Headlines around the world, again. The hunt for the Higgs boson

[2] In case you are wondering why the CERN experimentalist did not share the award, it is 'just' because, right or wrong, the Nobel rules forbid the Physics Prize to be awarded to more than three individuals, as well as to institutions.

Sailing the Ocean of Complexity. Sauro Succi, Oxford University Press.
 DOI: 10.1093/oso/9780192897893.003.0001

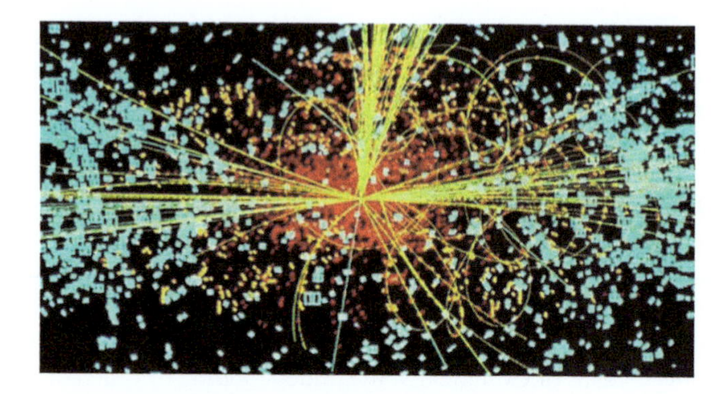

Figure 1.1 *Big science at the lower extreme of the Universe: high-energy collision revealing the Higgs Boson in a Large-Hadron-Collider experiment.*
Source: reprinted from commons.wikimedia.org.

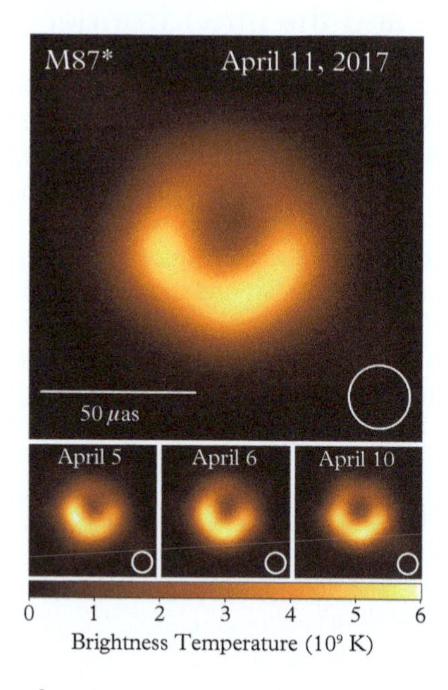

Figure 1.2 *Big science at the upper extreme of the Universe: the first image of a (supermassive) black hole.*
Source: reprinted from commons.wikimedia.org.

on the one hand and the detection of gravitational waves and black holes on the other, stand out as pinnacle icons of big science and its breath-taking inescapable fascination. It is indeed a widespread opinion that the most challenging and exciting science takes place at the extreme frontiers of size and energy: the ultra-small, elementary particle physics, and the extra-large, the Universe and the Cosmos (see Fig. 1.2).

As just as it is, this view hides a latent and often unspoken prejudice, namely that whatever lives away from the physical extremes is second tier, the more so if directly accessible to our senses, 'the things we can see', so to speak. In this book, I wish to offer my own small contribution to dispel this prejudice, by arguing instead that there is mystery and fascination, hence frontier science, at *all scales*, not just at the extremes. In other words there are 'inner' frontiers where science can be enthralling, rich, fascinating, and socially impactful, without being physically extreme.

In fact, just confining our attention the comparatively small window of the Universe, which goes from our Sun to the molecules we are made of, we find mind-boggling Oceans of Complexity, the understanding of which raises an intellectual challenge which has nothing to envy of the previous the grand extremes. After all, it is often heard that we humans, and particularly our brains, are the quintessence of Complexity in the physical Universe.[3] The fact remains that while high-energy physics and cosmology explore the lower and upper frontiers of physical space, respectively, the science of Complexity explores a precious inner one: in many respects, the limits of the human mind itself. If you see any reason to call this inner frontier less 'fundamental' than the outer ones, gimme a call, for I really don't. And I am surely not alone.

1.2 What is a complex system?

Sarah, oh Sarah, so easy to look at, so hard to define.
(Sarah, B. Dylan)

So much for the glory but, ... what is a complex system in the first place? Providing a definition of a complex system is not easy and with good reason: otherwise, it wouldn't be complex, correct? This is less facetious than it seems: we may legitimately define a complex system as one whose behaviour cannot be described in a few words. Example: take a calm, uniform river flow, versus ocean waves relentlessly breaking on the shore during a winter storm. There is hardly any question that it takes more words and information to describe the latter than the former: the stormy ocean is more complex than a quietly flowing river.

A subtler definition is the one given by the great Russian polymath Andrei Kolmogorov (1903–1987), namely the minimal length of a computer programme which describes = generates the object. For instance, the string 'xyzxyzxyzxyzxyzxyzxyzxyzxyz' is described by the 15-character rule 'write

[3] Interestingly enough, such an anthropocentric view is kind of supported by numbers, since the square root of the mass of the proton and the mass of the Sun gives about 55 kg. Of course, this is nothing more than an incidental observation (due to the Royal Astronomer Martin Rees), yet an amusing one [95].

xyz ten times'. On the other hand, to generate the string 'xlkoutrsryuowqdydcyb' there is no other way than stating 'write xlkoutrsryuowqdydcyb', which consists of 25 characters. No surprise, given that the first string is a ten-fold repetition of the same 'xyz' pattern, hence it exhibits a great deal of regularity, not to be found in the second one.

However, as we shall see in this book, it is often the case that fairly complex behaviour can be *generated* (but not described) by very short rules, often encrypted in just a few lines of computer programme. Hence, Kolmogorov's definition falls clearly short of capturing this crucial kind of Complexity, the one which sprawls from simple rules. This occurs whenever the complex system exhibits what physicists use to call *Universality*; that is, the behaviour of the system is dictated by a few universal organizing principles and proves largely insensitive to specific details.

The intellectual fascination of *Universality* can hardly be avoided, and indeed it sheds high hopes that the Theory of Complexity could make major strides across a variety of apparently disconnected disciplines, from physics to engineering, to biology, economy, and social sciences. However, at closer scrutiny, the capturing paradigm of complex behaviour from simple rules often proves too simplistic to provide a quantitative description of real-life complexity. Along this line, Giorgio Parisi, a towering figure in the field, and freshly awarded the 2021 Nobel Prize in Physics, defines a complex system as one 'that crucially depends on details' [82]. Indeed, most real-life systems show a very subtle blend of *Universality and Individuality*, namely the general organizing principles *and* specific details. In my view, this coexistence of *Universality and Individuality* is precisely one of the most profound and distinctive traits of Complexity. Another popular route to convey the point of complexity is to contrast *complex* versus *complicated*. The latter associates with a long list of intricated but consequential design rules, say a set of instructions to assemble a TV appliance, or to build a car or an airplane.

These are very complicated designs, in which each piece must fall gracefully in place in order for the overall device to deliver the function it was designed for. But a complicated system is supposed to passively obey the rules 'top-down'. A complex system, on the other hand, is *very* different: it is not passive, it takes an active part in the process, and displays behaviours that cannot be inferred from the basic rules of the game, a distinctive property called *Emergence*. Emergent behaviour cannot be inferred from the basic rules of the game, not even once these rules are specified to the very finest macroscopic detail [61, 73, 114, 48]. Albeit insightful, it seems to me that this picture still sounds a bit too abstract to succeed in the task of conveying a concrete idea of what a complex system really is. Therefore, given that complex systems abound around us, I prefer to proceed by example: the weather, the environment, the jungle, cities, traffic jams, the cell, the brain, the immune system, diseases, financial markets, our own society, and above all, life itself, are all inscribed under the same rubric of complex systems (see Figs 1.3 and 1.4).

Too much? Yes, but, no worries, we shall shortly get more specific.

Figure 1.3 *A very scenic natural complex system: waterfalls.*
Source: reprinted from pixabay.com.

Figure 1.4 *A man-made complex system: fancy skyscrapers.*
Source: reprinted from pixabay.com (carloyuen from Pixabay).

1.3 The science of sciences?

The idea of this book was prompted by a series of lectures which I have been giving over the years to high school students in various places around Italy. In the course of these lectures, right after having listed disparate examples of complex systems, I would pause and caution the kids as follows: if I were in your shoes, at this very point, I would look at me (the speaker) with a great deal of healthy scepticism; I would think to myself, here comes one more '*tuttologo*', in my native Italian, the man who is ready (and happy) to talk about everything and anything, in fact the bread and butter of any TV talk show.

How can one possibly address physics, biology, finance, society, psychology and more, all in a single mouthful? This sounds shallow in the first place, not to mention arrogant and in fact totally outrageous to the serious scholars who devote their entire life and career to the study of *sub-subjects* of each of these disciplines! Indeed, if the question is how one could possibly be an expert of all of these fields, the answer is very plain: no way! For sure Complexity scientists cannot replace a biologist's on cell issues, nor the economist on the financial markets, and so on. Yet, hopefully, they can bring *something* useful to the table for both.

Twofold question: first what and then how? The point is that, despite their vast disparity, *complex systems share a number of general distinctive features, which transcend the (crucial) details of each separate discipline.* To the extent that *Universality*, (read insensitivity to *Individual* details), and *Individuality* (sensitivity to individual details) manage to coexist, as they surely do in most complex systems, the Science of Complexity surely has her place in the dancefloor.

I have experienced this myself in my own work: methods we developed to study turbulent flows, sure enough a very complex natural system, have found profitable use in the study of deoxyribonucleic acid (DNA) translocation across membranes, again unquestionably a complex system, although totally different from turbulence! The Science of Complexity has been occasionally named the science of sciences [50], (shall we say super-science then?) and, even though I am not particularly fond of bold definitions, I sense some truth in this one, inasmuch as it reflects the inherently transversal nature of the Science of Complexity across disciplines[4] [99, 114, 61].

This is hardly a surprise, since the Science of Complexity deals mainly with *organizational principles*, rather than the (all-important) details which pertain to each specific discipline. In a way, more forest than the trees. This is a double-edged sword though: the risk of inconclusive shallowness is always lurking behind the corner. Indeed, generality and interdisciplinarity are the major blessings of the Science of Complexity and its main liabilities as well.

This is why its true impact will be measured by the extent to which it will prove capable of contributing to the solution of problems that mono-disciplinary research alone cannot tackle. There is no shortage of promising candidates, especially in the context of the so-called grand challenges to modern science and society: energy, environment, pandemics, medicine, finance, urban planning, are potentially poised to reap concrete benefits from the concurrent synergy between specialists, contributing deep and hard-won knowledge of the details of a given system, and Complexity scientists dealing with the quantitative analysis of such details, once embedded in the general framework of the overarching organizing principles [50, 73, 114]. Yet, as per the disclaimer in the Preface, this is not what this book is about: I'm rather going to stick to a small, yet representative, window

[4] Farrar, Strauss. and Giroux. New England Complex System Institute, https://necsi.edu, where we read that: The world is transforming. Why has the unexpected become commonplace? Because of complexity. Even with mountains of data available, neither decision makers nor traditional science is equipped to unravel collective behaviors with far-reaching consequences. Science is advancing. Using the next generation of mathematics, complex systems science helps us understand the dynamics of society and civilization by making more accurate predictions than either statistics or calculus.

of the physics-biology interface [48, 102]. Some of these organizing principles, the ones I perceive as most relevant to the aforementioned interface, will be covered in Chapter 2.

Next, we proceed with a few 'philosophical' considerations.

1.4 Universality and individuality

As the reader will appreciate shortly, organizing principles usually come with very far-reaching consequences, which typically go even beyond the realm of science, to impact epistemological and philosophical issues. As anticipated in the opening of this chapter, *complex systems are the rule and not the exception*, the fact that they live around and within us, that they very often belong to the realm of 'things that we can see', has for a long time sentenced them to second-tier scientific status, 'boring and often disturbing details', as reflected by the often-heard dismissive statement, 'the rest is detail'.

Like Dylan's wife Sarah, they are often 'so easy to look at, so hard to define'. Complexity has also been termed the tombstone of definitions [23], but some useful definitions can nevertheless be found. For instance, Giorgio Parisi writes: 'A system is complex if it crucially depends on the details of the system' [82], a definition which, in my view, does indeed capture an essential trait of complexity.

Indeed, complex systems inform us that, besides the overarching organizing principles, they always exhibit some *details that do matter*, no less than the informing principles they are embedded within. This is perhaps most clearly seen in the field of biology, but all of the aforementioned scientific and social grand challenges send out the same message. This means that finding the right blend of *Universality and Individuality* is one of the deepest-running intellectual challenges raised by complex systems.

In loose terms, it is fair to say that Complexity thrives on freedom in the first place: life gets complex when we are presented with many, often conflicting, options, not when we have none. No choice may be a miserable situation, but not a complex one. Indeed, complexity is often quantified in terms of what scientists use to call *degrees of freedom*, the variables which specify the state of the system. A point-like particle constrained to move along the horizontal axis of a plane features just one degree of freedom, the same particle free to move along both axes of the plane has two.[5] Of course, a particle free to move in the plane has more freedom to trace fancy trajectories than one constrained on a line. By the same token, two particles moving in the plane feature four degrees of freedom, and their interaction spawns further complexity in their joint trajectories.

[5] To be precise, a particle moving along the horizontal axis has *two* degrees of freedom, one for its position and one for its velocity. Indeed, given the position at a given time, without knowing the velocity as well, it would be impossible to predict the position at any subsequent time.

1.5 Relevant and irrelevant details

It is worth emphasizing that not all degrees of freedom are equally relevant: typically, some are more important than others, so that complexity depends critically on the divide between the two, sometimes called *relevant* and *irrelevant*, respectively. As a concrete example, take the case of fluid turbulence (Fig. 1.5), to be detailed in Chapter 7, in which large groups of molecules (easily many billions of billions) move in a highly coherent fashion, as if they were basically one: collective motion. That's how the beautiful vortices and curls we observe all around form in the first place. Beautiful vortices meet well to our eyes, much less to those of the car designer, who knows that they invite two highly undesired guests, *dissipation* and *noise*. Does the car designer need to focus on the motion of each and every molecule flowing around the car? Happily, the answer is a plain no!

Designers can safely focus their attention on higher level structures, sometimes even the vortices themselves, or tiny subregions thereof, but never down to the level of single molecules. This is the divide we were referring to, molecules are the irrelevant degrees of freedom and vortices are the relevant ones.

Don't miss a subtlety: vortices are made of molecules, hence molecules do matter, by definition! The point though, is that they do not matter individually because it is the *collective motion* of large groups of molecules which dictates the amount of energy dissipated by the car. The profound divide between individual versus collective behaviour goes way beyond fluids and represents a distinctive trait of most (should I say all?) complex systems at large (see Fig. 1.6). In broad strokes, *Individuality* fuels Complexity via a large number of degrees of freedom, while *Universality* usually works in the opposite direction, bringing Order

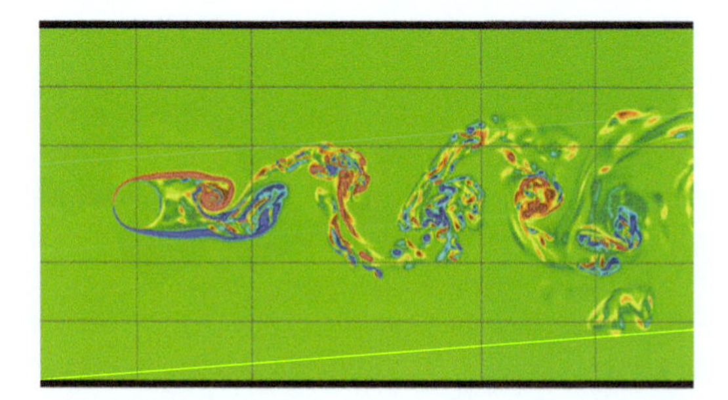

Figure 1.5 *A prototypical complex system: turbulent flow past a cylinder (projected onto a two-dimensional plane). The relevant degrees of freedom to describe the quantitative behaviour of this system are the fluid vortices, in blue and red colour for clockwise and counter-clockwise rotation, respectively. These vortices are the emergent structures which result from the coherent motion of zillions of underlying molecules moving along closed patterns. There are astronomically fewer vortices than underlying molecules, but the Complexity of the vortex motion still defies our full understanding.*
Source: courtesy of G. Amati.

Figure 1.6 *An enchanting example of man-made visual Complexity: Vincent van Gogh's* 'Starry Night'. *This too is a collection of molecules, but its inescapable beauty is not a property of the single molecules. It derives instead from the coherent patterns emerging from their arrangement in space, including starry vortices . . . .*
Source: reprinted from pixabay.com.

back into Disorder. But this is by no means the full story, as it would be a gross mistake to equate Simplicity with Order and Complexity with Disorder. The real challenge is hidden in the combination of the two. Each complex system, or, better said, each *class* of complex systems, carries its own peculiar mix of both.

The champion science of *Universality* is theoretical physics, with its ceaseless quest for simple and unifying principles, the Theory of Everything, possibly condensed in a single formula, short enough to fit on your T-shirt. Arguably, the most famous example of *Universality* in physics are Newton's laws of Universal Gravitation, stating the two bodies attract each other with a force proportional to the product of their masses and the inverse of their distances squared. If two 'bodies' sounds very anonymous, it's because anonymous it has to be indeed: ants, lions, humans, cars, buildings, we are all are subject to the same law, all that matters to gravity is mass, all the rest is an irrelevant detail. If Alice and Bob both weigh 60 kilos, they experience the same attraction to Earth, gravity is genuinely gender-blind! On the other hand, think of a different world, in which gravity would respond not just to our mass, but also to our height, or gender, or the colour of our eyes. No question this would be a rather more complex (or complicated) world indeed!

At the other extreme lies biology, with its painstaking attention and love for the diversity spawned by sensitivity to individual features. Cats and dogs are different animals, cats miaow and dogs bark, but if Jimmy the cat and Felix the dog happen to weight the same, Jimmy and Felix are the same to gravity! Hence, if you wish to know more about cats and dogs, the answers to your questions surely won't spring from Newton's theory of universal gravitation. Not that cats and dogs have nothing in common, they both come with four legs, two eyes and one tail, and indeed they both abide by the organizing principles of Thermodynamics. At the same time, though, they differ in many substantial – hence relevant – details, because, even once the organizing principles are fulfilled, there is still plenty of room left for different specific realizations. Keeping a long-sighted eye on universal principles, without losing the short-distance grip on the relevant details is the basic challenge of the Science of Complexity. This challenge runs all the way across science, some disciplines might be more or less dependent than others, but the border between universal/individual behaviour is the critical portal to the process of understanding how complex systems work.

1.6 The gifts we neither understand nor deserve

As I write, very few scientists would deny that the Science of Complexity stands on a par with the most challenging and glorified intellectual endeavours, be they fundamental particles, quantum gravity, cosmology, and similarly glorious disciplines. Even Steven Hawking (1942–2018), one of the main champions of big science and its glamorous Theory of Everything, once proclaimed that the present century will be the century of Complexity. And if there is a glorious name to Complexity, it is life itself. No doubt life is complex, not only in its biological sense but also in a very practical, social one, which means the science of complexity is not about thin air, it's actually about us, you and me.

The most fascinating and emotional part, the one I feel still remains underdisclosed to the public at large, is that we have been gifted with the wonderful chance of navigating the Oceans of Complexity around and within us. Differently restated, the gift is our ability to survive with the uncertainty which inevitably comes along with Complexity [80]. But how? In metaphorical language, I would say by sailing the rough and deep waters of Ocean Complexity, with two nimble ships called knowledge and insight. This is often (mis)taken for arrogance, but I feel like the opposite is true. While we are justly proud of our intellectual triumphs, we should never forget that the ability to make (some) sense of the world around us is by no means a given, but rather 'a gift we neither understand nor deserve', as vividly expressed by Eugene Wigner [120]. Wigner referred specifically to the 'unreasonable' power of mathematics to describe the physical world, but his statement may just as well be extended to the broader picture of the Science of Complexity. The great gift is that nature does not impose exhaustive knowledge on us as a conditio sine qua non for gleaning Knowledge and Insight. It allows us to survive with uncertainty.

Nature is generously redundant it provides irrelevant degrees of freedom galore which *can* be ignored; (on an individual basis) without compromising the basic understanding we need to survive and sometimes even prosper in this strange and often scary but breath-taking world. Thus, the gift is that, subtly interspased within the ocean of irrelevant degrees of freedom, there are magic islands of relevant ones, where things are 'just right' for complex behaviour to emerge out of Oceans of Complexity. And the second, even more important, gift is that Oceans of Complexity have been equipped with navigation routes which allow us to bring the ship to harbour in the magic islands. This is why modern formulations of free energy will take centre stage in this book. One may justly wonder why then are irrelevant degrees of freedom around in the first place, given that they beg to be ignored? My answer is very plain: I don't know.

But I do know, though, that this is precisely where I sense a most enchanting beauty springing from the ground. If irrelevant degrees of freedom were not around, the relevant ones would stand right in our face, making the sailing trivial and unengaging, hence boring: only highways, no scenic routes, a fall-asleep scenario. To the very-welcome contrary, the game is subtle, hence by no means one that comes for free: ingenuity, fantasy, and, of course, discipline as well must be invested in the enterprise of sailing Oceans of Complexity. The bar is not low, the *Universality/Individuality* interface is a 'long and winding road', which goes over high bars and through narrow doors, instead of low and wide ones. Another facet we shall cover in some detail in this book. This is again personal (well, after all, this is my book, isn't it?) but even back in my teens I never trusted invitations to the low and wide path, especially if heralded to make my life easier. I always felt, as I still do to this day, that such invitations often arise from sources that care far more for their agenda than they do for mine.

Sorry for the outpouring, back to facts: not all details can be ignored, some are as essential as the general principles informing universal behaviour and neglecting them comes at a price—in worst-case scenarios it might even sink the boat .... *nature plays hide and seek*, and we must find the right tiny corners where she likes to hide. As we shall see, such corners are often mind-bogglingly tiny, literal straws in a huge biological haystack. Searching for the magic islands in an ocean of irrelevant details, this is the art of sailing the Ocean of Complexity (see Fig. 1.7)! If we search with labour, passion, ingenuity, and great respect for the explored ground, we may actually dock the ship on the islands where the 'Impossible' comes true, the land where, to borrow from the Wizard of Oz (Over the Rainbow), 'skies are

Figure 1.7 *Sailing through rough waters. Source*: reprinted from pixabay.com.

blue', 'troubles melt like lemon drops', and 'dreams that you dare to dream really do come true'.

This is the land where we eventually learn to live with Uncertainty without (too much) anxiety, ultimately the art of living altogether. I find this subtle and beautiful; it instils humility instead of arrogance, precisely in the sense expressed by Saint Francis's 'Necessary, Possible, Impossible (NPI) principle' in the opening of this book: start by doing what is necessary, then do what is possible, and suddenly you will be surprised to find that you are doing the impossible.

If this book would succeed in conveying just a glimpse of this beauty, it will not have been written in vain.

1.7 Summary

It is a widespread opinion that the most challenging and exciting science takes place at the extreme physical frontiers in size and energy: the ultra-small, elementary particle physics and the extra-large, the Universe and the Cosmos. This is true, fair, and square. But it is equally true that there are also breath-taking mysteries away from the physical extremes. This is the playground of the Science of Complexity.

2

The Guiding Barriers

One of the core organizing principles of my life is that success comes through a delicate balance between making things happen and letting things happen.

(Robin S. Sharma)

2.1 The three barriers

In the Preface, we have hinted at the fact that the Science of Complexity explores the inner frontiers of human knowledge, typically away from the physical extremes of big science. In this chapter, we shall substantiate this statement by inspecting three prime Complexity-generating mechanisms more closely, namely *nonlinearity*, *nonlocality*, and *ultradimensions*. These mechanisms prove capable of building up formidable barriers to human understanding. However, they also provide powerful organizing principles to handle the barriers they have contributed to build in the first place. In brief, they inform the strategies to navigate the Oceans of Complexity.

2.2 The common threads of Complexity

Complex systems display a number of common underlying features, which serve as general organizing principles beneath the diversity of their specific natures. In the following, we provide a selective survey of such organizing principles, the idea being to confine our attention to the leit motives which are particularly relevant to the kind of Complexity discussed in this book, namely the interface between physics and biology. Here comes the list, three singles:

1. *Nonlinearity*
2. *Nonlocality*
3. *Ultradimensions*

and three duals:

1. *Order/Disorder*
2. *Equilibrium/nonequilibrium*
3. *Smoothness/roughness*

Sailing the Ocean of Complexity. Sauro Succi, Oxford University Press.
 DOI: 10.1093/oso/9780192897893.003.0002

This list leaves out several central topics in modern Science óf Complexity, particularly complex adaptive networks and their widespread use in science and society, for which several excellent books are available in the literature [48, 54, 52, 20, 15]. However, as discussed in the Preface, this is not a book on Complexity at large, but on a specific window at the frontier between physics and biology. For the sake of compactness, this chapter focusses on the first three items, leaving the remaining three to the next one.

2.3 Nonlinearity

If you drink a coffee and pay 1 Euro for it, upon drinking two you expect to pay 2 Euros, and that's exactly how it goes, you pay in direct proportion to how much you drink: this is a linear system, one in which the effect (pay the for the coffee) stands in direct proportion to the cause (drink the coffee). A non-linear system is not like that: two coffees may cost you less than 2 Euros or may be more, but definitely not 2, see Fig. 2.1. Weird as this sounds, this is the way most systems around us, including ourselves, function for real, except that we do not necessarily realize it because when the causes are in some sense 'small enough', so are the effects, and the nonlinearity remains silent. Take a piece of material, say rubber, and pull it from the two ends: as a result, the piece of rubber stretches out,

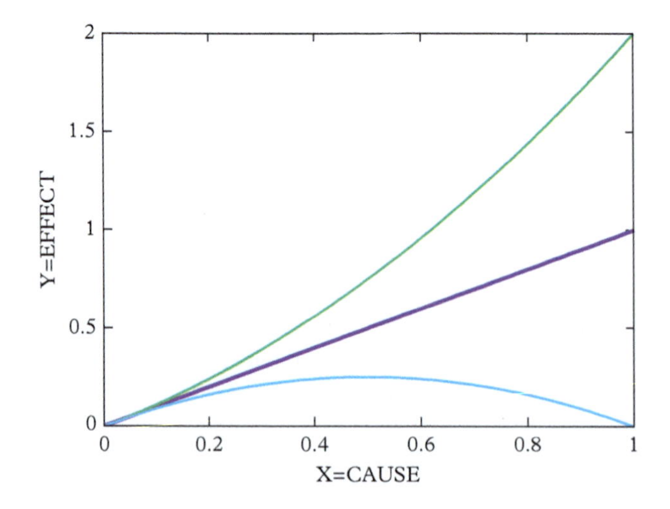

Figure 2.1 *Graph of a linear function (middle), sublinear (bottom), and superlinear (top). The linear function is a straight line, whose constant slope measures the ratio between cause and effect. The sublinear case is characterized by a decreasing slope as the cause is made larger, indicating that the material gets 'stronger' (less deformation) under increasing pressure. The superlinear case, on the contrary, shows an increasing slope at an increasing cause. For the case of materials, this corresponds to the material offering less and less resistance (larger deformation), until eventually rupture occurs.*

getting longer and thinner in the process. Rubber responds to the external load (cause) by a shape change (effect), typically by elongating. As long as you pull it softly enough, the deformation is in direct proportion to the intensity of the pull: double pull, double elongation. As you keep pulling harder and harder, though, the material starts to undergo larger and larger deformations, much larger than the incremental pull, until it possibly breaks down. Material scientists call this 'ductile rupture', to indicate the case when the system sends signals (precursors) that something weird is going to happen, *before* it does. Much worse is the case of 'fragile rupture', in which no such signals are dispatched to warn about the incumbent disaster.

Occasionally, nonlinearity may also act the other way round; by increasing the pull the material gets stiffer, a property sometimes referred to as 'anti-fragility'. Although to a different extent for each given material, every material displays its own response to stress (the pull), and sooner or later, this response is bound to become non-linear. And, if you think that this is exclusive to material science, a minute's thought about the human mind and psychology shows that this not the case. The slogan 'Don't crack under pressure' speaks clearly for the point .... Non-linearity is the prime engine of Complexity, one the chief reasons why complex systems are hard to predict, the weather being probably the most popular case in point. To speak in a metaphor, two and two does not always make four, and if it did, we would not be here to tell you so in the first place, since, life, as we know it, is quintessential nonlinearity.

So, if not four, how much is two plus two? The answer is that it depends: sometimes it is less, sometimes it is more, denoting negative (positive) interference, technically known as *sublinear* and *superlinear* behaviour, respectively. Interestingly, the same complex system can exhibit both kinds of behaviours, depending on the intensity of the solicitation.

Eventually, more is incommensurably more, an extreme behaviour epitomised by the famous *butterfly effect*: the wingbeat of a butterfly in Cuba causing a hurricane in Miami.[6]

Fragile rupture is a prototypical butterfly effect, past a given threshold, even a minute's increase of the load leads to rupture, and similar examples can be found everywhere, outside and beyond the realm of material science. Indeed, we don't need hurricanes to grasp the idea, we all know that a small change in many/most decisions in life may lead to completely different pathways, as nicely captured by a big advertisement I saw many years ago in Brussels Airport, which went like this: 'In life, the difference between failure and success is paper-thin'. We don't need to be told twice how true this is.

[6] I used to ask high school students about the butterfly effect, and they typically know about it, if only sometimes with a bit of an overstretch. One student once told me that the wingbeat in Cuba would cause a hurricane in Moscow. Too much! But I am told that the locations change very much from place to place ....

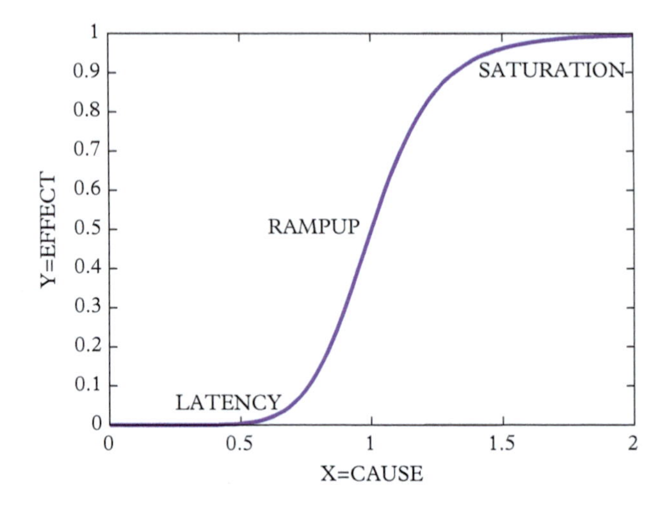

Figure 2.2 *Sketch of a paramount nonlinear function: the sigmoid, with the three typical stages: latency, ramp-up, and saturation (LRS).*

As mentioned previously, sometimes, more rarely though, the inverse is also true: a big change is absorbed without much of a consequence. To stick with animal analogies, we may call this the *elephant effect*: you push a lot and get little response in return. If you wish, this is once more anti-fragility, not only does the material not crack under pressure, it gets even stronger! And again, the idea can be easily applied to other contexts, such as psychology and social sciences. To go with the ancient Romans '*Per aspera ad astra*', (through adversity you reach the stars).

We have mentioned before that the same system can host multiple types of nonlinearities. A key example in point is the sigmoid function (see Fig. 2.2).

This function is characterized by three distinct regimes: *latency, ramp-up and saturation* (LRS). Latency is the stage where the system provides a very weak response to increasing solicitations, the system is silent, no signal, a potentially dangerous stage. Ramp-up is the regime in which the system wakes up and provides a very substantial response to increasing solicitations. Finally, saturation is the stage at which the response starts flagging down, until it eventually stops growing altogether, no matter how large the solicitation. Clearly, the key information is the threshold at which the ramp-up starts off and the one at which it goes away, for this is the range where we get a substantial response, and, of course, the extent of the ramp-up. 'Calm' systems exhibit a substantial separation between the two thresholds, 'nervous' ones don't: for these latter, a small change in the cause intensity leads to a sudden jump from the silent state to the saturated one. Physicists call this a *phase-transition*, to indicate a sharp transition between two qualitatively different states, say liquid water versus vapour (boiling) or vice versa (condensation).

The notion of phase-transition goes far beyond the realm of physical sciences and indeed the LRS behaviour is very widespread in complex systems, from biology to electronics, to computer science, and many other systems. For an enthusiastic description of its crucial importance, see the book by P. Domingos [30].

2.3.1 The superposition principle

The expression 'two plus two does not make four' is a purely metaphorical way of emphasizing that, in a complex system, the whole is more (or less) than the sum of its parts, simply because the parts interact/interfere with each other, and those interactions form an integral part of what the complex system is: no man is an island! The far-reaching consequence is that *complex systems cannot be broken in pieces*, each to be studied in isolation, and then reconstructed without losing information. In technical terms, they do not obey the *superposition principle.*

The importance of this observation cannot be overstated because the superposition principle lies at the very basis of the canonical route pursued by science from its beginnings to the present day, the glorious route of 'scientific reductionism'. That is: take a huge problem, too huge to be chewed in a single mouthful, and 1) break it into smaller pieces, 2) solve each piece separately and 3), recompose the global solution by summing up each single contribution. Ancient Romans (again!) knew it all too well, '*Divide et Impera*' (divide and conquer), being their (successful, judging by the size of their empire) motto.

If the system is linear, with no interference (positive or negative, does not matter) between its components, this is perfectly fine, in fact it is literally exact! If my task is to sum four numbers, say $1 + 5 + 3 + 9$, for a total of 18, I can certainly sum the first two $1 + 5 = 6$, then the second two, $3 + 9 = 12$, and sum the two partials, to get $6 + 12 = 18$, which is the -exact- result. Nothing was lost in the process, the superposition principle shines bright! This is because the sum is a quintessential linear operation. Now try the same trick with the *square* of the sum, namely $(1 + 5 + 3 + 9) \times (1 + 5 + 3 + 9)$, which is $18 \times 18 = 364$. The square of the two partials, though, gives $6 \times 6 + 12 \times 12 = 36 + 144 = 180$, which is less than half of the correct result! The trick of dividing into two subsystems, computing the sub-squares and summing them up, falls flat on its face, as more than 50 per cent of the result is lost along the way! This is because the square of the sum is more(less) than the sum of the squares: there is a positive(negative) interference between the numbers to be summed, which is completely overlooked by the process of breaking the system of four numbers into two separate pairs of two. This is neatly reflected by the elementary formula for the square of the sum of two numbers, say a and b, namely:

$$(a + b)^2 = a^2 + b^2 + 2ab \tag{2.1}$$

This is because the square of the sum means multiplying the sum by itself, i.e. $(a + b) \times (a + b)$ and the product consists of *four* terms $a \times a$, $b \times b$, $a \times b$, and

$b \times a$. The first two are the sum of the squares, but the second two provide an extra contribution, expressing the 'interaction', so to say, between a and b. If a and b have the same sign (positive interference), the square of the sum is more than the sum of the squares, whereas if the numbers are opposite in sign they interfere negatively and the square of the sum is less than the sum of the squares. If you think that this is dry math in thin air, please think twice, because that is precisely what our own life is made of, at all levels, biological, psychological, and social. And the reason is that we are not isolated systems, the 'bricks' we are made of are constantly interacting with each other and if their interaction was linear, they would never succeed in keeping us going. We shall hear more of this in this book, but here I just wanted to make the point that losing the superposition principle, the pillar of linear science, comes with pretty far-reaching consequences for our predictive capabilities.

So, does this mean that we are at a total loss with nonlinear systems? Happily enough, not at all: humans are smart, and they have invented plenty of ways around, rarely exact, granted, but pretty powerful and insightful approximations, nonetheless. If the interaction, the scientific name for interference, is 'weak', in some sense to be defined system by system, reductionistic practices still work pretty well, essentially because we are equipped with well-defined procedures to obtain controlled approximations of increasing accuracy. But, when the interaction is weak no more, the very notion of isolated system crumbles to dust, and new methods, ideas, and tools are needed. That is where Complexity makes its glorious entry in modern science.

2.3.2 Multiscale coupling

Another distinctive property of nonlinear systems is that they operate across multiple scales in space and time. Just look at our own body: quarks form nuclei, nuclei form atoms, atoms form molecules, molecules assemble in cells, cells grow into tissues, tissues form organs, and organs assemble into our own bodies. The whole process covers about fifteen decades in space, from the femtometre (one millionth of a billionth metre) to about a metre and even more in time. Indeed, the typical timescale for elementary chemistry is the femtosecond, while human beings live on average almost three billions seconds, for a total of about twenty-four decades in time. This is the spacetime region swept out on average by human beings over their lifetime. Clearly, such a huge and hierarchical multiscale chain must be sustained by a highly orchestrated flow of energy and information across these different scales. Here again, nonlinearity plays a crucial role.

Indeed, a basic property of nonlinear systems is their ability to *transfer mass, energy, and information across scales:* large to small and small to large and likewise. When two droplets melt into a single, larger droplet, energy and mass move from small scales, the size of the two merging droplets, to larger scales, the size of the single droplet. Similar phenomena occur at every scale in nature, galaxies and black holes making no exception. Likewise, when a single mother droplet breaks up into two smaller daughter droplets, energy and mass flow from large to small scales. *Coalescence* and *breakup*, two fundamental processes that occur in most natural and industrial activities, would not be possible if the energy involved in the process were a linear function of the droplet volume.

Why? The point is that droplets coalesce or break up whenever there is an energetic incentive to do so. Coalescence occurs whenever it takes less energy to sustain a single large droplet than two smaller ones. Likewise, breakup occurs precisely in the reciprocal scenario. For the record, this is the working principle of nuclear fusion and fission, respectively. If energy were a linear function of the droplet volume, such an energetic drive would vanish and neither coalescence nor breakup would occur spontaneously. The argument is simple. The energy of the droplet usually consists of a part proportional to its volume (bulk energy) and one to its area (surface energy). For a sphere of radius r, the former scales like the cubic power r^3 and latter like the square power r^2, prefactors being irrelevant at this stage. Upon merging two droplets of radius r into a single one of radius R, volume conservation gives

$$R^3 = 2r^3 \tag{2.2}$$

which delivers $R = 2^{1/3}r \sim 1.26r$. Note that this is much smaller than $2r$: the coalescence of two droplets leaves the volume unchanged and increases the radius only by about 25 per cent, not a factor two. What does the area do in the process? Since the area scales like the square of the radius, we obtain:

$$R^2 = (2^{1/3}r)^2 = 2^{2/3}r^2 \sim 1.6r^2 \tag{2.3}$$

which is less than total area of the two droplets before they merged, that is $2r^2$. the area decreases by roughly a factor $1.6/2 = 0.8$. This means that the bulk energy stays the same, while the surface energy decreases, hence the total energy decreases and the process of coalescence is energetically favoured.

It is intuitively clear that a segregated world where scales don't couple, small remains small and large remains large, an attitude that scientist call *scale separation*, is much easier to predict compared to one in which scales mix with each other. Much easier but also more boring and less inventive. We shall return in detail to these fundamental isues in Chapter 3, but it is hoped that by now the reader sees clearly why, in a world where two plus two is four, small remains small and large remains large, very few, if any, of the marvels and good and bad surprises we see around us all the time could ever stand a chance. This is the ultimate power of non-linearity.

2.4 Nonlocality

A pillar of modern theoretical physics is the *principle of locality*: events take place whenever agents meet next to each other in space and time. And the far-

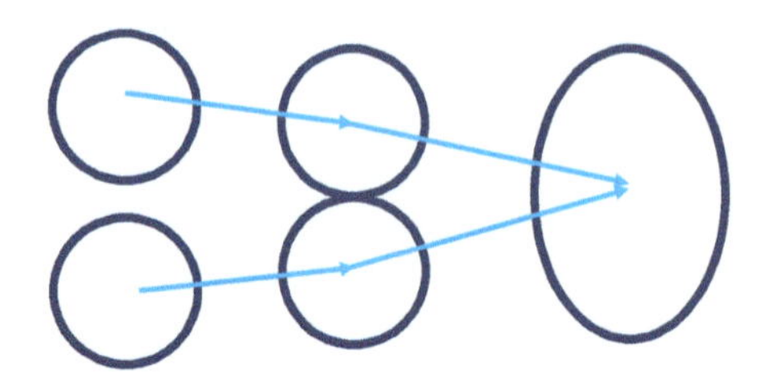

Figure 2.3 *Coalescence of two droplets into a bigger one, a process which moves mass and energy from small to large scales. Breakup is just the reciprocal process (right to left, reverted arrows).*

ther apart they lie, the weaker their interaction. In the world of particle physics, the interactions take place via carrier particles which travel from place to place in space and time, the prime case in point being an electron at position $x1$ and time $t1$, emitting a photon (light), travelling to position $x2$ to meet a second electron at a subsequent time $t2$. That's how non-local interactions take place, through particles that travel from place to place carrying the information to be exchanged between the interacting partners. Light messengers, the lightest one being the massless photon we call light, travel a long way: heavy messengers, the massive particles which mediate the interactions between protons in the nuclei, called gluons, travel much less, just about a millionth of a billionth of a metre, the size of the proton. So, the mass of the messenger and of course the degree of emptiness of the medium hosting the interaction, determine the way that the intensity of the interaction decays in space and time. These relations are magnificently captured by one of the most beautiful and powerful tools of theoretical physics, known as Feynman's diagrams, after the great American physicist Richard Feynman (1918–1988), one of the very few scientists known to the wider public, for his flamboyant personality, besides his exceptional talent.

But these are elementary particles, what do they have to do with us? Discounting the fact that they are our elementary constituents, it is indeed true that the way macroscopic objects interact with each other, namely the causality relation:

Effect(here,now) versus Cause(there,then)

can hardly be inferred by inspecting Feynman's diagrams! In other words, even though in the end everything traces back to elementary particles, Feynman's diagrams are of little avail here however, (see Fig. 2.4). Indeed, the Causality Relation

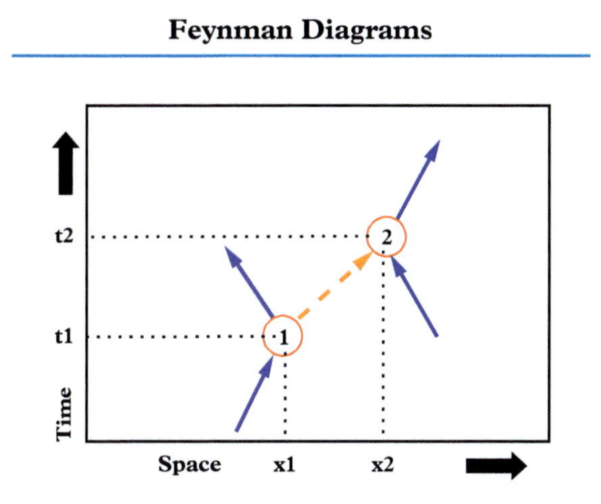

Figure 2.4 *Two electrons interacting at position x1 and x2 in space and instants t1 and t2 in time. The rightwards propagating electron at x1 emits a photon (dashed line), also propagating rightwards and gets deflected leftwards in the process (blue arrow at point 1 pointing up and to the left). The emitted photon travells rightwards until it meets the second electron at position x2 and time t2. The second electron absorbs the photon and get deflected rightwards (blue arrow up and to the right at point 2). The two electrons interact non-locally, but only because the 'invisible' photon connects them in space and time. The actual interactions between electron 1 and the photon, as well as electron 2 with the same photon are completely local. That's the locality principle in its full splendour!*

in complex systems can often turn out to be non-local in space and especially in time, a property that scientists call *memory*.[7]

2.4.1 Memory, history, and hierarchy

Complex systems often show a distinct sense of memory, they don't forget their past, which, as we shall see shortly, proves an effective tool to anticipate their future. They often display a strong sensitivity to their history, the future does not depend on the present alone, but also on the near, or maybe not so near, past. What happens '*here,now*', depends on what happened '*there,then*'. After all, as noted before, memory is precisely the mechanism by which past errors can be turned into lessons for a better future, whatever this may mean. It is no exaggeration to say that memory is a pre-requisite for learning, hence a primeval tool of survival.

[7] Non-locality has become a particularly hot subject in modern physics, especially in connection with the weirdest aspects of quantum mechanics, often condensed under the exotic term 'quantum entanglement'. For a lively discussion of this fascinating topic see [76].

In technical parlance, one speaks of *correlations*: a given event in the past can still make itself felt (co-related) today. Human societies are the prime arena of memory dependence: some events in the remote past of human history still affect our daily life in a very concrete sense. Industrial revolutions, wars, and similar major events make our present-day society look very different from the way it would had they not taken place. That's after all the very raison-d'etre of history as a learned discipline.

Once, a friend and colleague asked me: what would we look like today had Hitler never been born? Would somebody else have done the same, maybe earlier, maybe later, or maybe never? Obviously, I can improvise a sketchy answer, but the real fact is that I simply don't know, and I bet good money that even the most learned historian does not either. But certainly, the fact that Hitler existed still has an effect in our society. My parents used to report that the Germans soldiers they met during war time were never too hostile to them. Had they been, I would not be here to tell. So, the Germans soldiers not having been nasty to my parents, surely did have a big effect on me, as we speak!

Many (not all of them) complex systems do not forget, the past lives on. Because humans are (very) complex systems, they have a mind and they pass over their thoughts in a way (language) a simple system would never prove able to. But memory does not necessarily require such a level of sophistication as a thinking or a conscious mind, but just an accumulation of effects from the past. Most natural systems exhibit some form of impersonal memory which profoundly affects their behaviour in time, nonetheless.

2.4.2 The Deborah number

A useful indicator of the degree of memory of a given physical system is provided by the so-called Deborah number (see Fig. 2.5), defined as the ratio of the physical timescale of change and the time available to observe such change:

$$De = \frac{t_{change}}{t_{obs}} \tag{2.4}$$

Memory lasts as long as the system is in flux, the end of the journey marking the complete loss of information on the initial state. If such relaxation time is much longer than the observation time, the system is said to have a long-term memory, which translates into a large value of the Deborah number. If, on the contrary, the system relaxes very fast as compared to the observation time, then it can be treated as basically memory-free.

This shows that memory is a relative concept and also accounts for the historical origin of the name, which stems from the Biblical song of the prophetess

Figure 2.5 *The Prophetess Deborah, portrayed in Gustave Dorè illustrations for la Grande Bible de Tours (1865).*
Source: reprinted from en.wikipedia.org.

Deborah[8] (allegedly 1107–1067 BC), where we read that 'The mountains flowed before the Lord' (Judges 5:5). Mountains flow before the Lord, but not before the humans, because humans don't live long enough to see them flowing (and when they do, they'd rather not, because this is likely to spell natural catastrophe).

Leaving aside the Complexity of society, biology is also littered with *history-dependence*, often encoded in *hierarchical* structures. Genealogical trees are a prototypical example: children would not exist without their parents and there is a definite hierarchy to be followed. You would not be here unless your mother and father had been around before! The tree is a hierarchical structure frequently met in complex systems, one of the reasons being that it greatly facilitates information flow across the system. Among other things, it protects against collapse due to over communication (think of the most intrusive use of cell-phones and social media).

The rule is simple. Suppose you are in charge of a small company or organization with, say, 16 employees (why 16? We shall learn in a moment). If you have to dispatch a message to each of them, that makes 16 messages for you to deliver. Now suppose you decide to appoint four assistants, delivering four messages each, the whole set of 16 employees can now be reached with no individual member being loaded with more than four messages. Of course, the structure is more complex, since four assistants have to be appointed, but the gain is apparent, no individual in the organization needs to dispatch more than four messages. In plain maths, you have traded 1×16 for 4×4. One could regard this as a luxury of a lazy boss, but replace now 16 with 160, or maybe 160,000 and the luxury is readily seen to turn into a literal survival strategy.

As usual, numbers matter: *quantity is quality*! That's what most complex system do: they organize in hierarchical structures which optimize the flow of information across the entire system so as to guarantee survival and actual functional

[8] The Hebrew word, *Dvora,* means bee, the honey maker. Despite the sweet name, being cursed by Deborah meant being haunted forever, no memory loss . . .

performance. Physiology does the same; the cardiovascular or respiratory systems being outstanding examples of the case of hierarchical organization, and the environment does the same too, as a glance to hydrological networks rapidly demonstrates. Since Complexity is often associated with large collections of individuals, be they atoms, cells, or human beings, an efficient flow of information is absolutely vital to the proper functioning and survival of the overall system. For the power of this principle in natural systems, see Bejan's book [14]. Before closing this section, we wish to caution the reader that, while history matters a lot, there are complex systems for which history does not teach much at all: we cannot predict the long- or mid-term future by extrapolating the past, weather forecasting being an exemplar case in point.

2.4.3 Chicken-egg causality

Memory implies a clear distinction between past and future, hence it is strictly connected with the central notion of *causality*, see Fig. 2.6. The basic principle of causality states that causes precede their effects and, besides forming a pillar of natural philosophy, it is also deeply ingrained within our daily life experience. Note that this does not necessarily mean that every cause has an effect, or vice versa, but simply that causes precede their effects and effects follow their causes. You may wish to notice that this assumes that we are always in a position to tell which is the cause and which effect, unmistakably. But is it always like this? A minute's thought reveals that this is not necessarily the case and indeed complex

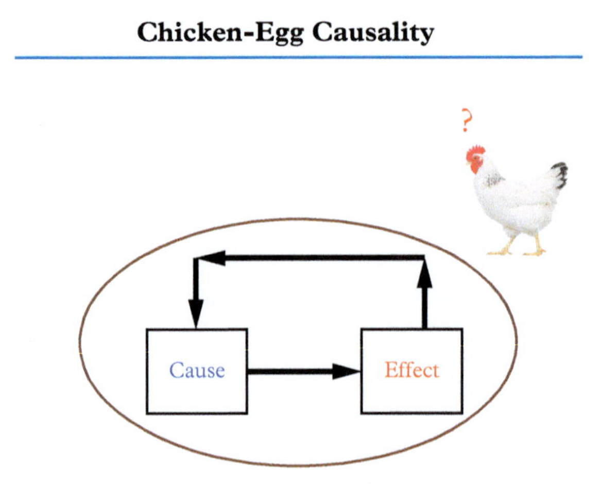

Figure 2.6 *Sketch of the chicken-egg loop expressing the principle of circular Causality. The cause produces an effect (causality), but the effect rebounds on the cause, leading to a circular structure whereby cause and effect can no longer be told apart. The chicken watches the egg and wonders . . ..*

systems often challenge this very fundamental principle. The point is as follows. Complex systems are typically *open*, i.e. they interact with their environment and affect it in return, often to their advantage. Briefly, they *learn, adapt, and react back*. If a change in the environment generates a response capable of modifying the environment itself, it is clear that the relation between cause and effect turns into a prototypical chicken-egg loop.

The classical sequential arrow cause-effect turning into a chicken-egg loop, also known as circular causality is another distinct hallmark of Complexity. The technical name is a feedback loop: actions are undertaken in the future on account of the lessons learned in the past. In many complex systems, chicken-egg causality is typically implemented via *error detection and correction*, a fundamental mechanism to turn learned lessons into convenient moves for the future.

Beyond a given level of Complexity, perfection, meaning literally *zero* errors, becomes a patent chimaera, just too costly, if at all possible. This is why complex systems must be *flexible*, i.e. actually develop tolerance to imperfections and errors. It is this very tolerance and flexibility that allows them to evolve by trial and error, always ready to learn and improvise unplanned solutions, a strategy often called *tinkering*. The ability to live with uncertainty and imperfection is another profound hallmark of Complexity and I would say of the human condition in general.

2.5 Hyperland and ultradimensions

Finally, our third item, hyperland. We are used to living in a three-dimensional space, in which objects have a length, a width, and height: one, two, and three, no more. Eventually, Albert Einstein (1879–1955) informed us that time should be treated as a factual fourth dimension, to be merged with space and form a four-dimensional entity called spacetime. Be as it may, if it is not three, it is four, just one more .... Many complex systems, however, inhabit hugely higher-dimensional spaces, and when I say 'hugely higher', I do not imply the ten or eleven dimensions invoked by string theory, but hundreds of them, or maybe thousands, or even much more. Biology is a goldmine of such ultradimensional hyperspaces.

Take a case we shall discuss in some length later in the book, protein folding, i.e. the fundamental process whereby the linear structure (primary) of the nascent protein turns into a nanometric stone-compact object, the so-called *native* form, the only one in which the protein can deliver its vital functions. Proteins are made of thousands of atoms, each of which comes with three coordinates (x, y, z) in space, plus additional coordinates, such as the angle of orientation between bonds, which we shall ignore for the sake of simplicity. Now, the precise state of the protein depends on -all- such coordinates, hence with, say, a thousand atoms, the protein lives literally in three-thousands dimensions: welcome to hyperland! And if instead of proteins we consider human deoxyribonucleic acid (DNA), with its three-billions base pairs, hyperland further expands into multibillion dimensions!

Given that three dimensions are enough to generate eminent Complexity, one may wonder how we could possibly compute anything useful at all in such a huge wasteland, the paradigm of which titles this book.

Well, the good news is that we can! The point cuts to the core of this book: the overwhelming majority of hyperland is a literal wasteland, in the sense that nothing particularly interesting happens there. The interesting variables, what we called relevant degrees of freedom in the previous chapter, are segregated within ultrasmall regions of ultralarge hyperland, the hide-and-seek attitude we alluded to in the previous chapter. Hence, the key question is how to find these crucial regions, the magic islands of the previous chapter. But before we do so, let's spell out a few practical figures which convey a better idea of how small small is here.

2.5.1 Lost in wasteland?

Consider a tiny piece of matter made of just a hundred water molecules, not much indeed considering that your average glass water contains billions of billions of them. Next, replace water molecules with simple spheres and place them in a *planar* box (think of a big egg carton) of side $10D$, D being the diameter of the sphere. Since the box is planar, we have $10 \times 10 = 100$ small boxes of side D, one per sphere. How many possible configurations can the hundred spheres take on? The calculation is simple: there are 100 boxes to choose from for the first, but only 99 for the second, 98 for the third and so on down line till the last molecule, numbered 100, is accommodated in the only box left free. The number of possible configurations is therefore $100 \times 99 \times 98 \ldots$, known as 100, hundred factorial in math language. This is a really huge number, and the math shows that it lands in the proximity of 10^{160}, way larger than the iconic *Googol*$=10^{100}$, whose resemblance with the high-tech giant of Silicon Valley is no coincidence.[9] This is the story: hyperland counts 300 dimensions, but the tiny corners where 'important' things happen, in a sense to be detailed in the following ideas, are usually mind-bogglingly small.

A few comments are in order. First, we notice that this monster number of configurations is, all the same, from a macroscopic perspective. For instance, the potential energy of the molecules depends only on their mutual distance, not on individual identity. The energy between two spheres, say 5 and 21, or 7 and 91 is just the same, as long as their mutual separation is the same. Energy-wise, particles are indistinguishable. This is a typical example of a multitude of microscopic states mapping into the very same macroscopic one, a crucial key to our chances of navigating across the wasteland of the Ocean of Complexity. The second crucial observation is that 100 molecules of diameter D in a planar box of side $10D$ have no freedom to move, they are frozen in. This is not what water molecules do in our

[9] For the initiated, there is a very handy approximation for $N!$, known as Stirling's formula, which reads $N! \sim \sqrt{2\pi N} \times (N/e)^N$, where $e = 2.718$ is the Neper number, the basis of natural logarithms.

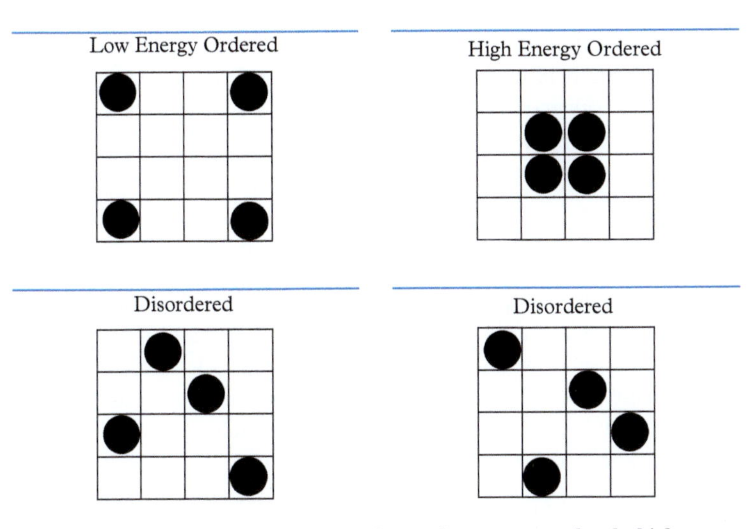

Figure 2.7 *Four spheres in sixteen boxes: low-energy ordered, high-energy ordered, and two intermediate-energy disordered ones. The energy between two spheres goes inversely with their distance.*

glass! At least, not unless we bring our glass into a super-fridge at absolute zero temperature, a mission impossible on planet Earth, as we shall see in Chapter 8. Given that at any finite temperature molecules must have freedom to move, let's grant them such freedom by making the box four times larger, so as to contain up to 400 spheres. Since we are still dealing with 100 spheres, though, it is clear that there are now many different ways to accommodate them in a box which can host up to four times as many, see Fig. 2.7. The amazing point is that many means really *very* many here! Indeed, the number of different configurations skyrockets from just 1 to $400!/(300! \times 100!) \sim 10^{100}$, i.e. 1 followed by 99 zeros before we meet 1![10]. Next, let us assume that the spheres repel each other with a force inversely proportional to some power of their mutual distance, which is indeed what real molecules in a glass of water do. Repulsion means increasing potential energy; as the pair of molecules approach each other, the energy slope they must climb to come close together. This means that not all the nearly 10^{100} configurations have the same energy: configurations with many pairs of close-by spheres (say closer than two diameters) possess more energy than those with less of such pairs.

[10] The number of configurations of P indistinguishable particles distributed in B boxes is $B!/(P! \times (B-P)!$. To convince yourself take the toy example $B = 3$ and $P = 2$, two particles in three boxes. The set of possible configurations is $\{012, 021, 102, 120, 201, 210\}$, where 0 denotes an empty box. This totals to $3! = 6$. Now make the particles indistinguishable, that is take $1 = 2$. The six previous configurations become is $\{011, 011, 101, 110, 101, 110\}$, of which only three are different. The math confirms, $3!/(2! \times 1!) = 3$, *QED*.

Here comes a crucial point: the probability of any given configuration $\mathcal{C}$ decreases exponentially with the ratio $E/k_B T$, where E is the energy, T is the temperature, and k_B a universal constant, named after the Austrian scientist Ludwig Boltzmann (1844–1906), a giant figure to whom we devote the full Chapter 9. In formulas:

$$p(\mathcal{C}) \sim e^{-E(\mathcal{C})/k_B T} \tag{2.5}$$

This formula holds the key to the exploration of hyperland. Indeed, the formula instructs us that configurations with energies much above the thermal one, $k_B T$, are exponentially rare, hence they contribute very little to the global, read thermodynamic, properties of the ensemble of hundred molecules. In other words, at *equilibrium*, such high-energy configurations are largely improbable, hence they contribute very little to the equilibrium thermodynamics of the system. The result is that thermodynamic properties, such as the average energy, temperature, pressure, and so on, are largely determined by just a very few *crucial* configurations which maximise their probability by attaining the minimum energy. It is precisely *these* very few configurations which occupy the tiny corners of hyperland! But how tiny are these precious corners? A realistic estimate for hundred spheres at the freezing point shows that basically only one out of 10^{260} contributes significantly to the thermodynamic properties [36]. This means a probability of 0 followed by 259 zeroes before we meet 1!

Here we go again with the explicit notation for 10^{-259}:

```
0.0000000000 0000000000 0000000000 0000000000 0000000000
  0000000000 0000000000 0000000000 0000000000 0000000000
  0000000000 0000000000 0000000000 0000000000 0000000000
  0000000000 0000000000 0000000000 0000000000 0000000000
  0000000000 0000000000 0000000000 0000000000 0000000000
  0000000001
```

And, if this is not mind-boggling enough, let us further observe that the typical number of molecules in a macroscopic piece of a matter is of the order of $Av \sim 6 \times 10^{23}$, i.e. six hundred thousand billions of billions, also known as *Avogadro number*, after the Italian chemist Count Amedeo Avogadro (1776–1856) (see Appendix 19.1 on Numbers). In explicit notation, 6 followed by 23 zeros:

```
6000 0000000000 000000000.
```

We shall come back to these matters in more detail in Chapter 13, but here we wish nonetheless to call the reader's attention to the fact that with an Avogadro numbers of molecules, the size of the 'relevant' sectors of hyperland take another mind-boggling leap, roughly speaking from 10^{100} to $10^{10^{23}}$, i.e. 1 followed by hundred thousand billions of billions of zeros! To the best of my knowledge, 10^{Av} carries no official name, so I decided to call it *avopex*, to indicate that it stands between the googol, i.e. 10^{100} and the googolpex, i.e. $10^{10^{100}}$.

The hierarchy is then avogadro, googol, avopex, and googolpex. These hypernumbers, especially the last two, escape not only our common sense, but they also defy our very imagination. Yet, these are the numbers that, *in principle*, rule hyperland. As a result, in order to make our way in hyperland, we must spot its hypertiny corners of the order of one part in an avopex. The obvious question being: how? This is one of the main triumphs of modern theoretical physics, most notably of its highly interdisciplinary branch known as statistical mechanics. Many ingenious techniques to identify the hotspots of hyperland, leaving wasteland alone, have been developed in the last decades, many of which have found a broader use in the context of complex system research, which is often confronted with monster-dimensional spaces aside from physics. These methods are most beautiful but pretty technical, hence beyond scope here. Yet, the general idea is worth exploring. Essentially, instead of groping in the dark, i.e. searching at random, hyperland can be explored under the guidance of so-called 'importance functions', sort of lamps or Global Positioning Systems (GPS), which inform us where to look. It is these guiding lights which permit us to turn around the otherwise unsailable wasteland, a main theme to which we shall return in later chapters, and most notably the one devoted to free energy Chapter 12, the main purpose of this book.

2.5.2 The hyperNile

We conclude with a geographical analogy which helps account for the monster numbers of hyperland from a geometrical perspective rather than a combinatorial one. The river Nile is a most precious and vital presence for the Egyptian nation and North Africa in general: a literal source of life, see Fig. 2.8. Hence, with a little stretch of imagination, we may parallel it to the 'important' part of the overall Nile basin. The question then is: what is the fraction of the basin area covered by the Nile? Or, perhaps more interestingly what is the probability of landing on the Nile if parachuted from an airplane flying in the Egyptian skyspace? The caption returns the answer: just about six parts in thousands. In scientific notation, 0.006, zero followed by just two zeros before we meet the first non-zero, that is 6. Now compare this with the billions of billions of zeros you must go through before meeting one in the hyperland of just your humble glass of water. Yet, this is less surprising than it seems. Now we will get to the point. The Nile leaves a two-dimensional geometry on Earth's surface. Now, instead of two dimensions take two hundred: the hyper-area covered by the hyperNile would be 0.006 raised to the power $200/2 = 100$. This gives 0.006^{100}, corresponding to about 10^{-220}, i.e. 0 followed by 219 zeroes, before meeting the first nonzero digit. This exponential suppression is an exquisite result of the exponential dependence of the volume on the number of dimensions, also known as *dimensional curse* in technical jargon. Finding the hyperNile, the Nile of hyperland is a heroic struggle against the dimensional curse. This is the formidable task that must be accomplished in order to keep our bodies going in the face of wasteland.

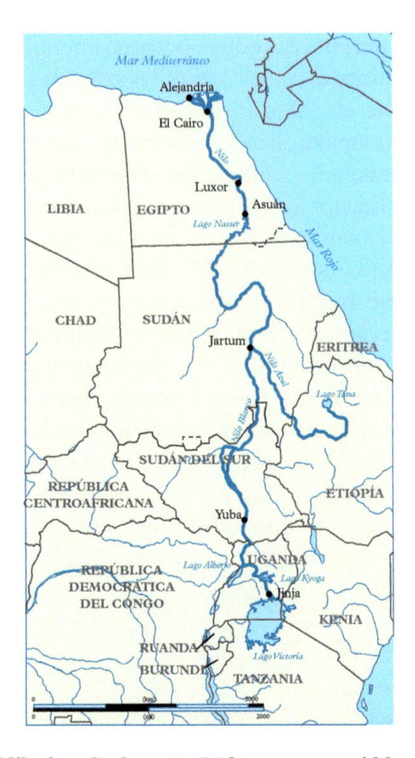

Figure 2.8 *The River Nile, length about 6,650 km, average width about 3 km, for an area of about 20,000 square km, against a basin area of about 3,400000 square km. If you land on the basin of the Nile from the sky, the chance of hitting water is 20,000/3,400000, i.e. about six parts in a thousand, 0.006. And if, instead of the Nile basin, you take the whole African continent, roughly 30 million km squared, the chance is a bit less than one part in a thousand, 0.0006.*
Source: reprinted from commons.wikimedia.org.

2.6 Summary

Nonlinearity, nonlocality, and ultradimensions are three powerful generators of Complexity, hence notorious barriers to human understanding. Unsurprisingly, they characterize a vast family of complex systems. Nonlinear systems do not respond in proportion to the solicitations they receive; small changes can lead to big effects and vice versa. Obviously, this hinders their predictability, and it also undermines the most powerful superposition principle, preventing us from piecing it back together, once we have dissected a given system in the search for ultimate simplicity. But, nonlinearity is also the source of immense freedom; a nonlinear system can function in many different states and actually jump from one to another without much warning. No surprise then that such immense freedom occasionally turns to our advantage as well, the constructive power of nonlinearity in biology being a sterling example in point, although certainly not the only one.

Nonlocality prevents different regions of spacetime being handled independently, all-to-all spatial connection. It impairs another basic principle of deductionism: divide and conquer. The effects of non-locality in time (memory) are possibly even more far-reaching, as they imply a strong incidence of history on the evolution of complex systems, once again highly visible in the biological realm. And, just as in the case of nonlinearity, a tremendous source of opportunities as well, the world-wide web being just one popular example in point. Finally, ultradimensions. They are totally foreign to our visual mind and brain: one dimension is easy, two are still fairly comfortable, but three is already not quite a walk in the park. Beyond three, visualization fades quickly away, leaving intuition, imagination, and math abstraction as the main (only?) guides. The great gift, as discussed in Chapter 1, is that these guides do work! And they do because hyperland has been equipped with beacons which guide the search towards the hyper-tiny corners where the 'important things' happen. Stay with us, the trip goes on!

3

Competition and Cooperation

Science is organized knowledge, wisdom is organized life.
(Immanuel Kant)

3.1 The two big C's

As we shall see time and again in this book, Complexity thrives on the coexistence between the two big C's: Competition and Cooperation. And, usually, the tighter competition and cooperation, the higher complexity is, for there is hardly any complexity without tension. The dynamic coexistence resulting from the ever-shifting battle between the two big C's, is a hallmark of Complexity in general, and it is particularly apparent in fields where survivals takes central stage, like biology, economy, and finance. In this book, we shall mostly be concerned with the competition and cooperation between two paramount thermodynamic quantities, energy and entropy, which make the object of a separate forthcoming chapter. In this chapter, we discuss three basic mechanisms which drive competition and cooperation in complex systems, namely:

1. *Order&Disorder*
2. *Equilibrium&Nonequilibrium*
3. *Smoothness&Roughness*

3.2 Order&Disorder

Two dangers constantly threaten the world: order and disorder.
(P. Valery)

The distinction between Order and Disorder is innate in our brains: a beautifully arrayed table before dinner and the same table after, surely conveys the point . . . It is a familiar fact of life that we are generally more at ease with Order than Disorder, as the former comes with a comfortable sense of being in control, while the latter conveys a disturbing and often alarming lack thereof. In some sense, it is therefore natural and intuitive to equate Order with Simplicity and predictability and Disorder with Complexity and uncertainty, (see Fig. 3.1).

To wax lyrical, Cosmos versus chaos.

Sailing the Ocean of Complexity. Sauro Succi, Oxford University Press.
 DOI: 10.1093/oso/9780192897893.003.0003

Figure 3.1 *An Ordered spiral (right) turning into a Disordered tangle (left). Two fairly different arrangements of the same material line. According to Paul Valery, they both threaten the world, but as a matter of fact, the combination of the two makes the world go around.*
Source: redrawn from 123rf.com.

Intuitive and natural as they might seem, the twin 'equations' do not do justice to the real essence of Complexity, which is rather a subtle blend and inextricable intertwining of the two. As we shall see, most complex systems thrive on the subtle and often slippery interface between Order and Disorder, in fact two faces of the same coin. In fact, Complexity attains its zenith when the two come to a close tie, any hands-down win of either resulting in a loss of Complexity, whichever the winning side. Interestingly, these cosmic but rather fuzzy notions map out onto very concrete scientific quantities, which hold centre stage within the realm of two paramount areas of physics, thermodynamics and statistical physics.

These central characters are *energy* and *entropy*, the two cosmic primadonnas of the natural world [6]. Energy is the more popular of the two and it enjoys a decidedly better reputation: apart from nuclear, perhaps, most forms of energy come with a positive press image: energy is the grand-constructor, the fuel of life, '*res viva*' (the living thing) for ancient Romans, Brahama for the Hindu. More down to earth, energy is the 'sine qua non' (the Latin for '*absolutely necessary*') for getting anything done. Construction generally implies some form of Order and, energy feeds the growth of orderly and organized structures, which deliver purposeful functions. In a nutshell, energy makes the world go around: no energy, no party!

Entropy, on the other hand, is almost a byword for the opposite: the destructive and degrading action of chaos, to continue with Hindu analogies.[11]In a more dramatic vein, energy is the sister of life and entropy is the sister of death.[12] Happily enough, the picture is neither so bleak nor so simply black or white; the

[11] For the sake of completeness, Hinduism caters for a third supreme deity, Vishnu, the Preserver, whose role is to protect humans and *restore Order* against the destructive power of Shiva. But again, the story is more complex than this, as Shiva can also destroy negative powers . . . . As we shall see, in thermodynamics the competition/cooperation between Order and Disorder is controlled by a well-defined physical quantity, the free energy, defined as energy, minus temperature, times entropy (see the relation (3.1) in the main text). From this relation, one may argue that temperature plays the role of tuning the balance between Order and Disorder.

[12] Perhaps, the most captivating metaphor is Shakespeare's famous 'walking shadow'.

two prima donnas compete but also cooperate in a sort of fascinating and subtle duet/duel, which lies at the heart of most crucial aspects of complex systems, biological ones in the first place. As with aesthetic pleasure in the arts, interesting things tend to happen at the fuzzy border between dull regularity (Order) and mind-shattering irregularity (Disorder). In the words of George Bernard Shaw, (1856–1950) 'Consistency is the enemy of enterprise, just as symmetry is the enemy of art'.

Coming back to earth, it turns out that the cooperation/competition between the two can be expressed through a simple mathematical combination of the two, that reads as follows:

$$F = E - TS \tag{3.1}$$

where E is the energy, S is the entropy, and T is the temperature. The quantity F at the left side is called *Free-Energy* and makes the central character of this book, as we shall detail in the next chapters.

3.3 Symmetry and broken symmetry

Order is the same as Disorder, with less fantasy
(sign in a roman cafè)

At a first glance, it would seem intuitive that Order should match with symmetry and Disorder or with lack thereof. A closer inspection reveals that the opposite is often true. Consider the process of solidification, i.e. a liquid turning into a solid upon being cooled below a given critical temperature. Atoms in the liquid can take virtually any position in space, while in the solid they are frozen around specific locations dictated by the crystallographic symmetry of the solid. Such symmetries come in many families, depending on the specific substance, but the essential point is that while the liquid can be rotated by any angle and still look the same (rotational invariance, also known as isotropy), the solid looks the same only upon specific rotations that leave its shape invariant. In two dimensions, for simplicity, a square is invariant upon rotations of 90 degrees and multiples thereof, while for hexagons the angle is 60 degrees (see Fig. 3.2).

That's how a disordered liquid exhibits *more* symmetry than the ordered solid!

The loss of symmetry in the transition from a disordered to an ordered phase is measured by a so-called *Order Parameter*, a physical quantity which is zero by definition in the disordered phase and acquires a non-zero value in the ordered one. For the case of solidification, the Order parameter is typically defined in terms of density difference between solid and liquid (usually a positive quantity since solids are denser than liquids, a property -not- shared by water . . .). But, the notion of Order Parameter runs far beyond the solidification process, to impact a broad array of phenomena in physics, engineering, and the life sciences. More precisely, the notion of *spontaneous symmetry breaking*, the loss of symmetry due to the change of an external parameter, say the temperature or pressure, is possibly

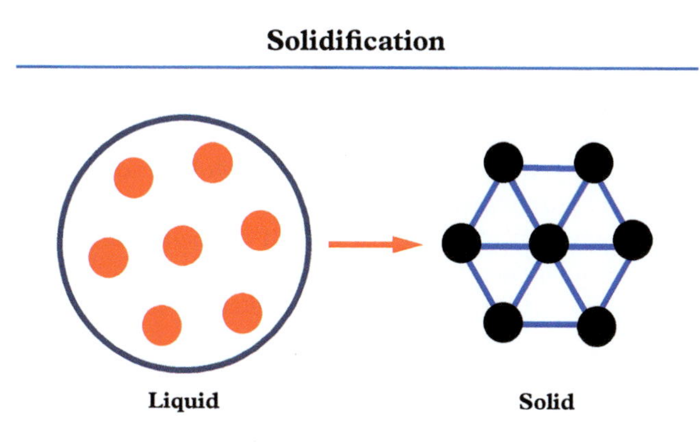

Figure 3.2 *Cartoon of a two-dimensional liquid turning into a solid with hexagonal symmetry upon cooling down below its solidification temperature. The liquid (left) has the symmetry of a circle; hence it looks the same upon rotations of any angle. The solid (right) has the symmetry of a hexagon, hence it looks the same only upon rotating it by multiples of 60 degrees. The disordered liquid has more symmetry than the solid.*

one of the most profound and impactful concepts of modern physics. In fact, in the words of P. W. Anderson (1924–2020), one of towering figures of modern physics and theory of Complexity, 'the theory of *broken symmetry* is the most rigorously based, physics-oriented description of the growth of complexity out of simplicity' [4].

3.4 Equilibrium&non-equilibrium

At equilibrium things work, out of equilibrium they discover new worlds.

As we have seen in the previous sections, Complexity has many, often contradictory, faces, but a key one is certainly the way in which complex systems change in time, i.e. their dynamics. Simple systems are generally smooth, i.e. they show fewer abrupt changes than complex ones, hence they are more predictable, almost a definition of Simplicity.

3.4.1 The ever-shifting battle

But what is the cause of such changes, and why do some systems change smoothly while others do not? In pretty general terms, change in time develops through the 'ever-shifting battle' (competition) between the various mechanism in action in a given system, some of which promote growth (gain), some others inhibiting it (loss). So, change is the outcome of the battle between promoters and inhibitors, gain and loss. Symbolically:

$$\textit{Change Rate} = \textit{Gain} - \textit{Loss} \tag{3.2}$$

In the following, I use a bare 'financial' analogy, but you may replace 'wealth' with whatever property you value in life, not necessarily a material one. Here we go with the analogy. If you earn 2,000 Euro/month and spend 1,500, your wealth grows at a rate of 500 Euro/month. If you swap these figures, earn 1,500 and spend 2,000, your wealth changes at the same rate, but on the negative slope (which is less pleasing . . .). Equilibrium is defined as that special condition whereby gain and loss come to a perfect balance, so that the system does not change anymore. Symbolically again:

$$\textit{Gain} = \textit{Loss}.$$

Within this example, you earn 2,000 and you spend 2,000, but any other number would do, as long as they are the same . . . However, life is generally way more complicated than that, gain and loss may depend on a variety of factors and in fact they may even fail altogether to reach any equilibrium at all. But, let's stay with stable ones: even leaving aside the many vagaries of daily life (fines, unexpected bills, or, more cheerfully, unexpected winnings), gain and loss may actually depend simply on your exact state, and particularly on the value of your current wealth. When your wealth increases, you might be inclined to be a bit more spendthrift than when you're on a shoestring budget, which means that when your funds are low, you make them higher by spending less. Likewise, if by some lucky circumstance, maybe an unexpected gift, your wealth goes up, you're likely to let your spending go a bit higher. Suppose you have been saving 500 Euros per month over five years, accumulating 30,000 Euro in the process. You may then decide that, at least for a while, you need no more savings, and you start spending exactly what you earn, namely 2,000 Euro instead of 1,500, your lifestyle improves, and your wealth stays unchanged. Gain equals loss, future wealth equals past wealth, wealth-wise you are at equilibrium: no change, time is erased from the picture. I see your reaction coming: life cannot be that simple, there are always unexpected events which prevent you from keeping your spending *exactly* in balance with your income, to the very last cent. The point is absolutely well taken, the scientific name for such 'unexpected events' being *perturbations*. And indeed, the way that equilibria react to perturbations opens up a fundamental distinction between them. Those which remain basically unaffected are called *stable*, while those that succumb to perturbations are called *unstable* (see Fig. 3.3). Let us illustrate the point by sticking again to our financial analogy.

3.4.2 Stable and unstable equilibria

You are at financial equilibrium when a lucky strike occurs, you win a lottery, say a substantial extra 20,000, raising your wealth to 50,000; how do you react that this lucky strike? If you're happy with your current life, you may opt for a really nice gift for your beloved, or perhaps a donation, and then go back to your previous

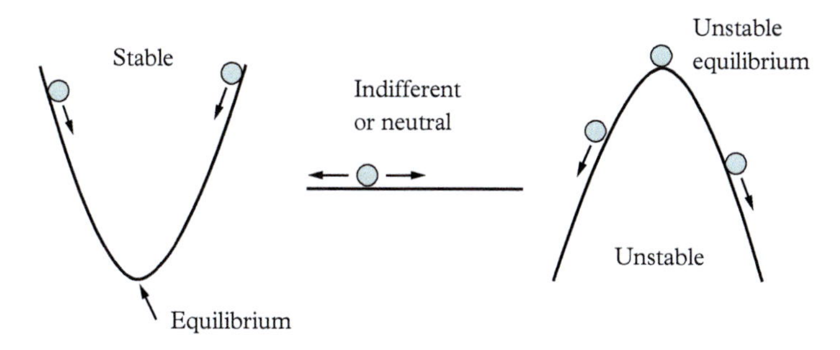

Figure 3.3 *A stable equilibria (valley) on the left, an unstable one (tip) on the right, with a neutral equilibrium (mesa) in between.*
Source: redrawn from quora.com.

lifestyle. The extra 20k was a pretty nice 'perturbation' which did not change your financial equilibrium, leaving no trace on your wealth once the extra 20k was spent. Manifestly, you were in stable equilibrium towards this perturbation. Now, suppose that you decide instead that 20k is enough for you to make a new investment (I know it's naive, but Steve Jobs didn't start with much more than that, I guess his first loan was around 50k dollars). Suppose further, that your investment takes off, and if it really flies high, you eventually make it to the rich. Your previous lifestyle was an unstable equilibrium towards this perturbation, the 20k win was large enough to take you away from your previous wealth. At this point, if asked which is better, stable or unstable equilibria, I guess every reader would go for the latter. Well, not so fast. Indeed, good for you that the perturbation took you to a better situation, but, beware, not all outcomes are this rosy; if your investment turns into failure, you may well end up loosing not only the extra 20k but even your initial 30 k. This was the same equilibrium as before, but you simply took the wrong path out of it, the result being financial catastrophe.

Same question again: which one is better, stable or unstable? The answer is that you can't tell, for the question is ill-posed. The fact is that stable equilibria allow the system to function on a long-term schedule, which is essential to deliver useful functions. Quitting a stable equilibrium comes with inevitable chances and uncertainty: you may end up in far better positions, or you may just ruin yourself. Yet, one thing you know for sure: if you never quit, you're guaranteed to miss any chance to discover anything better than what you have. As I like it to put it: *at equilibrium things work, out of equilibrium they discover new worlds.*

3.4.3 Conditional equilibrium

The inquisitive reader may note that the picture I portrayed may change drastically if instead of 20 k, the lottery win was to be just a modest 2 k. With only 2 k extra, you would hardly go for a new investment! The point is again well taken, in that I should have spoken of 'marginally' stable and unstable equilibria, meaning by this that the stability/instability depends on the intensity of the perturbation. Strictly speaking, *absolutely stable* equilibria are those which resist any perturbation, no matter how intense, while *absolutely unstable* ones are those which evaporate away, no matter how small the perturbation (the infamous butterfly of chaos theory, to which we shall return in Chapter 5). For a useful visual metaphor, think of an absolutely stable ball at the bottom of a valley and the same ball sitting on the top of a hill, respectively (see Fig. 3.3). An important and general feature of stable equilibria is that under small perturbations the system experiences restoring forces which are *linearly* proportional to the intensity of the perturbation, but opposite in sign, whence their restoring effect. As an example, think of a row of beads kept together by springs all at the same distance from each other. If, by chance, one such bead moves slightly towards its left neighbour, the pull from the right neighbour immediately exceeds the one from the left, thus pulling the bead back to its equilibrium place. If the displacement is small, in some suitable sense, the restoring force is linearly proportional to the displacement, and everything is fine and stable.

However, if the displacement is huge, the linear proportionality between the displacement and the restoring force is lost. The system enters a nonlinear regime, where many other options open up, including the possibility that the restoring force changes its sign, so that, instead of attracting the bead back to its place, it just pushes away from it. The system has gone unstable. Although oversimplified, this simple analogy tells a realistic tale: near equilibrium, the relation between cause (the displacement) and the effect (the restoring force) is typically linear. *Far from equilibrium, however, nonlinearity takes stage.* This is a very general organizing principle across virtually any complex system, be it physics, biology, finance, or society [50].

3.4.4 One, none, or many equilibria?

The next crucial point is to realize is that in a complex system there are usually many equilibria, oftentimes astronomically many, some of which are stable and some not. They are typically arranged in the form of dynamic networks, whose nodes are the equilibria themselves, while the links represent the communication channels through which the system jumps from one equilibrium to another. Stable equilibria are attractive, unstable ones are repulsive. In a landscape analogy that we shall often meet along this book, stable equilibria are valleys and unstable ones are the peaks. The system tends to sit on the valleys, where it remains until a strong enough perturbation (a not-so-little butterfly) kicks it off, towards a different stable equilibrium. In order to do so, however, a barrier must be passed, which, coming back to the previous financial analogy, corresponds to the amount of the lottery win: on a win below the threshold you stay, above the threshold you go for another destination. The routes connecting the various stable equilibria can be overly complicated, and the time it takes to complete the transition between them may vary wildly across the network, as we shall detail in Chapter 12 devoted to free energy.

The main lesson here is that, almost by definition, a *complex system can function in many states*, typically arranged on a network topology, and *can jump from one to another without much warning*. This is another distinctive feature of Complexity. Simple systems don't have such freedom, in many cases they may exhibit just a single stable equilibrium, and not much can happen. Stable equilibria are key, because they are the states where the system enjoys enough 'peace' to deliver its functions on a long-term basis.[13] Unstable equilibria, on the other hand, are ephemeral places, where life is maybe bright but short. Yet, they are crucial, because they encode the freedom to move from one stable equilibrium to another. It is only through them that the system can take 'jumps into the unknown' and cross the portals of new discoveries.

The main take-home is as follows: at equilibrium the system has enough time to accomplish useful functions, subject to the constraints which stem from the defining condition gain = loss. This resilience to change is a curse and a blessing at the same time, for it secures stability but at the price of frustrating inventiveness. Out of equilibrium, such constraints are relaxed, and the system enjoys more freedom, hence more room for Complexity, be it for good or for ill.

That sounds pretty much like real life, doesn't it?

3.5 Smoothness/roughness

We just asserted that complex systems are hard to predict, which means that it is hard to compute their future state based on the past. In brief, they are hard to predict. Computing is indeed a central aspect of the theory of Complexity and a key property of complex systems altogether. In the following, we shall focus on the former, i.e. the mathematical tools which help us in the task of computing(predicting) the behaviour of complex systems.

Much of the most powerful mathematics to this day relies upon a formidable tool called *infinitesimal calculus*, tracing back to Isaac Newton (1642–1727) and Gottfried Leibniz (1646–1716), who were less than gracious about each other's role in its paternity. This rests upon a very strong assumption; namely that space and time are infinitely divisible. Choose a point in space, or an instant time, you can always find another point in space and another instant time arbitrarily close to the initial ones, no matter how small the stipulated distance. This assumption is reflected in Leibniz's sentence '*Natura non facit saltus*', nature does not jump. In those days this was indeed an undisputed tenet. No more: by now, we know that not only matter, but quite likely also space and time, are 'atomic', i.e. they cannot be divided indefinitely, there comes a point where divisibility is no longer possible. Matter, space, and time are discrete. If you try to make the distance zero, i.e. take the continuum limit in technical parlance, many complex systems react very badly, returning infinity, a result that physics is hardly willing to take for

[13] Sometimes 'peace' is a bit too much, in that it lasts indefinitely, a condition which, at least in biology, can only occur when the system is dead . . ..

an answer. This takes us to intriguing paradoxes, the most popular being perhaps the one due to the Greek philosopher of Elea, fifth century BC.

3.5.1 Zeno's paradox: Achilles and the tortoise

Here goes Zeno: the grand-hero of the time, Achilles, stands behind the tortoise, but being much faster ('piè veloce' in Italian, whose literal translation is 'fast foot'!) he expects to take over easily, and so does anybody else. But here comes Zeno on his way, arguing that by the time Achilles moves up the position that the tortoise was at a given time t, the tortoise, albeit much more slowly, has nonetheless moved a small bit ahead, so that Achilles is still left precisely *that* bit behind (see Fig. 3.4).While taken aback, Achilles promptly reaches to the new position, only to realize that meanwhile the (naughty) tortoise has moved another tiny bit, thus leaving Achilles still behind. Much to Achilles's despair, the story repeats *ad libitum*, virtually an infinite number of times, the upshot being that the fast-footed Achilles will never take over the slow-footed tortoise! So argues Zeno. We all 'know' this can't be true, but ... what's wrong with it? (Incidentally, let's concede that, at this point, we are all rooting for Achilles's side ...).

The solution is plain yet subtle, that's Zeno's specialty after all! The tiny bit which allows the tortoise to stay ahead of Achilles becomes increasingly small at each stage of the paradox and literally zero upon an infinite number of iterations. And, as Achilles keeps trying, the tiny bit gets tinier, down to length zero. The subtle, and yet very practical, Achilles's delay decreases so fast that even an infinite sequence of such small bits sums up to a finite time lapse! Which means that after a -finite- amount of time Achilles finally manages to catch the naughty tortoise!

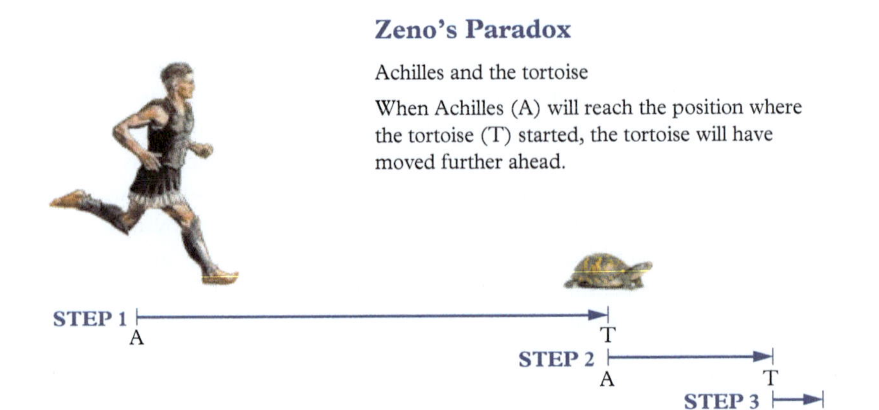

Figure 3.4 *Illustration of the Zeno paradox.*
Source: redrawn from medium.com.

As a result, just wait longer than this finite time lapse takes Achilles ahead of the tortoise, and all falls back in place; common sense is restored.

Zeno's trap lies in the over-contrived schedule of Achilles's moves, and in this respect, it is readily disposed of by practical common sense: just give Achilles enough time *per move* and he will overtake the tortoise, no doubt. But, if you fall in the trap of the over-contrived schedule, then the way out is not obvious; it takes the non-trivial realization that the *sum of an infinite number of tiny bits is not necessarily infinite*, it can converge to a finite number, provided the bits decrease fast enough. Thus, even in the case of an infinite number of infinitely small steps, Achilles would reach the tortoise in a finite amount of time (assuming no nervous breakdown in the meantime!). Such smooth phenomena form the joy and the triumph of infinitesimal calculus, an immensely powerful predictive tool to this day. Note that smooth here means that space and time go together and by making the time lapse zero, the corresponding space interval is also forced to zero. But how about systems for which nature '*facit saltus*', i.e. she does take jumps? A minute's thought reveals that this unleashes the most infamous taboo of physicists: infinity. Why? Because if you move a non-zero distance in space in a zero time interval, the ratio of the two (what we call 'velocity') is infinity. Scientists dub such kind of jumpy behaviour 'singular', as opposed to 'regular', which indicates 'regular' phenomena abiding by the Leibniz 'no-jump' paradigm.

With this in mind, it should come as a little surprise that complex sytems are often not smooth, but rough instead, and in some instances, they may even take jumps. And I am likewise confident that at this point, the reader sees clearly why rough and jumpy systems are harder to predict than smooth ones. A word of precision is warranted: rough systems are less regular than smooth ones but more regular than jumpy ones. Take the curves represented in Fig. 3.5. The values of the smooth curve change gently from place to place in the sense that not only its value but also its slope changes gently from place to place. The rough curve still doesn't take jumps, but its slope does. So the chain of ascending irregularity is: smooth, rough, jumpy.

3.5.2 A new beauty in town: Fractals

Being abhorred by infinitesimal calculus, roughness has been swept under the carpet for about four centuries. It took the genius of the Polish-born French-American mathematician Benoit Mandelbrot (1924–2010) to put it on the map of mainstream science, through the development of *fractal geometry*. This subject is worth many books on its own, hence there is no point of delving further here. We simply quote from Mandelbrot's book *The Fractal Geometry of Nature* [71]:

> Why is geometry often described as 'cold' and 'dry'? One reason lies in its inability to describe the shape of a cloud, a mountain, a coastline, or a tree. Clouds are not

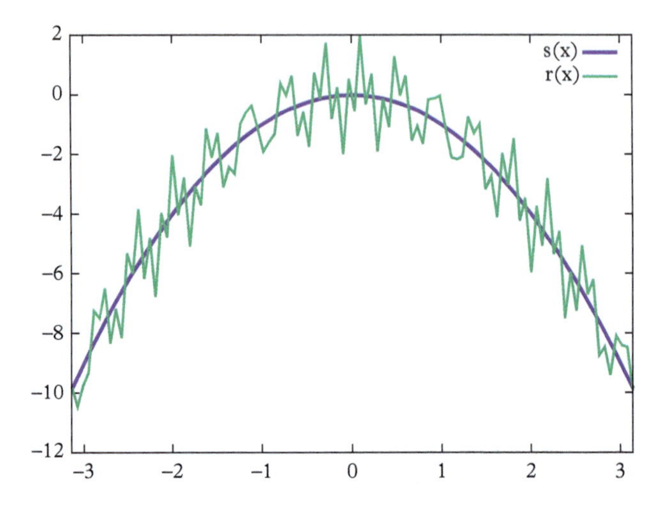

Figure 3.5 *Graph of a smooth function* ($s(x)$) *and the same function with some roughness on top* ($r(x)$). *To be noted is that both curves are continuous, i.e. jump free. However, while the slope of the smooth curve changes gently from place to place, the same is not true for the rough one. The jumpy quantity is not the curve itself but its slope. Rough curves are less regular than smooth ones, but more regular than jumpy ones.*

> spheres, mountains are not cones, coast-lines are not circles, and bark is not smooth, nor does lightning travel on a straight line. More generally, I claim that many patterns of Nature are so irregular and fragmented, that, compared with Euclid—a term used in this work to denote all of standard geometry—Nature exhibits not simply a higher degree but an altogether different level of complexity.

Fractals are objects which never become smooth, no matter how close you look at them. More precisely, upon looking at them closer and closer, they reveal a sequence of self-repeating patterns, a property referred to as *scale invariance*. Snowflakes make a good example in point: a very organized form of roughness. Fractals have made it a long way in many disciplines beyond geometry, from physics to material science, all the way to biology and finance. Whether they contributed new fundamental discoveries in any of these fields remains open to debate, but it is hard to question that they did open up an entirely new and fresh eye towards the geometrical Complexity of nature. No longer a disturbing imperfection, a departure from the idealized beauty of Euclidean geometry, but a genuine source of a new beauty itself. A beauty which reflects the *vitality* of complex systems, i.e. their ability to express change even at the shortest scales in space and time. A vitality hardly shared by smooth systems.

3.5.3 Computer simulation

Roughness has a profound impact on the mathematical description of complex systems: for one, it sets a strong bias in favour of discrete versus infinitesimal calculus. In discrete calculus, infinities are tamed from scratch because, by rule, zero plays no part in the game, hence ruling out its dreaded inverse as well, infinity. Not that this would not make life any easier; to the contrary, this leads to a proliferation of calculations which would be automatically zeroed in the continuum. The key point, though, is that while these calculations are a nightmare for our calculational brain, they turn out to be bread and butter for digital computers, which famously win hands down over the human brain at the so called brute-force computing.[14] This has spawned an entirely new branch of modern science, known as computer simulation, incidentally this author's main occupation, which has literally transformed every other discipline.

The power of infinitesimal calculus is placed in jeopardy by rough and jumpy systems and a different math is needed, in particular, one that endorses discreteness at the outset and refrains from assuming the continuum limit. A typical example in point are the cellular automata (CA) schemes to be discussed in Chapter 7. In a nutshell, these are discrete dynamical systems which live at the discrete nodes of a regular lattice, and evolve in time according to simple rules; the same at every lattice site and the same at every instant in time. Even though such rules look innocently simple, their repeated application in space and time proves capable of spawning spatio-temporal patterns of breathtaking Complexity (see Fig. 3.6). Turbulent flows, also to be covered in Chapter 7, provide one of the most spectacular examples in point. Another major example are evolutionary games [79]. Note that we have spoken of rules, precisely to indicate that at no point does the CA game assume the continuum limit. Some elegance may be lost along the way, but the predictive power of discrete calculus is incommensurably broader, once digital computers are endorsed. Indeed, by their very nature, (electronic) computers deal with radically discrete (digital) information, 0 or 1, nothing in between. This explains why computer simulation is playing such a unique role in unveiling the mysteries of Complexity, by no means just a number-crunching exercise, but a strategy to explore nature with the aid of computers as genuine tools of discovery.

To quote Frank Wilczek, Physics Nobel 2004

[14] Just to give a sense, most powerful current-day computers are nearing the exascale performance, namely they can multiply a billion of billions fourteen digit numbers every second! Compare with a standard human, who, in the same time, can barely multiply a three-digit pair .... Try yourself on, say, 1.356×3.543 and if you can do it in less than a second (no hand-pocket calculator), please, let me know! For the record, as of 2020, the fastest human calculator, allegedly Bhanu Prakash, a 20-years-old !ndian boy, can multiply $86946385 \times 73 = 63470861269$, in just 26 seconds. Mind-boggling as this is, it's nowhere any near to electronic computers.

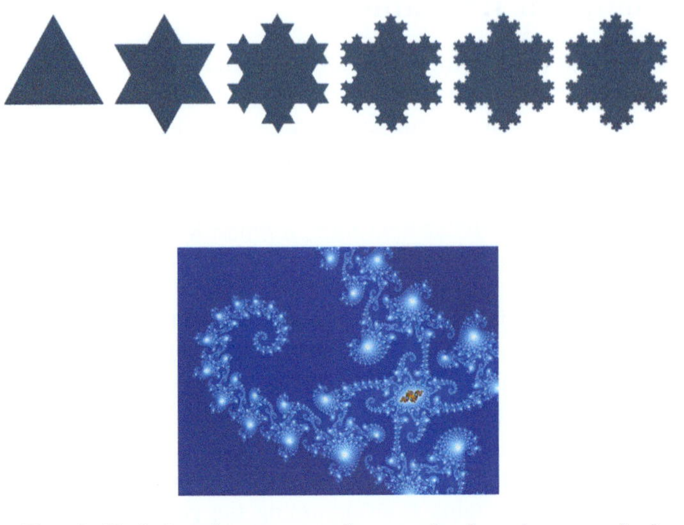

Figure 3.6 *Top: the Koch fractal curve; start from a triangle and generate further triangles, one on each of the three sides. The procedure repeats indefinitely, thus giving rise to a curve of increasing roughness at smaller and smaller scales. Bottom: a scenic fractal.*
Source: reprinted from commons.wikimedia.org.

> '... Concepts and equations that computers can run will be powerfully leveraged, concepts and equations that cannot be turned into algorithms will be regarded as deficient' [123].

Even though Wilczek would not be described as a computational physicist, it's hard to think of a more apt and cogent description of what computer simulation really is.

3.6 Summary

Summarizing, complex systems thrive on the subtle coexistence between Order and Disorder, Order keeps the organization going, Disorder may disturb organization, but sometimes it does it in an inventive way that triggers innovation. They also host the coexistence of equilibrium, the condition ensuring long-term operational stability, and non-equilibrium, the regime where both the destructive and constructive powers of nonlinearity are fully unleashed. Finally, complex systems are best explored via computer simulation, which is better suited to handle their often highly irregular and non-smooth behaviour in space and time. It is no exaggeration to state that our understanding of Complexity would be nowhere close to its present state were it not for the outstanding advances of computer simulation in the last few decades. A statement which applies to the whole body of science in general.

3.7 Appendix 3.1: Summing series

Let us consider the series obtained by summing the number 1 n times. For instance, with $n = 10$

$$1 + 1 + 1 + 1 + 1 + 1 + 1 + 1 + 1 + 1 \tag{3.3}$$

which is obviously equal to 10. By summing n times, with n an arbitrary integer, the sum is obviously n. Hence, by send n to infinity, the series returns infinity. Next consider the following series

$$1 + 1/2 + 1/3 + \ldots 1/n \tag{3.4}$$

In other words, the n-th term is $1/n$, meaning that the terms in the series become increasingly small, in inverse proportion to their place in the series. By sending n to infinity, the terms become increasingly close to zero, so that the series grows by increasingly smaller amounts. Does the infinite series still return infinity? Unlike the previous case, the answer is not immediate. The explicit calculation shows that, yes, the result is still infinity, although a much 'weaker' infinite than in the previous case. Finally, consider the following series

$$1 + 1/4 + 1/9 + \ldots 1/n^2 \tag{3.5}$$

That is, the summands decrease with the square of their position, n, which is a much faster decrease than in the previous case. Hence, this series is certainly smaller than the one in (3.4). Is it still infinite, though? The answer this time is no: this infinite summation converges to a finite number, and more precisely, the exact result is $\pi^2/6 \sim 1.5$, where $\pi \sim 3.1415 \ldots$ is the famous 'Greek pi'. If you don't see where this strange result comes from, don't feel bad, as it took the genius of one of the greatest mathematicians ever, the Swiss, Leonhard Euler (1707–1783), to figure it out. The key point is plain: the sum of an infinite series of terms needs not to return infinity: it all depends on the decrease rate of the subsequent terms. This is Zeno's paradox unveiled.

4

Nonlinearity, the Mother of Complexity

Using a term like nonlinear science is like referring to the bulk of zoology as the study of non-elephant animals.

(Stan Ulam)

4.1 *Anna Karenina*

In the previous chapters, we have discussed (some of) the organizational principles which subtend the behaviour of complex systems. All of them are key to Complexity, but if I had to cherry pick just one, I would single out nonlinearity, to which we devote the next three chapters. Tolstoy's famous opening line of *Anna Karenina*, 'All happy families are alike, each unhappy family is unhappy in its own way', fits well with nonlinearity: there is only one way of being linear and literally uncountable ways of being not! So, let us begin by discussing first what nonlinearity is not, i.e. linearland, the country of 'happy families' where two plus two is always four.[15]

4.2 Linear math

In ordinary life, when we wish to point out a plain consequence of some facts which leave little to be disputed (the nightmare of lawyers), we often say 'two plus two makes four'. At least, this is how we say it in Italy. Indeed, in ordinary calculus, two plus two makes four $2 + 2 = 4$, i.e. the first 2 sums up to the second 2, to form their *linear superposition* $2 + 2 = 4$. Linear superposition means that by the time you add the second '2', the first '2' stays unchanged. The two numbers to be added do not 'affect' each other, they are indifferent to each other, they do not interact.

In general mathematical terms

[15] It should be clearly understood that, at least in ordinary math, two plus two definitely *is* always four, so the expression 'two plus two does not make four', is a pure metaphor. More in the Appendix 4.2.

Sailing the Ocean of Complexity. Sauro Succi, Oxford University Press.
 DOI: 10.1093/oso/9780192897893.003.0004

$$s = a + b \tag{4.1}$$

where a and b are any two numbers and s is their sum.

Symbols are readily substantiated with concrete numbers. If you spend $a = 1$ Euros for your coffee and $b = 1.50$ Euros on an accompanying croissant, your total spending is $s = 1 + 1.50 = 2.50$ Euros (if you sit in a classy bar, you'd better prepare yourself for steeper prices). Another prototypical linear operation is the multiplication of a number, say a, by a constant, say c, to give the product p. In mathematical terms

$$p = c \times a \tag{4.2}$$

This relation is geometrically represented by a straight line, whose slope gives the proportionality ratio between the two, the constant c in our case. Here again, if you double a, since c is constant, you double p as well.

To keep going with our breakfast analogy, if you buy $c = 2$ two coffees, you expect to pay $p = 2 \times 1$, 2 Euros, and if the bartender asks you for more, you have a solid ground to argue, wouldn't you? Well, that is linearity in action, you have just applied equation (4.2)! The two mathematical pillars of linearity given previously nicely combine with each other. Continuing on the breakfast analogy, assume you are with a good friend of yours, and you offer them coffee plus croissant. Your spending is readily calculated: two coffees, one Euro each, makes two Euros, and two croissants, 1.50 Euro each, makes 3 Euros, for a total of $2 \times 1 + 2 \times 1.50 = 2 + 3 = 5$ Euros, no rocket science needed here. What you did in the process of computing your bill was to multiply each unitary cost, 1 Euro for the coffee and 1.50 for the croissant, respectively, by the corresponding number of consumers, and sum them up. Believe it or not, in this very act of summing them up, you have applied a very powerful rule, gloriously known as the superposition principle (see Chapter 2): the total cost is the plain sum of the partial ones. This is sometimes expressed by the fuzzy but evocative metaphor *the whole is the sum of its parts*, whose precise meaning is exactly the procedure you used to compute the bill for your breakfast. Multiply and sum according to the two basic linear relations (4.1) and (4.2). So much for the math.

At this point, I bet that my reader is fairly unimpressed, since what we did so far appears to be just plain common sense, backed by elementary algebra. And if this sounds very plain to you, it's because it is plain indeed. But please, now pause for a moment and ponder the assumptions beneath this elementary calculus. A linear system is one which returns a response (effect) in direct proportion to the stimulus (cause) it receives: double the stimulus, double the response, half the stimulus, half the response, as simple as that. In our case, the cause was the purchase, and the effect was the payment: you buy two coffees, one for you and one for your friend, you buy two, you pay twice, as simple as that: that is how things work in linearland.

4.2.1 Entry discount

Next, pause for another moment and ask yourself: is it going to be the same no matter how many friends you offer breakfast to? In a linear world, this is what would happen regardless of the number of friends you invite for breakfast. If you take three, instead of one, for a total of four, you would definitely pay twice as much, (I assume nobody declines coffee, not to mention croissants . . .). But does this keep going indefinitely? Well, as long as we are talking a comparatively small number, say fewer than ten, probably yes. But if the number goes up, you would likely be tempted to ask for a discount, with a good chance of success; the larger the number, the higher the chance. Again plain common sense, but beware, the very minute you start thinking about a discount and actually get it, you have crossed the fence of the nonlinear world! Why? Because, as soon as discount enters the scene, the prices no longer add up in linear proportion: you paid 5 Euros for two, you and your friend, but if instead of one friend you had, say, nine for a total of ten, you are very likely to pay less than ten times your own 2.5 Euros, namely 25 Euros. I bet you could easily save at least of couple of Euros, making perhaps 23, which is less than ten times 2.5 (to be sure, an 8 per cent discount). The discount broke the rules of the linear world. The moment the price of any given item depends on the number of items you buy, you leave linearland behind; welcome to the nonlinear world!

Trivial as it may seem, this mundane example illustrates a number of far-reaching points: in the linear world, effects are proportional to the cause, *no matter how large the cause*, and multiple effects just add up on top of each other, *no matter how many*, this is what the superposition principle does. Natural as they seem at a first glance, it is precisely these *no matter* conditions which prove unrealistically restrictive when confronted with the real world. That's exactly where nonlinearity kicks in massively!

4.3 Nonlinear math

We have just illustrated the sum of two numbers and multiplication by a constant as a mathematical paradigm of linearity. Let us now move on to non-linear analogues. Suppose that instead of summing two numbers, the task is now to sum their *squares* instead. For the sake of concreteness, let us take $a = 3$ and $b = 2$, whose squares are $a^2 = 3 \times 3 = 9$ and $b^2 = 2 \times 2 = 4$, respectively, summing up to $a^2 + b^2 = 9 + 4 = 13$. Now, instead of the sum of squares, let us consider the square of the sum, that is $(a + b)^2$. This gives $(3 + 2) \times (3 + 2) = 25$, nearly twice as much! The systems respond more than linearly, a property also called super linearity. The superposition principle falls flat on its face; the square of the sum is nearly twice larger than the sum of the squares. Why? Because, squaring the sum entails a constructive interference between the two numbers. This is the elementary rule that states 'the square of the sum is the sum of the squares, plus twice their product'. In math formulas

$$(a+b)^2 = (a+b) \times (a+b) = a^2 + b^2 + 2a \times b \tag{4.3}$$

As anticipated in Chapter 2, the new entry is the double product $2ab = ab + ba$, standing precisely for the interaction/coupling between a and b. By the time you square their sum, you have to multiply the numbers by each other. The reason is that, by the very action of squaring, there are four products to compute a^2, b^2, ab, and ba, hence the previous formula.

This is a quadratic nonlinearity, as it involves squares and products. One can easily generalize to higher nonlinearities. For instance the cubic analogue of (4.3) is

$$(a+b)^3 = a^3 + 3a^2b + 3ab^2 + b^3$$

In words, the cube of the sum is the sum of the cubes plus three times the square of the first times the second, plus three times the square of the second times the first. The geometrical interpretation is a three-dimensional generalization of Fig. 4.1 to a cube of side $a + b$. This splits into eight cubes of sides (a, a, a) and (b, b, b) plus the remaining six cross-coupling terms, corresponding to parallelepipeds of sides (a, a, b), (a, b, a), (a, b, b), (b, a, a), (b, a, b), (b, b, a). This cubic nonlinearity is slighthly more laborious that the quadratic one, and the number of cross-couplings increases at increasing degrees of nonlinearity, but the principle stays the same. These cross-couplings are the reason why a non-linear system is more (or less) than the sum of its parts. We wish to note that many nonlinearities do not stop at any finite polynomial degree! One such case is the familiar square root. The function $\sqrt{a+b}$ consists of an *infinite* sequence of powers of a and b!

Square of the Sum

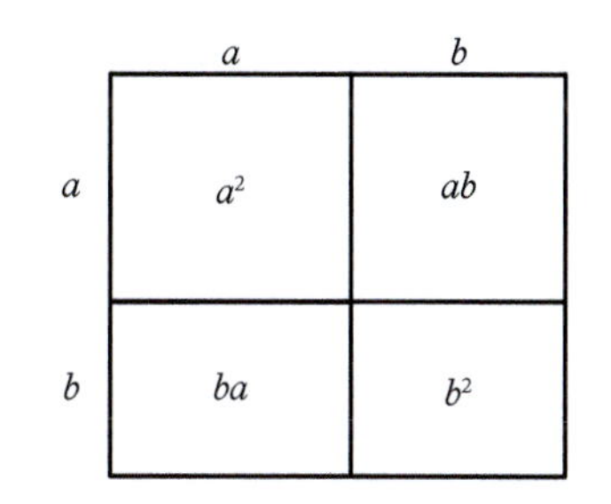

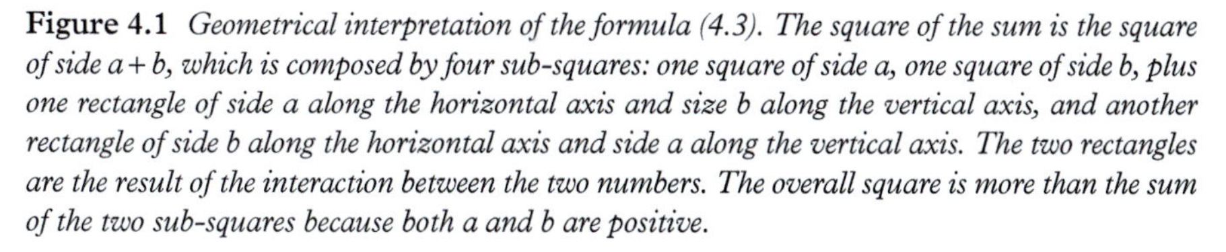

Figure 4.1 *Geometrical interpretation of the formula (4.3). The square of the sum is the square of side a + b, which is composed by four sub-squares: one square of side a, one square of side b, plus one rectangle of side a along the horizontal axis and size b along the vertical axis, and another rectangle of side b along the horizontal axis and side a along the vertical axis. The two rectangles are the result of the interaction between the two numbers. The overall square is more than the sum of the two sub-squares because both a and b are positive.*

4.3.1 More or less?

The reader may justly ask under what conditions would the system be *less* than the sum of its parts. For the example in point, the answer is easy, since 3 and 2 are both positive, their product is positive and therefore the square of the sum is more than the sum of the squares. Now make the two numbers of the opposite sign, one positive and the other negative, for instance $a = 3$ and $b = -2$. Summing 3 and $b = -2$ gives $3 + (-2) = 3 - 2 = 1$, which is just the same as subtracting 2 from 3, hence their sum is $3 - 2 = 1$, whose square is also 1. But the sum of the squares is still 13 because the square of a negative number is a positive one, i.e. $(-2) \times (-2) = 4$, hence, in this case we have lost 12 units instead of gaining them. If the numbers have opposite sign, one positive and the other negative, the square of the sum is less than the sum of their squares. The upshot is clear: if the two numbers carry the same sign, both positive or both negative, they interfere positively, they *cooperate*, namely their product is positive and makes the square of their sum larger than the sum of the squares. But, if their sign is opposite, they hinder each other, the product is negative, and it subtracts from the sum of the squares. The numbers interfere negatively, hence they *compete* destructively.

Although highly simplified, this simple example conveys a very general trend which goes far beyond the elementary algebra we have been discussing here. If the elementary units which compose a given system interfere positively (cooperation) the system is more than the sum of its parts. If, on the other hand, they interfere negatively (destructive competition) then the system is less than the sum of its parts. Even though nonlinear systems are the rule in life, they have long been treated as the exception. Not without good reasons, the first being that under small loads, most systems react indeed linearly. It is only under sufficiently large perturbations that nonlinearity is unveiled. The second is that nonlinear system are much harder to model and compute, hence, to predict [108].

4.4 Pitagora's theorem and beyond

The notion of linear superposition becomes even more transparent for the case of geometrical extensions of standard numbers which go by the name of break *vectors*. For our purposes, we can think of them as arrows with a head and tail, living in the two-dimensional plane. Let us now consider two vectors $\vec{a}$ and $\vec{b}$, the upper arrow denoting them as vectors, and form their vectorial sum $\vec{s} = \vec{a} + \vec{b}$. Taking for simplicity the case in which the tails of both $\vec{a}$ and $\vec{b}$ are located at the origin of the plane, and further assuming without loss of generality that $\vec{a}$ is aligned with the horizontal axis, the sum vector $\vec{s}$ is nothing but the vector whose head lies at the upper vertex of the parallelelogram (see Fig. 4.2) formed by $\vec{a}$ and $\vec{b}$.

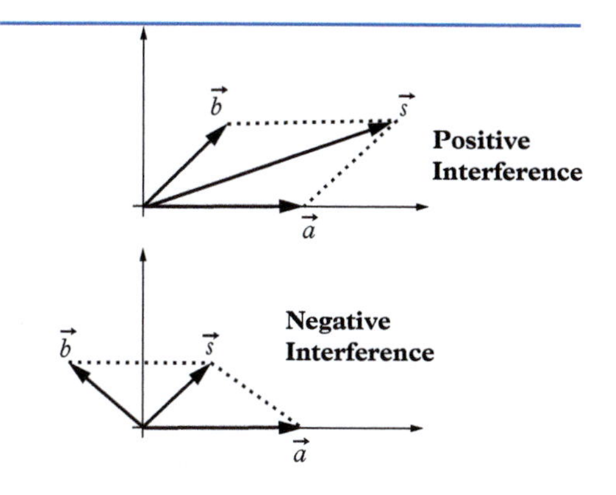

Figure 4.2 *Geometrical meaning of the vectorial sum. Positive (top) and negative (bottom) interference.*

The notion of positive or negative interference is entirely encoded in the angle between $\vec{a}$ and $\vec{b}$, let's call it θ. Let us see why. Elementary trigonometry informs us that the square of the sum vector is given by the formula

$$s^2 = a^2 + b^2 + 2ab\, cos(\theta) \tag{4.4}$$

In, a and b denote the length of the two vectors. The same elementary trigonometry also tells us that the factor $cos(\theta)$ varies between 1 and 0 in the range $\theta = [0, 90]$ degrees (the first quadrant of the plane) and between 0 and -1 in the range $\theta = [90, 180]$ degrees (the second quadrant of the plane moving counterclockwise).

The case $\theta = 0$ means that $\vec{b}$ is perfectly aligned with $\vec{a}$, indicating maximum cooperation. In this case the formula (4.4) gives $s^2 = a^2 + b^2 + 2ab$, and since both a and b are positive by construction (the length of the arrows) the square of the sum is more than the sum of the squares. The case $\theta = 180$ sits just at the opposite end: the vector $\vec{b}$ points in the opposite direction of $\vec{a}$, indicating maximum competition. The formula (4.4) now gives $s^2 = a^2 + b^2 - 2ab$, and the square of the sum is more than the sum of the squares. Finally, the intermediate case $\theta = 90$ is kind of special. In this case the vector $\vec{b}$ is neither aligned nor counter aligned with $\vec{a}$, in fact, it is perfectly neutral as it projects zero upon $\vec{a}$. In technical parlance, the two vectors are said to be orthogonal, they basically ignore each other. In this case the formula (4.4) gives $s^2 = a^2 + b^2$, and the square of the sum is exactly equal to the sum of the squares, something that cannot happen with standard numbers. The attentive reader may notice that is nothing but the famous Pitagoras's theorem!

4.5 Nonlinear materials

So far, we have stayed with pure elementary math, but how about the physical world? Needless to say, although we may not perceive it immediately, nonlinearity is all around (and within) us! A common example of nonlinear system is provided by ordinary materials: upon pulling them apart, material objects respond by deforming in proportion to the applied stress: pulled twice as strongly, they respond with twice as larg deformations. Some, for instance rubber, deform more and some, such as iron, deform much less, to the point that we don't notice by eye, but they all do to some extent, as per the famous πάντα ῥεῖ (everything flows) by the Greek philosopher Eraklitus of Ephesus (575–435 BC).[16] This simple proportionality rule cannot last forever though. Beyond a certain threshold, the deformation usually becomes disproportionately large, to the point that even a tiny amount of extra pull leads to a catastrophic breakdown of the material (see Fig. 4.3). Above this critical threshold, any extra stress, no matter how small, is never small enough to prevent the catastrophic response we call 'rupture'.

We can get a bit more specific, hopefully in the interest of clarity. Suppose we represent a material as a series of springs (dear reader, wait before you laugh, the spring model is pretty useful and popular among professional scientists!) which respond to an external pull by returning a resisting force linearly proportional to their elongation. For the sake of the argument, suppose we have 100 springs, each

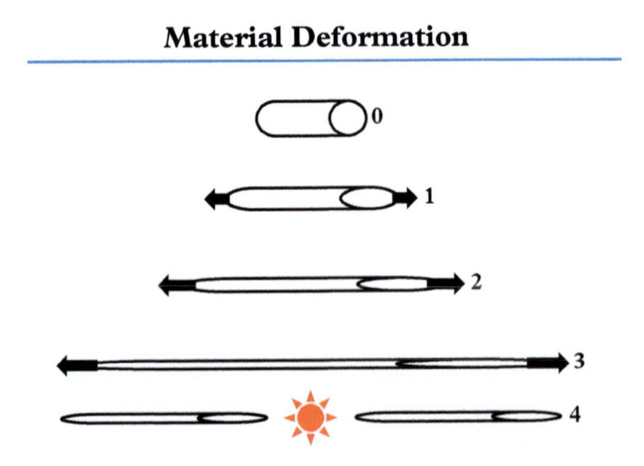

Figure 4.3 *A sample of material under pull. The undeformed material (label 0) stretches upon being pulled at its ends. When the pull is small, the material deforms in linear proportion with the intensity of the pull (labels 1 and 2). Upon increasing the pull further, the material deforms in more than direct proportion (label 3) until it breaks down (label 4). The figure depicts the case of a ductile rupture, i.e. the one preceded by anomalous deformations. Fragile materials go straight from small deformations to rupture. In other words, they don't stand large pulls. The analogy with psychological behaviour is compelling.*

[16] As a matter of fact, such literal statement does not appear anywhere in any of his writings. Nevertheless, it summarizes his philosophy well.

capable of elongating at most one millimetre before breaking down. Let us however assume that we have some statistical spread in their resistance: some would break down as soon as this threshold is passed, some others, the strong ones, are capable of withstanding larger deformations before they break down. As long as the external force can be balanced by an elongation below one millimetre, raising the force generates a linearly proportional increase in the spring deformation. When the pull reaches a critical value, forcing each spring to elongate by exactly one millimetre, a qualitatively new scenario unfolds. As soon as the force is raised above the critical threshold, *some* springs start breaking, leaving the burden of balancing the external pull to the smaller number left that prove capable of elongating more than 1 millimetre.

Suppose these 'survivors' are 90 out of the original 100: it is then clear that in order to compensate for the missing 10, they have to elongate more than proportionally in line with the previous size to the increase of the pull, to be sure $10/9 \sim 1.1$ millimetre. The nonlinear regime begins. This story continues: by further increasing the pull, maybe another 20 break down, and only 70 are left to resist. And these 70 left must now elongate up to $10/7 \sim 1.43$ millimetres. The picture is now clear, since fewer and fewer springs are in charge of withstanding an increasing force: sooner or later, the moment comes when even a tiny increase of the force will overcome all survivors, causing breakdown; this is the point of rupture of the material, the crack under pressure. This scenario is by no means confined to materials, the principle 'no small force is small enough' is in action also, and we would say especially, in human endeavours, such as the social, the financial, and more generally whenever psychological variables come into play. In fact, revolutions are literal forms of social rupture.

The opposite behaviour is also observed, although less frequently. Under mounting stress the material shows less and less incremental deformation, i.e. it develops increasing strength, a property called *anti-fragility*, a fancier word for sub-linearity.

Both hinder predictability since, whether bad or good, they spell surprises lurking behind the corner. In material science this happens whenever molecules manage to group in tangles which offer higher resistance to the external drive, ordinary paints being an example in point. In social systems, this is the force of cooperation and mutual help which typically develops after major catastrophes, such as hurricanes, earthquakes, and similar extreme events. Be that as it may, the bottom line is that under sufficiently large loads, far enough from equilibrium, sooner or later nonlinearity kicks in. What happens then? Let's take a look.

4.6 Butterflies and elephants

By now, the reader is acquainted with the fact that in linearland, small changes of the cause are accompanied by small changes in the effect. Obviously, if small returns small, we know beforehand that, as long as we go softly enough on it, the system will not respond with wild changes. Once we know that small returns small and large returns large, we are in pretty good shape. The system is *predictable*, we have a firm handle on it, we are in control. If I know that a coffee costs me an Euro and its cost is not going to change no matter how many of them I decide to drink, I know exactly how much I would pay for any number I wish to buy. That is, I am in perfect control. The point is that while the 'small returns small' side of the coin is usually fine (without butterflies around, though . . .), the 'large returns large' side is generally much less safe because, if the system is nonlinear, it is hard to guess how large the second large (effect) really is.

Even if small changes produce small effects, it is everybody's intuition that this by no means implies that large should also produce large. Consequences of big changes are much harder to predict, because they launch the system in a distant *terra incognita*, where it might well behave in a totally different way. If the system is linear, we can project, hence predict, the effects of arbitrarily large changes, no matter how large, because they *always* go in proportion with the cause; we have telescopes for any distance, no terra incognita is too far for them. With non-linear systems such powerful telescopes become much more short-sighted, and the only option is to proceed in little steps. But the quirkiest case comes when 'small returns large', which can be taken as a byword for instability, since then even a small perturbation can generate a huge response. The opposite, 'large returns small' is also quirky, although less worrisome, as it generally speaks for stability; you push hard but get little motion in return. For reasons shortly to be apparent, we shall refer to the previous ideas as the butterfly and the elephant effects, respectively (see Figs. 4.4 and 4.5).

As observed before, both butterflies and elephants hinder predictability since, whether bad or good, surprises are lurking behind the corner anyway, and complex systems are littered with such corners. Is this bad or is it good? It's neither and both: butterflies, otherwise popular creatures, play villain because of chaos: a wingbeat in Cuba triggers a hurricane in Miami, but chaos may eventually play in our hands, as it is the case for the fortunate few who manage to win a million-dollar lottery upon buying a single one dollar ticket . . . .

Likewise, elephants are ordinarily perceived as good, because they imply forgiveness: you make a big mistake, like crossing the road without watching, and nothing happens, your potentially life-costing mistake is forgiven. But elephants can be pretty frustrating as well, for instance, when you put in a lot of work preparing for a major product to be launched on the market, and the market responds with just a faint 'blip', no return on investment. The fact is that nonlinearity is neither bad nor good: as we discussed in Chapter 3, what it really does in the end, it spawns nearly unlimited freedom. Therefore, like freedom, nonlinearity is good, but if you abuse it, it can turn against you, 'a friend made enemy', to borrow from the U2 band. This said, in Chapter 5 we are going to take a closer look at butterflies, and more precisely to the 'dark side' of nonlinearity: chaos.

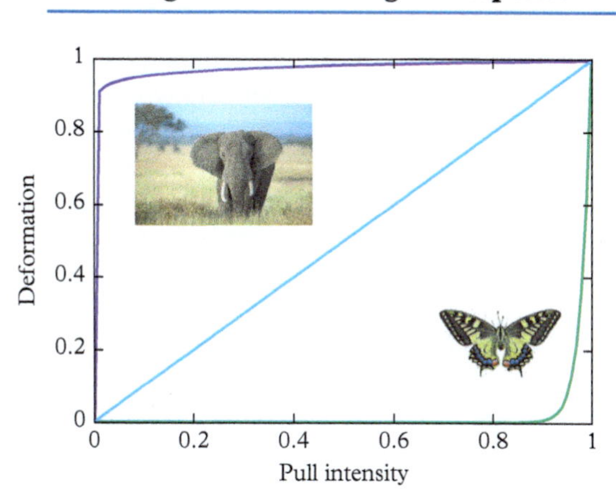

Figure 4.4 *Linear (middle line), fragile (lower curve), and antifragile (upper curve) response. In the fragile case, the material deforms very little, until a threshold is reached, beyond which the deformation grows disproportionately with the increase of the pull intensity. The system goes from a 'large gives small' to a 'small gives large' (butterfly) regime. The antifragile case shows the opposite behaviour, a 'small gives large' initial regime followed by a 'large gives small' (elephant) regime.*

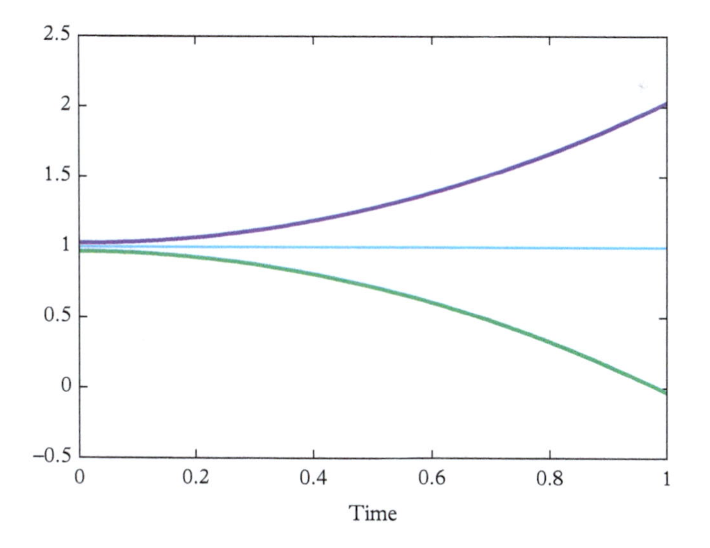

Figure 4.5 *A divergent (butterfly) trajectory. If the trajectory starts slightly above 1, it grows indefinitely. If it starts below, it decreases indefinitely. If it starts exactly at 1 it stays at 1 all the time (neutral). The elephant case is just the opposite, initially distant points get closer and closer in time (convergent behaviour). The elephant trajectory is like the butterfly one in reverse time.*

4.7 Summary

Summarizing, a nonlinear system is more or less than the sum of its parts because the parts interact. If these interactions are cooperative, the system is more than the sum of its parts and if they are competitive, it is less. How much more or less, strictly depends on the specific nonlinearity. And since there is only one way of being linear and literally infinitely many of being nonlinear, it is clear that nonlinearity breathes a prodigious freedom, hence Complexity, into our world.

4.8 Appendix 4.1: Nonlinear functions

In mathematics, a linear function is defined as one obeying the following equality:

$$f(x+y) = f(x) + f(y) \tag{4.5}$$

with x and y both positive for convenience. What this means is that the output of the sum (left-hand side) equals the sum of the outputs (right-hand side), an operational definition of the superposition principle. One can readily check that the only solution of the the previous relation is the linear function:

$$f(x) = ax$$

where a is an arbitrary constant.

Geometrically, this is a straight line of slope a. Any departure from the expression (4.5) defines a non-linear function, which says it all about the very peculiar and restrictive case of linearity. *Anything* else is nonlinear! Here, a distinction can be made between sub and superlinear functions. Let us define the Interference as the difference between the two sides of the equation, namely:

$$I(x,y) = f(x+y) - f(x) - f(y)$$

Sub(super) linear behaviour is defined by negative(positive) sign of the Interference term. By taking $y = x$, the previous definition simplifies to:

$$I(2x) = f(2x) - 2f(x)$$

This shows that $f(x) = x^2$ is superlinear, since indeed $(2x)^2 - x^2 - x^2 = 4x^2 - 2x^2 = 2x^2 > 0$. Likewise, one can readily check that its inverse, the square root function $f(x) = \sqrt{x}$ $(x > 0)$ is sublinear. More generally the p-th power x^p, where p is a positive number, is superlinear for $p > 1$ and sublinear for $p < 1$.

4.9 Appendix 4.2: How much is it two plus two?

The curious reader may want to know that there are perfectly viable algebras in which it is literally true that 'one plus one is not always two'. To this end, consider the binary algebra used by our electronic computers. Such algebra consists of just two numbers, 0 and 1 (bits, for binary digits), so that 1 + 1 cannot make 2, simply because 2 doesn't exist! Binary algebra informs us that, $0 + 0 = 0$, $0 + 1 = 1$, $1 + 0 = 1$, and ... $1 + 1 = 0$. The first three make perfect common sense, the fourth one does not. The mathematicians call this 'sum modulo 2', indicating that whenever the sum exceeds 1, you must subtract 2. That is: $1 + 1 = 2 - 2 = 0$! Incidentally, we note that subtracting 2 is precisely what makes the sum modulo 2 a (very) nonlinear operation .... Should we then conclude that 1 + 1 does not make 2? The answer is: yes and no, as it depends on the representation. In the binary one, it surely doesn't but in the decimal one, the one we are familiar with, it does. How come?

The point is that the 2 we have subtracted in sum modulo 2 does not get lost but carried to the next position (rightwards) in the binary representation. In detail: 1 in decimal is written 10 in binary notation, that is $1 \times 2^0 + 0 \times 2^1 = 1$ (we remind that any number raised to power 0 returns 1). In decimal, $1 \times 10^0 = 1$. Summing 1 + 1 digit-by-digit gives $10 + 10 = 01$, $1 + 1 = 0$, with carry 1 and $0 + 0 = 0$ which becomes 1 after summing the carry. The result is a binary 01, namely $0 \times 2^0 + 1 \times 2^1 = 2$ decimal. That's it! Bottomline, 1 + 1 is always 2, in decimal notation, but not in the binary one! If we use binary notation (that's what computers like best) to express decimal numbers, then we have to supplement the sum modulo 2 with the shift of the 1 from the first to second position (counting left to right). That's how 1 + 1 keeps making 2 in decimal notation. Sorry for all these complications, but this was just to make sure that the metaphor 'two plus two does not make four', does not sound misleading or superficial, hence irritating, to the attentive reader, as it apparently was to a valued referee of this book.

5

The Dark Side of Nonlinearity

Chaos: When the present determines the future, but the approximate present does not approximately determine the future.

(E. Lorenz)

5.1 Introducing chaos

In Chapter 4, we have noted that small changes may lead to huge consequences, a fact we know all too well in ordinary life, where some decisions can change the course of an entire lifetime (the movie *Sliding Doors* builds around this very premise). We have further observed that the reverse is also true, although, it seems to me, to a lesser extent; big mistakes rarely get forgiven in full. Be that as it may, extreme sensitivity to small perturbations is best known to the public under the somewhat threatening name of chaos. Chaos has ancient roots in history, as the epitome of the supernatural power beyond our control [108]. In more modern terms, it goes under the much less dreadful name of the *butterfly effect*, a colourful metaphor for the *extreme sensitivity to small perturbations* which undermines our ability to predict the future [49, 97] (see Fig. 5.1).

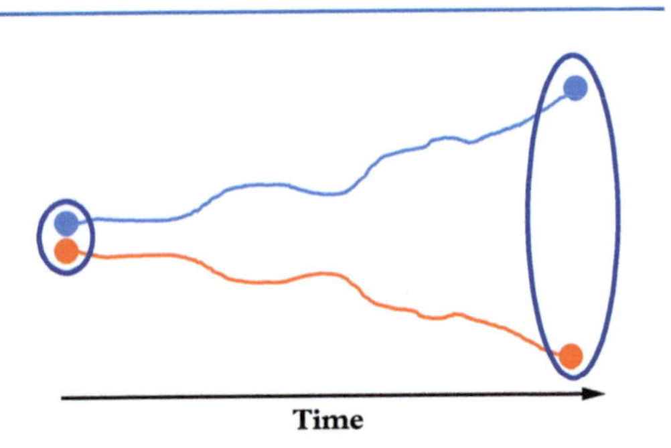

Figure 5.1 *Sensitivity to initial conditions: the blue and red trajectories start very close nearby but end up at very distant positions as time unfolds.*

Sailing the Ocean of Complexity. Sauro Succi, Oxford University Press.
 DOI: 10.1093/oso/9780192897893.003.0005

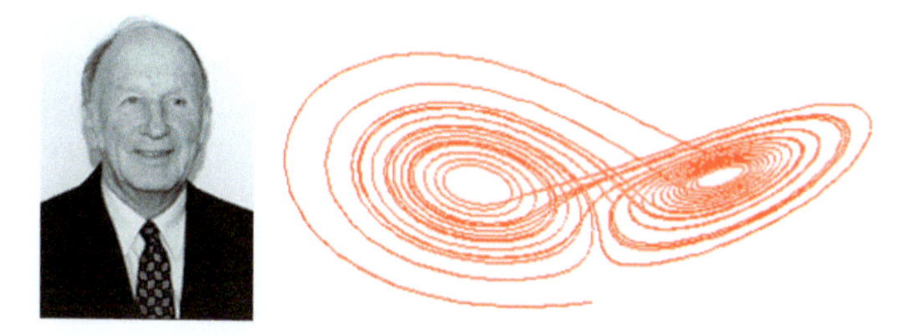

Figure 5.2 *Edward Lorenz, the father of chaos theory, and his eponymous 'strange attractor', the subset of three dimensional space where the Lorenz trajectories settle in the long-time limit. The butterfly-looking shape of the attractor is striking ....*
Source: reprinted from en.wikipedia.org and commons.wikimedia.org.

According to this metaphor, the wing beat of a butterfly in Cuba can trigger a hurricane in Miami (there are assorted geographical versions depending on where you ask the question ...). Meteorology has provided the historical cradle of modern chaos theory and still serves as a paradigm of unpredictability. As Bob Dylan sings, 'the answer, my friend, is blowing in the wind'. More down to earth, 'Climate is what you expect, weather is what you get' (Robert Heinlein, *Time Enough for Love* (1974).[17] The scientific roots of chaos theory were laid down in the nineteenth century by the French polymath Henri Poincaré (1854–1912), who first discovered extreme sensitivity to initial conditions in the study of celestial mechanics [57]. However, the birth of modern chaos theory is usually associated with the Massachusetts Institute of Technology (MIT) meteorologist Edward Lorenz (1917–2008), as we are going to discuss next (see Fig. 5.2).

5.1.1 Surprises over lunchtime

In the early 60s the MIT meteorologist Edward Lorenz was busy trying to predict the weather. The equations that govern atmospheric phenomena are known in principle, but were impractically hard to solve in those days, and, notwithstanding enormous progress, they still are a 'tough cookie' to this day. So, Lorenz did what theoretical scientists are used to do: he worked out a simplified model which would relinquish most math difficulties while hopefully retaining the physical essence of the phenomenon. In this specific case, he boiled millions of equations down to just *three*, describing a layer heated from below through the rate of heat flow, the change of temperature along the plate, and across it, for a total of three. Just three, yes, but ... nonlinear.

Three is a small number, but with nonlinearity in the room, three equations can be far too many to solve with traditional 'paper and pencil'. In fact, they cannot, which is why Lorenz used his computer (since we are talking the 60s, though,

[17] In order not to alienate the meteorologists in the audience, I hasten to pay due homage to the major progress they have made in the last decades, current 3-day weather forecast being much better than the one day it used to be a couple of decades ago.

much less powerful than present-day ones, they were already far beyond the calculational abilities of humans). The usual practice here is to start from some initial values and see how they change in time under the drive of the equations. If you want to develop a sense of your model, what you usually do is to try out many different initial data and see how they evolve in time, namely the trajectory of the system. Under ordinary circumstances, you would expect that two trajectories emanating from nearby initial data will remain comparatively close all along. And so did Lorenz, but as mentioned previously, a big surprise was lurking behind the corner. At some point in his toil, Lorenz went out for lunch (yes, occasionally, even scientists get lunch, defying the famous 'Lunch is for wimps' of the legendary Wall-Street geek Gordon Gekko). Much to his surprise, when he got back, he found results which had apparently nothing to do with those he left on his table before going to lunch. All he did, in the process, was to slightly change the initial data. He reacted in the most natural way, i.e. chase a mistake in the computer programme. But, after what I may imagine as an intensive and extensive search,[18] none showed up.

'Mistakes are the portal of discovery', James Joyce (1882–1941) informed us, but sometimes great discoveries come without any mistake, they are in fact the fruit of mere serendipity, 'Lady Luck' if you wish, provided you do your best to assist her with your own hard work. This was the case with Lorenz; by serendipitous chance, he stumbled upon chaos: his three innocent-looking nonlinear equations exhibited a disproportionate sensitivity to small changes in the initial data. Of course, examples of extreme sensitivity abound in life: if you are on the thin ridge of a glacier (for as long as they still exist), take the wrong step sideways, you may end up very far down the ravine. These dangers have been known for as long as we humans have been around. But, the point with the ridge is that you *do see* that it is dangerous. In the case of Lorenz, the appearance was totally innocent! The fact that such ridges could be hidden within three innocent-looking equations was a true scientific and *philosophical* revelation, the birth of modern chaos theory. Chaos in hundreds of millions equations would be no surprise, but that it could hide under just *three*, was mind-boggling indeed. The conceptual import of this discovery is hard to overestimate and even though Lorenz himself was not honored by Nobel Prize, the 2021 Nobel in physics to Syukuro Manabe and Klaus Hasselmann "for the physical modelling of Earth's climate, quantifying variability and reliably predicting global warming", along with Giorgio Parisi, "for the discovery of the interplay of disorder and fluctuations in physical systems from atomic to planetary scales" is a wonderful and long overdue vindication of his work and legacy.

To this regard, let me quote verbatim from the close of chapter 11 of Leonard Smith's book [108]:

> Prophecy is difficult; it is never clear which context science will adopt, but the fact that chaos has changed the goal posts may well be its most enduring impact on science. . . . . chaos has forced to rethink what it means to approximate nature.

[18] I have experienced this many times myself, although with less spectacular outcomes.

5.2 A closer look at chaos

Caveat: the following section is a bit denser in math than the rest of this book. The reader not willing to take the challenge can safely skip it with no serious damage. The ones willing to go over, are expected to enjoy a sharper view of the subject.

In the previous section, we have been speaking of chaos in pure words, which is good enough to convey the philosophical sense of it. However, we can do better than that. Just a little bit of quantitative analysis discloses a more solid appreciation, and, hopefully, more pleasure as well. The main point we shall make in Section 5.2.1 is that simple (nonlinear) rules can generate fairly chaotic behaviour, and we are going to see the explicit face of these rules, for they are simple enough to be formulated within a few lines.

5.2.1 Population dynamics

A rich pedagogical example is provided by the dynamics of populations which we now discuss in some detail. Let us consider a population of individuals, could be humans, could be lions, bacteria, or cells, the essential point being that the species is capable of reproducing and, no surprise, in order to do so, it needs resources, primarily space and food. Let us denote by r the replication rate of a given individual per cycle, meaning by this that each individual gives birth to $r - 1$ offsprings at every cycle. With $r = 1$, no new individual is born, which means that growth implies r larger than one, $r > 1$ in math notation, while less than one ($r < 1$) indicates decay. Note that replication here does not necessarily mean sexual-reproduction, a cell splitting into two cells (mitosis) every twelve hours would account for a rate $r = 4$ per day. It is readily appreciated that if each individual gives rise to $r > 1$ individuals per cycle, starting with one in the beginning (cycle zero), at the first cycle we have r, at the second, we have $r \times r = r^2$, at the third $r \times r \times r = r^3$ and so on down the line. Clearly, any r larger than 1 results in explosive (exponential in math terms) growth. The math is simple enough; with $r = 2$, ten cycles turn the initial single individual into a population over thousand, 1 at cycle 0, 2, at cycle 1, 4 at cycle 2, 8, at cycle 3, 16, at cycle 4, keep going and what you find at cycle 10 is 2 multiplied by itself ten times, also known as $2^{10} = 1,024$.

This is called exponential growth, hardly a champion of sustainability.[19] Of course, the actual (logistic) impact of such exponential growth depends on the

[19] Exponential growth is nicely exposed by the oft-told story of the ancient king of Persia who wanted to reward the inventor of the chess game, named Sissa Ben Dahir. Sissa's request was deceivingly modest; a chessboard is made of eight by eight squares, and all he asked for was a grain of rice in the first square, two in the second, four in the third, and so on, until the 64th is reached. The king was baffled at such an apparently modest request, until he realized that in order to meet it, he would have had to offer his inventor the overall (actual) output of rice for the next two millennia! The number is indeed $1 + 2 + 4 + \ldots 2^{63}$, which, even if you approximate with the very last entry, gives you something like nine billion billions rice grains. To be sure, the exact number is $18,446,744,073,709,551,615$ grains, a bit over eighteen billion billion. Now compare with the number when grains are simply added one by one, namely $1 + 2 + \ldots 64$, the result being a tiny 2,040.

time duration of a single cycle. Doubling every hour means sixteen million new individuals in just one day, a demographic nightmare! Doubling every day gives one billion in a month, still a nightmare! Doubling every year results in eight billion, slightly short of the current worldwide population, in just 33 years. The bottom line is that growth requires values of r above 1, but sustainability commands that r be only *slightly* above 1, for otherwise the result are literal demographic bombs. For the record, the actual growth rate of the world population is about $r = 1.011$, namely about 1 per cent growth per year. If this sounds like a small number, please consider that the very same number means about 82 million newborn people per year. Or, perhaps even more impressively, it means that the worldwide population in 1970 was only half of what it is today . . . .

The attentive reader must have noticed that we have made a very strong assumption here, namely that no individual dies off during the entire replication process. This is obviously unrealistic as applied to long periods as compared with the lifetime of single individuals. If we are talking humans, it is certainly unrealistic to expect zero deaths over the fertile population in 30 years. This means that death rates, besides birth rates, must be accounted for, making the whole scenario much less explosive. In actual fact, the replication rate r is to be intended as the *difference* between the birth and death rates. Still, if we talk resource planning, space (and food) must definitely enter the equation besides time, and even assuming that time can be treated as infinite (the Sun will burn for a few more billion years), space most certainly cannot. At least, not until Musk, Bezos, and other Silicon Valley billionaires manage to take us to outer space (at an affordable cost) thereby opening up the new era of multi-planetary life (weekend on Mars?). That is to say, since each individual needs some space and some food, and since neither of the two commodities is unlimited, sooner or later the time comes when some of the individuals start to compete for one or the other, most likely both. That marks a dramatic change in the growth trend, let us see how.

5.3 Entry competition: The logistic map

Here comes the second act after replication, that is competition and the subsequent aftermath, starvation for the losing side. There is nothing graceful about starvation, the losing end ultimately dies out for lack of food. Clearly this puts exponential growth to a halt, and the next question is: does the number of individuals ever stop growing in time and settle to a finite number? And if so, what is this number, and how does it depend on the parameters of the problem, growth rate and competition strength, in the first place? Again a very practical question, and urban planners surely know what we are talking about. The answer to this question is readily found by solving a very popular equation, known as the logistic map. Here it comes:

$$N(t+1) = rN(t) - cN^2(t) \tag{5.1}$$

Don't be afraid of the math, the meaning is simple. The left-hand side is the number of individuals at cycle $t + 1$. The first term at the right-hand side is the growth rate term and the new entry, the second term at the right-hand side, stands for the losses due to competition, whose strength is measured by the parameter c. Note that the competition term depends on the *product* of the number of individuals because, in order to compete for food, at least two individuals must meet at the same time at the same place. If they were three, we would have a triple product, namely a cubic power, $N \times N \times N = N^3$. Hence, competition is the name of nonlinearity in this picture. Now, back to the question: does the number of individuals ever stop growing? The answer is: yes, it can. Which does not mean that it necessarily does, as it all depends on the specific value of the replication coefficient r.

Let us begin by inspecting the simpler case when it does. If the number of individuals stops growing in time, it means that their number at time $t + 1$ stays exactly the same as it was at time t, namely:

$$N(t+1) = N(t)$$

Inserting this equality in the logistic equation, since the left-hand side is zero by definition, the right-hand side must be zero too. That is:

$$(r-1)N - cN^2 = 0$$

This is a simple quadratic equation which delivers two solutions.

The first is a plain zero:

$$N_e = 0,$$

corresponding to *extinction* of the species (whence the subscript e).

The second and more interesting one is:

$$N_c = \frac{r-1}{c}$$

which corresponds to *saturation* of the population, see Fig. 5.3.

The number N_c is also called 'carrying-capacity', for it corresponds to the maximum number of individuals which can coexist indefinitely in time for a given value of the replication coefficient r and competition coefficient c. The interpretation is clear: first, we note that r must be greater than 1, otherwise the carrying-capacity is negative, which means extinction. Next, we note that the carrying-capacity goes in inverse proportion to the competition rate, which is also very intuitive. From a different angle, one could also interpret c as a measure of consumerism, individuals that can live on short supply (small c) can obviously

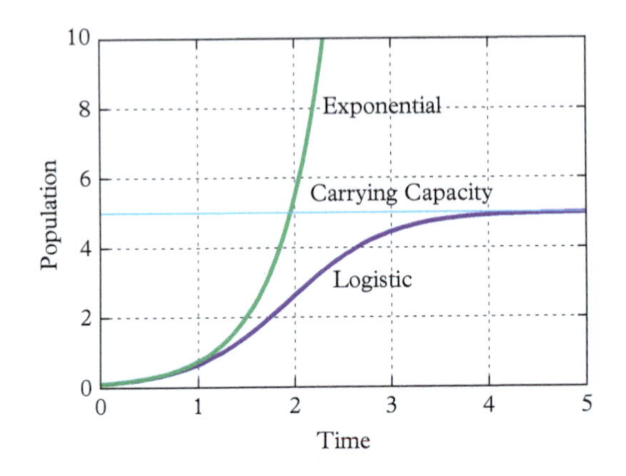

Figure 5.3 *The logistic curve versus the exponential one. In the initial stage, the system grows exponentially so that the two curves can hardly be told apart. As time unfolds, the exponential grows untamed while the logistic curve starts to slow down until saturation, reaching its carrying capacity. Note the typical sigmoidal shape of the logistic curve.*

reach higher capacities than those who demand a lot of resources on an individual basis. Large c makes me think of big Smart Utility Vehicles (SUVs) in an overcrowded city like Rome . . . .

In the limit of no competition at all, $c = 0$, the capacity is infinite, which is the way math informs us that the population never stops growing in time, no saturation. It may take a short time, or it may take a long time, but in the end, only an unlimited world, *in space, time, and food*, is compatible with this scenario. No need of Greta Thunberg's fans to realize that the name of this imaginary world is not planet Earth . . . .

5.4 Chaos: Lost and found

At this point we have exposed another feature of nonlinearity: nonlinear saturation. Owing to nonlinear competition, unbounded growth is tamed, which is good. But we started with chaos, and we ended up with a pretty rational and well-behaved situation. So, where did chaos disappear in this story? To answer this question, let us refer to the logistic map, in a slightly different form (for details see the Appendix):

$$x(t+1) = rx(t) \times (1 - x(t)) \tag{5.2}$$

where x lives in the interval $[0, 1]$.

The steady-state condition $x(t+1) = x(t)$ now yields a carrying capacity

$$K = \frac{r-1}{r},$$

a number always smaller than 1.

As for most maps, the verbal formulation of the logistic rule is utterly simple: 'Take a number, multiply by r and multiply again by one minus this number; the result is the number at the next step'. Despite this utter simplicity, weird things can and do happen.

For instance, the previous formula predicts that the carrying capacity should slowly increase towards to the value 1, full saturation of the system, upon sending r to infinity. But reality tells another story.

In particular, four distinct asymptotic (long-term) regimes can be identified depending on the value of the replication rate r. For r between 0 and 1, we have *Extinction*: the system dies out, reaching the value $x = 0$ regardless of the initial condition. This is very natural since $r < 1$ implies less individuals from one generation to the next, hence in the end the overall population must decay to zero. For r between 1 and 2, we have *Monotonic Saturation*: the population grows monotonically to full capacity, regardless of the initial condition. No chaos. For r between 2 and 3, we have *Oscillatory Saturation*: the population eventually exceeds capacity and whenever this happens the nonlinear term $-rx^2$ prevails over the linear one, so that the population at the next cycle must decrease. Then linear growth prevails again, dragging back the population below capacity again, whence the oscillating trend. For r between 3 and 4, we finally enter the *Chaotic regime*. The oscillations become irregular and larger in amplitude: the system has entered the chaotic regime where even slight changes of the initial condition may result in a very different trajectories. Chaos has finally trickled in (see Figs. 5.4 and 5.5). To be

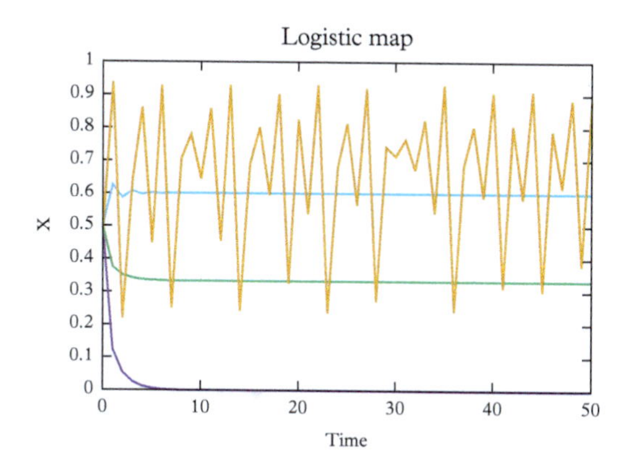

Figure 5.4 *Trajectory of the logistic map. From the bottom upwards: r = 0.5 (extinction) r = 1.5 (monotonic saturation), r = 2.5 (oscillatory saturation), r = 3.75 (chaos). All trajectories a start at x = 0.5 at time t = 0.*

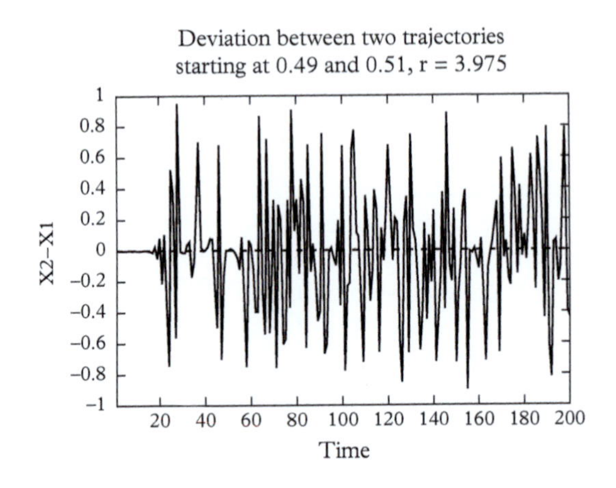

Figure 5.5 *Departure between two trajectories of the logistic map starting at $x(0) = 0.49$ and $x(0) = 0.51$ with $r = 3.975$.*

more precise, with r between 3 and 3.44949, almost any initial condition leads to periodic oscillation around two values. For r between 3.44949 and approximately 3.54409, the population will approach permanent oscillations among four values. With r increasing beyond 3.54409, from almost all initial conditions the population will approach oscillations among eight values, then 16, 32, and so on: the so-called *Period-Doubling Cascade.* At $r \sim 3.56995$ the onset of chaos is observed, at the end of the Period-Doubling Cascade. From almost all initial conditions, we no longer see oscillations of finite period. Slight variations in the initial population yield dramatically different results over time, that is chaos. Most values (but not all of them) of r beyond 3.56995 exhibit chaotic behaviour, not without isolated ranges of r, sometimes called *islands of stability* islands of stability, without chaoticity.

For all its simplicity, the logistic map highlights a far-reaching property of chaos, namely that the border between chaos and Order (periodicity) is not a neat and sharp one, but the two are intertwined, with islands of regularity within chaotic regions and vice versa (see Fig. 5.6).[20]

[20] The inquisitive reader may wonder what happens for r above 4. This is a perfectly reasonable question and the answer is *NaN*, for 'Not a Number', the dreaded message every programmer gets from its computer whenever the computation goes bust! The point is that for $r > 4$ the right-hand side of the logistic equation is no longer bounded between 0 and 1, as one can readily appreciate by noting that the product $x(1-x)$ attains its maximum, $1/4$ at $x = 1/2$, so that, if $r > 4$, the right-hand side exceeds 1. In this case, the logistic map becomes unstable, and it produces numbers which escape the representation capabilities of digital computers. Sometimes, I feel like 'NaN' should be promoted to the status of a fifth regime of the logistic map, but this is a fascinating story that we must leave for another book . . . .

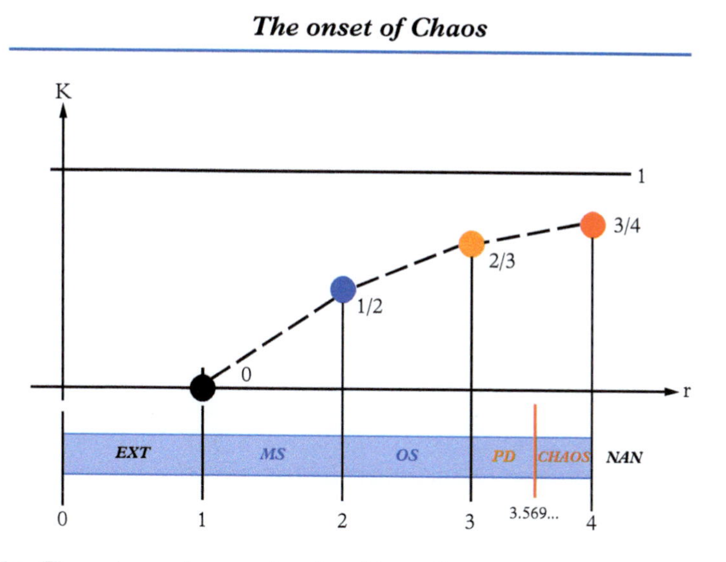

Figure 5.6 *The various regimes as a function of the replication parameter r. Extinction (EXT, $0 < r < 1$); monotonic saturation (MS, $1 < r < 2$), oscillatory saturation (OS, $2 < r < 3$), Period Doubling (PD, $3 < r < 3.56995$) and finally Chaos ($3.56995 < r < 4$). For $r > 4$ the mapping is divergent, and your computer would respond NAN, for 'Not a Number'.*

5.4.1 Are maps realistic?

If all this seems abstract and far apart from real life, you may be pleased to learn that many animal species show precisely this kind of behaviour, the typical case being known as the predator-prey mechanism. The predator thrives upon eating the prey, but if it eats too many, the preys become extinct and the predator becomes extinct as well for lack of food! If you are in the business of ecosystems planning, you'd better be informed of logistic-like maps and ensuing chaotic behaviour! For all its simplicity, the logistic map provides a sensible model for a broad variety of systems with competing interactions, the gain-loss mechanism described in Chapter 3. Another remarkable nonlinear map is the so called Lotka–Volterra model, describing the time evolution of a two-species ecosystem, typically the prey and predator system mentioned previously. And the Lorenz attractor also belongs to the same family, if slightly more complicated, with three populations instead of one. Maps provide the time sequence of a single or multiple variables, all sitting on the same location in space. Despite such limitation, they offer a wealth of insight into real-life systems. Yet, most real-life systems extend in space as well, hence the natural question arises: what happens to Chaos when space joins the party? This takes us into a very broad territory, which goes far beyond the scope of this book. Yet we shall touch on two major phenomena, *turbulence* and *morphogenesis*. Let's proceed with the former first, leaving the latter to Chapter 6.

5.5 From chaos to turbulence

'Please, fasten your seat belts, as we are entering an area of turbulence'. How many times have we heard this announcement on a plane, sometimes not without a bit of accompanying anxiety . . . . Turbulence means random buffets up and down, rough navigation in general. And we are talking huge airliners, several tens of tons in weight, buffeted like mosquitos, all this by a thin substance most often than not associated with weightless appearance. Yes, we are talking thin air. Thin air can toss around the tons of metal of advanced materials which make modern airliners. This is the power of turbulence (see Fig. 5.7). Familiar as it is in our daily life, turbulence is an excruciatingly difficult topic in science, one that kept scientists busy for more than two centuries and still represents one of the most extreme frontiers of modern science. Ironically, all this stems from a comparatively innocent-looking set of equations known for about two centuries, after the seminal work of the French engineer Louis Navier (1785–1836) and the British mathematician Gabriel Stokes (1819–1903). The Navier–Stokes equations, conceptually nothing but the celebrated Newton's law $F = ma$, force equals mass times acceleration, as applied to a finite volume of fluid. These equations are quadratic, simply because the kinetic energy of the fluid is given by the mass, times the *square* of its velocity: $E_{kin} = \frac{mv^2}{2}$. Besides being quadratic, the Navier–Stokes equations are also non-local because the change in time of the velocity $v(x, t)$ at a given position x and time t depends on the difference between the kinetic energy and pressure at neighbouring sites.

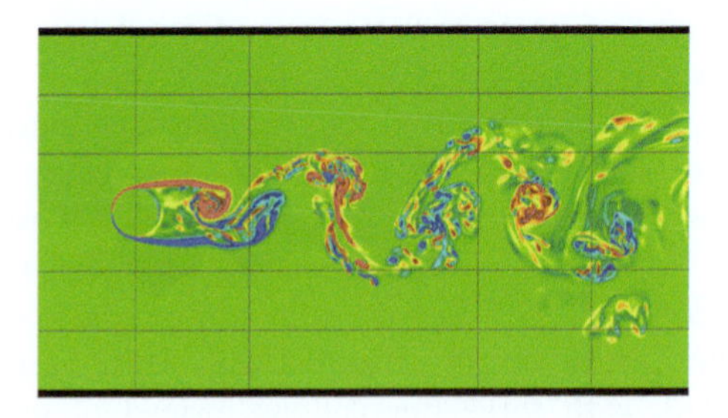

Figure 5.7 *Snapshot of a turbulent flow past a cylinder, resulting from a computer simulation from the author's team. The Reynolds number is modestly around 1,000, yet visibly sufficient to generate a lot of whirls and fancy structures all around. The simulation was performed using the Lattice Boltzmann method (see Chapter 6).*
Source: courtesy of G. Amati.

5.5.1 The energy cascade

The combination of this 'simple' quadratic nonlinearity with nonlocality, gives rise to a crucial phenomenon: energy flows across scales, typically from large to small, a process aptly called *energy-cascade* by professional in the fields. This is the familiar phenomenon by which large vortices break up into smaller ones, then into still smaller ones and then again down the line of the energy cascade (see Fig. 5.8).

The immediate question is: does the cascade ever stop?

And, if so, when? The answer, a definite yes, traces to the pioneering work of the Russian polymath Andrej Kolmogorov, who provided fundamental insights into turbulence in the early 40s. And the answer is that the cascade ends when the eddies (vortices) are small enough for dissipation to take over. But this is like answering with another question: why would dissipation take over when eddies are 'small enough'? A bit of math would offer a quick and lucid appreciation of the point, but we can get away with a qualitative statement, namely that below a given size, the molecules which form the vortices (zillions of them) are *no longer capable of sustaining collective behaviour* which *defines* fluid motion. They start to go on their own, and by doing so, fluid motion is dissipated in molecular-like disorganized motion. Incidentally, as we shall see in chapter 8, such disorganized motion is precisely what we call heat.

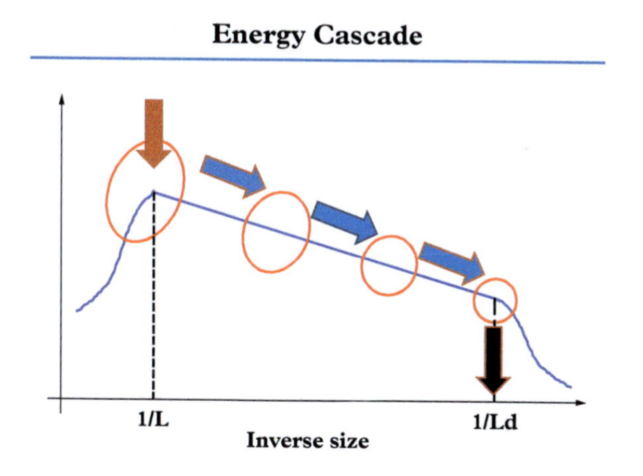

Figure 5.8 *The turbulent energy cascade. The energy is injected at the macroscopic scale L, and large-scale structures break up into daughter structures of smaller sizes, which in turn decay further into third generation daughter structures of still smaller scales. The hierarchical cascade from large to small structures proceeds until the last-generation structures become too small to sustain coherence against dissipation, thus terminating the cascade at the Kolmogorov or dissipative scale Ld.*

The eddy size at which this happen is (no surprise) called *Kolmogorov length*, and it is usually a tiny fraction of the macroscopic size of the object under study. For instance, in the case of our ordinary car, say 3 metres long and proceeding at some 100 km/h, the Kolmogorov length is in the order of a few tens of microns, i.e. about a hundred thousand times shorter than the car! These numbers alone convey a pretty clear idea of the spatial span of the cascade, the vortices can break up and transfer energy to smaller and smaller scales over about six decades in space! Not hard to accept that they are given a lot of freedom in the process, and it is precisely this freedom which makes turbulence hard to compute and predict.

5.5.2 The Reynolds number

We cannot talk turbulence without mentioning another major hero, the British Osborne Reynolds (1842–1912), who lends his name to possibly the most important parameter in the field, the eponymous Reynolds number, defined as the product of the object size, L, times its speed, V, divided by the fluid kinematic viscosity, ν. In math terms:

$$Re = \frac{VL}{\nu} \tag{5.3}$$

The Reynolds number measures the relative strength of nonlinearity, which fuels the energy cascade to dissipation, which stops it. The key point is that the numerator and denominators in the previous expression are largely out of balance, and not by coincidence. The car is a macroscopic object and so its size is measured in metres and its velocity in (tens of) metres per second, while the kinematic viscosity is a *molecular* property, and since molecules are tiny in car units, the kinematic viscosity scores one part in hundred thousands (10^{-6}) square metres per second. The end result is that the numerator is easily in the order of ten millions times larger than the denominator, which precisely the order of magnitude of the Reynolds number of the car. Based on its definition, the relation (5.3), it is readily appreciated that the Reynolds number can escalate to much larger numbers than hundreds of millions, take geophysics or meteorology, let alone astrophysics and cosmology!

This provides a timely opportunity to return in way more specific terms to the relation between Complexity and relevant Degrees of Freedom, which we have qualitatively introduced in Chapter 1. It goes like this. Based on comparatively simple arguments, essentially dimensional analysis, back in 1941 Kolmogorov was able to show that the smallest active scale (dissipative scale) in a turbulent flow at a given Reynolds number Re, is approximately given by $l_d = L/Re^{3/4}$. The number of active eddies, ('vortices' for convenience), in such a flow, a cube of side L for simplicity, is therefore:

$$DoF \sim (L/L_d)^3 = Re^{9/4} \tag{5.4}$$

This means that the number of relevant degrees of freedom which affect the motion of our car scores in the order of $10^7 \times 9/4 \sim 10^{16}$, also known ten millions of billions! This is much smaller than the number of molecules in the same cube, which counts tens of Avogadros of them, i.e. 10^{25}, a hundred billion million times more.

This means that each vortex contains approximately hundred millions molecules, a dramatic compression of information. But not dramatic enough to quench complexity, first because a billion billion vortices is still a large number and second because these vortices interact with each other in a more orchestrated way than molecules they are made of.

Indeed, these numbers are more than enough is more than enough to unleash a literal Pandora's box of Complexity. Amazingly complex structures can pour out of it, which fill up space and time in a most pervasive and inventive way, over many scales of motion, as we have just seen even in the humble case of an ordinary car. Interstellar winds (turbulence plays a key role in the formation of structure in our Universe, including galaxies, stars, and planets), tornadoes, hurricanes, gusts of wind, but also smoke and the circular rings in your coffee as you mix sugar in, all fall under the huge umbrella of turbulence, sometimes also called the 'fourth' state of matter.

So, is Turbulence the extreme face of Chaos?

Actually, no: it isn't! turbulence is more complex than that, it is a sort of subtly organized chaos in space and time, where 'subtly organized' means that the multitude of vortices and structures which make a turbulent flow so hard to predict (again, weather forecast is the premier example) are not random at all. They are *correlated* in both space and time; the way one moves affects the way others do, which is where complexity arises from. In qualitative terms, it is as if we would have a quadratic map for the flow velocity, similar to the logistic one, at each point in space except that the map depends not only on the fluid velocity and pressure (four variables) at that given point, but also on the velocities and pressure at nearby positions. This spatial coupling in is essential to the development of turbulent structures. As a result, upon iterating in time, the information propagates from place to place in space as well (gossip is a good analogy) until it invades the entire system. This highly complex spacetime structure is the literal fabric of turbulence and the reason why turbulence, at variance with chaotic systems, does not seem to settle into any low-dimensional attractor such as the Lorenz one.

The reader can now appreciate, if only on qualitative grounds, the physical and mathematical origin of the Pandora's box of complexity called Turbulence. In Chapter 1, we wrote that the science of complexity explores the frontiers of the human mind and, notwithstanding the fact that it is two-centuries old, purely Newtonian physics, with no quantum physics and no relativity, turbulence still stands out as one the most challenging problems of modern science.

5.6 The pleasure of being unpredictable

In this chapter we have illustrated the 'dark side' of nonlinearity and its direct connection with the threatening notion of Chaos, as the quintessential epitome of power beyond our control. Chaos theory provides us with a less-bleak metaphor, the naughty butterfly standing for hypersensitivity to fluke changes.[21] We have also touched briefly on turbulence and mentioned the reasons why it cannot be reduced to chaos. Yet, the identification of Chaos or turbulence with purely destructive power is a very poor caricature of reality, as both can be put at our service to enhance the performance of a number of devices in physics, engineering, and biology. In Chapter 6 we shall dig deeper into the shiny side of the coin, namely the constructive power of nonlinearity.

5.7 Appendix 5.1: More on the logistic map

Consider again the logistic map:

$$N(t+1) = rN(t) - cN^2(t) \tag{5.5}$$

and define the reduced population as $x = N/N_c$, i.e. the number of individuals divided by the capacity $N_c = (r-1)/c$. The logistic map takes then the form

$$x(t+1) = rx(t) \times (1 - \frac{r-1}{r}x(t)) \tag{5.6}$$

The steady-state condition $x(t+1) = x(t)$ defines the capacity for the reduced population as $x = 1$, as it should be. This is very similar to the logistic map, except for the prefactor $(r-1)/r$ at the right-hand side, which is precisely the capacity of the logistic model (always smaller than 1). Repeating this map from cycle to cycle, delivers the time sequence

$$x(0), x(1) \dots x(t)$$

namely the trajectory of the system. The exercise is simple because each step involves just two subsequent cycles, which means that $x(1)$ is a quadratic function of $x(0)$, $x(2)$ is a quadratic function of $x(1)$ and so on down the line. But now let us take another perspective, namely express $x(t)$ as a function of the initial value $x(0)$. This is precisely what we mean by 'dependence on initial conditions'. Since the mapping is quadratic, $x(2)$ depends on the -fourth- power of $x(0)$, and likewise $x(3)$ depends on the-sixth-power and so on (remember the grain of rice story earlier on in this chapter). Now imagine performing 100 cycles, you end up with

[21] I doubt that a fully predictable world is anything we would subscribe to. Quoting verbatim from chapter II of Stefan Klein's *How to Love the Universe* [58]:

> Nature allows us deep insights into its regulatory system while at the same time preventing us from seeing what's really up to. Some scientists might regret that, while others among us will be relieved to note that our emotional life remains incalculable. . . . .. Seems to me that it is precisely this unpredictability that marks the boundary between life and death

Death is way more predictable than Life: no surprise that we like to be unpredictable!

a polynomial of order 200! Here, you start to smell where complexity (and chaos) may hide behind. Please note another crucial point, if the mapping were linear, which is the case without competition, it would remain linear all the time, because any power of 1 returns 1 itself (mathematicians call this property idempotence, which means leave things as they are). Now you see the potential of nonlinearity in action: each single step is simple, just an innocent quadratic equation, but each iteration doubles the degree of the polynomial, as applied to the initial condition, and this is why the solution may depend wildly on the initial condition. Incidentally, this also shows why there is no need of high-order nonlinearities to generate complex behaviour.

6

The Bright Side of Nonlinearity

Criticism is always easier than constructive solutions.
(Jaron Lanier)

6.1 Constructive chaos

In Chapter 5 we have painted a rather bleak portrait of nonlinearity: chaos hidden beneath innocent-looking maps, naughty butterflies undermining our ability to predict the future or make sense of the past. Fortunately, the picture is less bleak than this: equating nonlinearity with chaos and chaos with unpredictability would be a parody of justice to the great richness of nonlinearity, which features plenty of positive sides as well. After all, it is thanks to turbulence that we can drink sweetened coffee without waiting the long hours it would take for sugar to mix in, if it would rely entirely on molecular collisions. While it is true that lack of predictability is a major inconvenience, I would not want to leave the reader with the impression that chaos is a purely destructive mechanism. Modern research in nonlinear systems has brought to light a number of remarkable instances in which chaos can be put at work to achieve goals that would be unattainable without it.

A most important case in point is the notion of *controlling chaos*, as first discussed in so called OGY model, named after the American physicists Edward Ott, Celso Grebogi, and James Yorke [83]. These authors have shown that the dynamics of a chaotic system can be controlled by applying carefully chosen small perturbations to it. The key idea is that a chaotic system typically exhibits an infinite number of unstable periodic orbits, any one of which can be stabilized by a small control. Hence, by properly selecting the orbit to be stabilized, enhanced performance can be achieved. Another related example of constructive chaos is the so-called *Stochastic Resonance* mechanism, first introduced by the Italian physicist Roberto Benzi, Alfonso Sutera, and Angelo Vulpiani [12], in the framework of meteorology. This is the mechanism whereby the addition of a properly chosen amount of randomness to a bistable system (a system with two stable minima and an unstable one inbetween) can again significantly enhance its systematic response and performance. Both examples of constructive chaos have found extensive applications in a broad variety in phenomena in physics, engineering, and biology. A detailed description of such phenomena goes far beyond the scope

Sailing the Ocean of Complexity. Sauro Succi, Oxford University Press.
 DOI: 10.1093/oso/9780192897893.003.0006

of this book, but I felt they ought to be mentioned in order to forestall the false picture of chaos as a purely destructive mechanism.

6.2 Nonlinear cooperation

I hope by now the reader agrees that non-linearity is not necessarily bad or threatening, in fact it provides several enchanting examples of *constructive power*, the power of cooperation. To put it in glorious ancient terms, nonlinearity, hence Complexity, is not only chaos but also Cosmos. So, time to illustrate the 'Cosmos' side of nonlinearity.

6.3 Moving information across scales

In Chapter 5, we commented on the ability of nonlinear systems to amplify small disturbances/inaccuracies/errors or any other form of uncertainty and eventually suppress large ones. There is another distinctive mark of nonlinear systems which, albeit less popular, plays a no less of a role in shaping up the world as we know it: *scale coupling*. By this, we mean the ability of nonlinear interactions to transfer information, mass and energy across different scales, from large to small and vice versa. This is pretty much like the 'small yields large' and 'large yields small' discussed in Chapter 5, with a crucial twist, though: instead of referring to the *amplitude* of a given phenomenon, it applies to its *size* in space. This property is no less essential: in a segregated world where large stays large and small stays small, there would be no growth of macroscale structures from molecules, nor would small structures stand any chance of being fed by the energy flowing from the larger ones, another process no less crucial to life. It would be a mostly sterile and uninspiring world without any crosstalk between structures of different sizes, in fact, a lifeless desert. Scientists have coined the world *cascade* to denote this kind of scale mobility, which conveys well the idea of flow of mass/energy/information from large to small (direct cascade) and from small to large (inverse cascade).

In the following, we shall discuss these cascades a little more.

6.4 From large to small: Breakup

We are all familiar with the beautiful scenery of ocean waves getting taller and steeper as they approach the shore, until they break out in a myriad of picturesque foamy shapes. The details of such process are excruciatingly hard, but the basic mechanism is pretty simple, in fact it is almost pure kinematics. If the velocity profile takes the shape of a wave, meaning by this that the top of the wave moves faster than its front and its rear side, it is readily understood that in time,

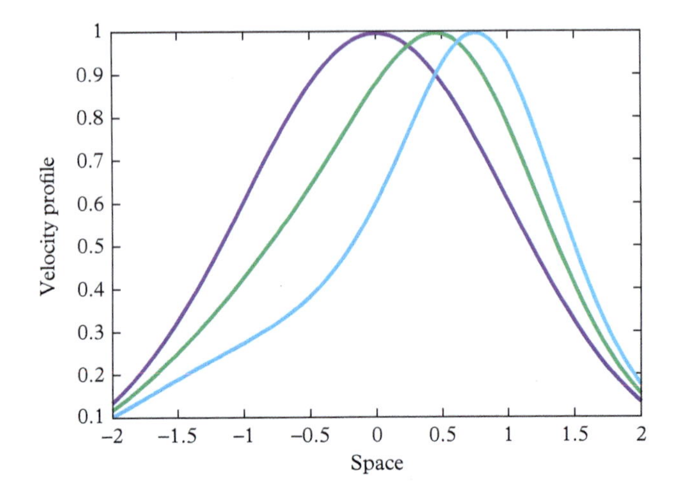

Figure 6.1 *The steepening of a velocity wavefront. Particles at the top travel at the highest speed, hence they get increasingly closer to the one ahead and increasingly ahead of the ones behind. The result is a steepening of the velocity profile which leads to a higher concentration of energy in a smaller region of space. As time goes on, particles on top overcome those ahead and the profile breaks down.*

the fluid molecules on the crest get closer and closer to the front ones and farther and farther from the rear ones. In other words, the wave gets steeper (see Fig. 6.1). The steepening comes to an end at some critical point, when the top molecules overtake the front ones. The very moment they do, they are left suspended in the air, since the formerly front molecules are left behind and cannot sustain the top molecules any longer, the result being that the wave breaks down under the effect of its own weight (see Fig. 6.2). To the detriment of poetry (perhaps), the attentive eye will notice that steeper profiles means that the energy contained in a smooth and long wave far out in the ocean, is transferred almost loss-free into a steeper and shorter wave near the shore. In other words, steeper here means *same velocity change confined within a smaller distance.*

This energy transfer from large to small is quintessential non-linearity in action; it is only possible because the equations which govern the dynamics of the water fluid are non-linear. To be more precise, as we already commented on in the case of turbulence, this instability derives from the change in space of the kinetic energy of the wave, which is the square, hence a nonlinear function, of its velocity. The picture evokes a sense of poetic chaos, the elegant and smooth wave crest breaking up in small pieces, all carrying a partial glimpse of the original beauty, which is nevertheless lost forever into a myriad of shiny droplets. Nonlinearity feeds shape destruction at large scales, but a very fertile destruction indeed, as it gives birth to new forms at small scales.

Figure 6.2 *A spectacular instance of nonlinearity: breaking waves.*
Source: reprinted from en.wikipedia.org.

There is a very positive side to this story: by promoting breakup, non-linearity manages to feed energy into small-scale structures, which would otherwise be left starving. To wax it lyrically, the death of large structures gives life to small ones, an essential process to keep the circle of life going. This is literally true, beyond metaphor: just think of the numerous multiscale networks that keep our body going, cardiovascular and respiratory in the first place. They are in charge of transporting small pieces of matter, blood cells in the former and oxygen molecules in the latter, down to the cellular level, where the vital intake processes take place. This is part of an even more general process, that Adrian Bejan has promoted to the status of 'Constructal Law', according to which nature has been designed in such a way to optimize the flow of matter across its structures, typically in the form of multiscale networks [14]. And, in the typical feedback loop of complex adaptive systems, such networks would evolve in such a way to optimize the traffic they support. Whether or not one may want to embrace Bejan's constructal law, the fact remains that even before you worry about the best way of moving matter from one place to another across a multiscale network, there must ways of *generating* matter at all such scales.

Which is exactly what breakup and coalescence do.

6.5 From small to large: Coalescence

The reciprocal process, from small to large, is called coalescence. Many phenomena in nature are characterized by the growth of shapes out of the merger of smaller constituents: molecules from atoms, cells from molecules, organs from cells, and so on up the ladder taking from the microscopic world of the 'things we cannot see' to the macroscopic or even megascopic objects which abound in our Universe. The overall history of the Universe, since its inception to the current days, is a tale of structure formation from the Big Bang on. This upward climbing of structures, the inverse cascade in physics parlance, is again quintessential nonlinearity in action, and one whose constructive power is hardly in need of any comment. But again, the positive face can be flipped upside down; we are all too familiar with the fatal aftermath of uncontrolled growth in biology and virtually all other ecological endeavours. Be as it may, the coalescence of structures is often, if not always, the result of an energetic bargain.

Consider liquid droplets in a vapour atmosphere: it takes energy to keep them apart from the surrounding vapour, and that energy is proportional to the area of the droplet surface. As a result, by coalescing two small droplets into a single and larger one with the same volume, the area is reduced, and so is the energy cost to sustain the newly formed droplet. This is basically the mechanism of liquid condensation and, in the case of nuclear energy, of nuclear fusion as well, the process by which we receive light (and life) from the Sun. Those willing to do the elementary math are mostly welcome to walk through the Appendix 6.1. For the others, the argument is that the volume is a cubic power of the radius while the area goes quadratically only. The result is that droplet mergers reduce the surface/volume ratio, hence the energy cost of the configuration. As a general rule, coalescence results whenever merging of multiple substructures into a single one is energetically convenient, a process which is generally driven by attractive forces at a microscopic level. The opposite holds true for breakup, which results from an energetic gain in splitting, as typically occurs in the presence of microscopic repulsion (see Figs. 6.3 and 6.4).

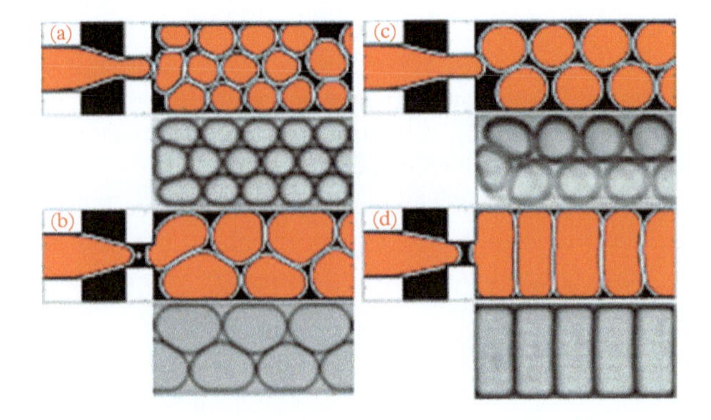

Figure 6.3 *Breakup of a liquid jet into droplets in a microfluidic device, simulation(top) and experiment (top). From A. Montessori et al. (2019), Physics Review of Fluids. 4: 072201(R). This process is key for the design of new families of droplet-based materials.*
Source: reprinted with permission of aps.org.

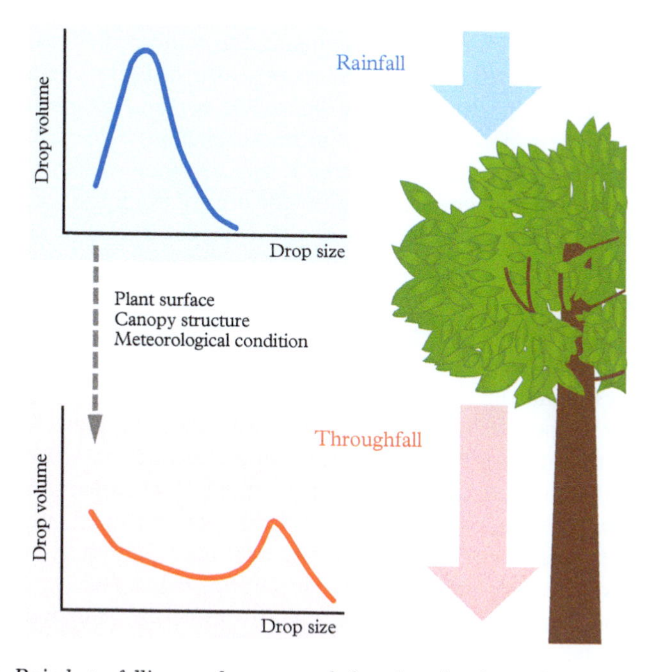

Figure 6.4 *Raindrops falling on the trees, and changing size due to breakup and coalescence processes triggered by their impact on the leaves. As the figure shows, the overall effect is to enhance the volume of larger droplets falling on the soil.*
Source: redrawn from onlinelibrary.wiley.com.

In physics, chemistry, and biology, both processes occur in a variety of phenomena: molecules group together to form small droplets and other supramolecular structures essential to biological functions. For instance, cells need protection from the vagaries of the external environment, in order to carry out their job in due peace, which is where segregation proves instrumental. At the same time, cells also need to split to generate new cells (mitosis) which is the basic act of replication. The details can be painfully hairy, but the organizing principles are adamant: segregation buys protection and peace to carry out the useful work, while fragmentation buys the capability to replicate and expand. The study of these fascinating phenomena, *where nothing is fundamental in the sense of elementary particles or cosmology, yet no less important*, forms a very active front of modern science, going by the name of *soft matter*. The name is apt, since soft matter deals with those states of matter most relevant to biological structures, which are typically soft and flexible. We shall have more to say about this subject in Chapter 15.

The reader may wonder what the role of nonlinearity is in all this. The point is that, as mentioned previously, the energy cost of forming a droplet of radius R consists of two terms, one proportional to its volume, hence to the cubic power of the radius, R^3, and one proportional to the area of its surface, hence proportional to the square of the radius, R^2.[22] The former is due to the attractive forces between molecules which, by convention, corresponds to a negative energy. Under the effect of pure attraction, coalescence would never come to a stop because energy keeps getting more and more negative upon aggregation. The surface term however brings this untamed growth to an end. Indeed, since sustaining an interface comes with an energy cost proportional to the area of the interface, the surface

When do the two competing processes come into a balance, namely equilibrium?

This happens at a critical value of the radius, known as nucleation radius, such that a small change to it, either positive or negative, leaves the energy unchanged. Below this value, aggregation is energetically favoured by the attractive forces, while above it is disfavoured by the cost of building the interface. Again a story of competition, and the nonlinearity here is purely geometrical: cubic for the volume, quadratic for the surface. If the two terms depended both on volume or both on surface only, the geometrical competition would dissolve and so would the whole process of condensation with it. To be sure, the picture is a bit subtler than this, as it involves both attractive and repulsive forces acting a molecular level, which depend very nonlinearly on the mutual distance of the molecules. We shall have more to say on this in Chapter 15. For now, it suffices to retain that coalescence and aggregation are driven by energy budget considerations, and such budget is governed by terms which depend nonlinearly on the size of the droplets.

6.6 Morphogenesis: The Turing model

One of the most important instances of nonlinear cooperation is the phenomenon of morphogenesis, that is, structure formation. The theoretical foundations of morphogenesis were laid down in the epoch-making paper 'The chemical basis of morphogenesis' published in 1952 by the British mathematician Alan Turing (1912–1954) [121] (see Fig. 6.5). Besides being one of the most impactful figures of modern science, for many the founder of computer science, Turing is also known to the public for his tragic life, as portrayed in the movie 'The Imitation Game', describing his heroic effort to automate the decryption of German secret codes during World War Two.

Be that as it may, Turing devised the following simple model. He imagined a regular array of 'chemical reactors', each sitting in the node of a regular grid. In each 'reactor' two species, say X and Y, transform into each other according to some prescribed nonlinear chemical rules. If each reactor would proceed on its own, each would reach the corresponding steady-state (if any) depending on

[22] To be sure, we should talk of *free-energy* cost, which is a combination of energy and entropy, as discussed in Chapter 8. For the sake of the points to be made in this chapter, we can safely ignore this distinction.

Figure 6.5 *A young Alan Turing (1912–1954), a founding father of modern computer science.*
Source: reprinted from en.wikipedia.org.

the initial conditions, regardless of the other reactors. Things take a much more interesting twist by assuming, as Turing did, that both species can migrate, or hop, from one reactor to its nearest neighbours, thereby generating an effective coupling between the whole string of reactors.

Spatial communication, combined with the local nonlinearity of the chemical reactions, is the key to morphogenesis. Indeed, upon solving the equations of his model, Turing discovered the emergence of various sorts of fascinating patterns, such stripes, fronts, hexagons, spirals, and so on, now justly known as *Turing patterns*. The association with real patterns observed in nature, like fish skin pigmentations, zebra and giraffe skin textures, and similar, is compelling (see Fig. 6.6). Quoting Turing himself: 'It is suggested that this might account, for instance, for the tentacle patterns on Hydra and for whorled leaves'.

The underlying principle is as subtle as it is powerful: the combined effect of nonlinear chemistry and spatial communication (the technical name being *reaction-diffusion*), gives rise to a selective mechanism for the size of the structures which can survive. In broad strokes, the small ones die out and the large ones can grow and prosper. Or, in more precise language, below a critical size they die out, and above it they grow till saturation (remember the logistic map . . .).

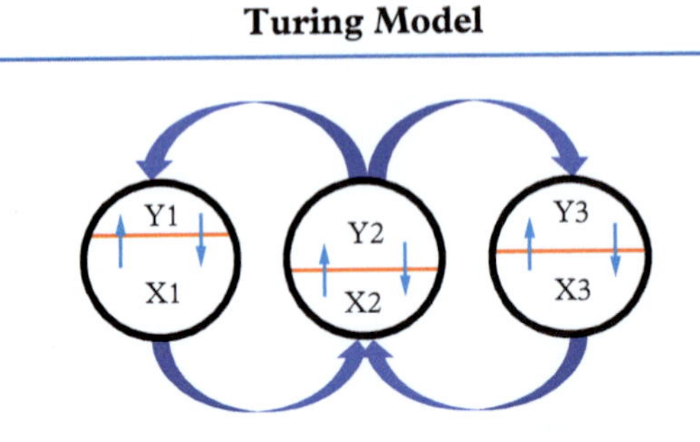

Figure 6.6 *Sketch of the Turing model. Three 'reactors' host two species, X and Y, which transform into each other due to the chemical reactions occurring within each reactor (vertical arrows). The three reactors are spatially coupled, so that any excess of species X or Y is transferred to the neighbouring reactors and vice versa. For instance, in the case of the figure, the central reactor has less amount of X than its neighbours, hence both neighbours transfer some amount of species X to it (bottom arrows) to compensate for the deficit. Conversely, having a higher content of species Y, the central reactor transfers a given amount of Y to both left and right neighbours. Chemistry and diffusion constantly compete and cooperate, and the Turing patterns are precisely the result of such competition and cooperation.*

Figure 6.7 *Typical Turing patterns, resulting from the combined action of chemical reactions and spatial diffusion.*
Source: reprinted from en.wikipedia.org.

Note that the conspiracy of the two, reactions and diffusion, is absolutely key: neither of the two alone would be able to achieve structure formation (see Fig. 6.7). The attentive reader may argue, how can large structures grow out of small ones if the small ones die out in the first place (See Fig. 6.8)? This is where the nonlinear cascade makes all the difference and provides an escape route from death: if the small scales stay small, they die, but if they merge and manage to coalesce into larger ones, above a critical size, they will survive, in fact grow and prosper! The cascade is a literal wave of life across scales! This is the paradigm of a *constructive instability*, the physico-chemical substrate of controlled growth. Turing considered a specific set of chemical reactions, but the paradigm holds for an extremely broad class of them, whence its crucial importance. This is why the so called reaction-diffusion models pioneered by Turing have taken central stage across

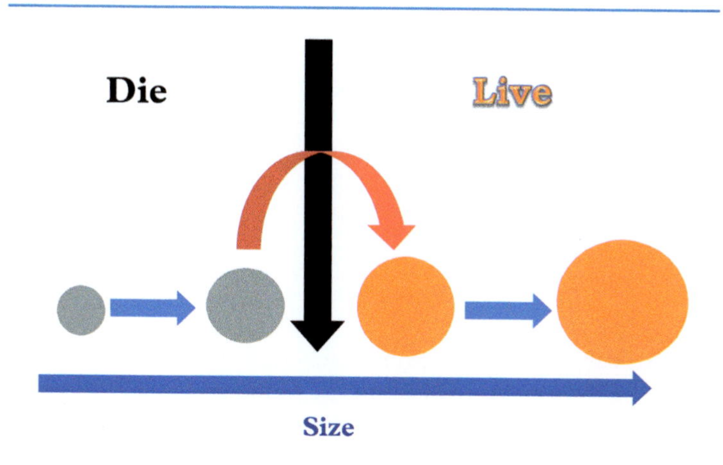

Figure 6.8 *The reaction-diffusion cascade: coalescence of small-scale structures in to larger ones above a critical threshold permits their survival. That's how coherent pattern forms out of small-scale noise.*

all disciplines dealing with structure formation, evolutionary and developmental biology at the forefront.

6.7 Summary

Nonlinear systems turn small inputs into large outputs (butterflies) and occasionally they do the opposite too (elephants). They turn small structures into large ones (coalescence) and occasionally they do the opposite too (breakup). In fact, this could almost be taken as an operational definition: a complex system is one that can do something and its opposite too. In a word, they do 'whatever they are allowed to do', since they enjoy a spectacular and sometimes frightening freedom. Nonlinearity is not necessarily bad, as the naughty butterfly would imply, nor necessarily good either, as the morphogenetic picture would suggest. In fact, it is neither bad or good, it simply is: an immensely powerful 'force' shaping the complex Universe around and within us (our brain is utterly nonlinear . . .). Such an immensely powerful force breathes enormous freedom into our Universe, and the very freedom that leaves us enchanted in front of a breath-taking sunset or delights our ear and heart upon to listening Beethoven's Moonlight Sonata, is also the same freedom which triggers devastating hurricanes before we can see them coming. It is the freedom which feeds both the anxiety about and the fascination for the unknown, two ancestral sentiments which have been with us, and probably driven us to this point, since the dawn of humankind. That is why Nonlinearity, hence Complexity, are deeply enmeshed with the human condition, a theme we shall return to in the final part of this book.

6.8 Appendix 6.1 Breakup and coalescence

Consider two droplets of radius r, merging into a single droplet of radius $R > r$ with the same volume. Since the volume is unchanged, we have:

$$V = \frac{4}{3}\pi R^3 = 2 \times \frac{4}{3}\pi r^3$$

which delivers

$$R = 2^{1/3} r$$

where $2^{1/3}$ is the cubic root of 2, approximately 1.26. Thus the radius of the single droplet is much less than the twice the radius of the two parent droplets. This is because in three spatial dimensions, the volume goes with the third power of the radius, a decidedly superlinear relation.

Now to areas. The area of the two parent droplets is $A_2 = 2 \times 4\pi r^2$, while the area of the single daughter droplet is $A_1 = 4\pi R^2 = 2^{2/3} 4\pi r^2$. Since 2 is greater than $2^{2/3} \sim 1.59$, the single droplet features less area for the same volume, or if you wish, a smaller surface to volume ratio. The surface to volume ratio is a key quantity in all processes associated with breakup and coalescence. Since it is precisely this ratio which controls the energy cost of keeping the liquid droplet intact in a vapour atmosphere, merging is a way of minimizing energy. Note that the sphere is the geometrical shape that minimizes this ratio. Indeed, for a sphere of diameter D, we have $S/V = \frac{\pi D^2/4}{\pi D^3/6} = \frac{3/2}{D}$. For a cube of side D, we have instead $6D^2/D^3 = 6/D$, which is four times larger. Thus, is takes six-times more surface energy to form a cubic droplet instead of a spherical one. And pretty much happily so, just think of the nightmare of cubic raindrops! It turns out that the energies in point depend on the spatial distribution of the droplet versus vapour densities, and this dependence is indeed a nonlinear one. If it were linear, there would be no driving forces for the coalescence process. Breakup works just the same way, but in reverse.

6.9 Appendix 6.2 Wave steepening

So far, we have discussed nonlinearity using numbers, now let us consider *sequences* of numbers, known as functions. Functions describe how a system, say the temperature in this room, changes from point to point in space. One says that temperature is a function of space. The curious reader may justly wonder what the basic mechanism is by which nonlinearity manages to turn 'small' things into 'large' ones and vice versa. Let's illustrate the point with some basic math. Two basic periodic functions are the sine $sin(x)$ and cosine $cos(x)$. The first starts from zero and oscillates between +1 and −1 in periods of 2π. The second does the same, except that it starts from 1 at $x = 0$, or, it lags behind by a phase $\pi/2$. They are complementary, in the sense that where one is zero, the other takes its positive

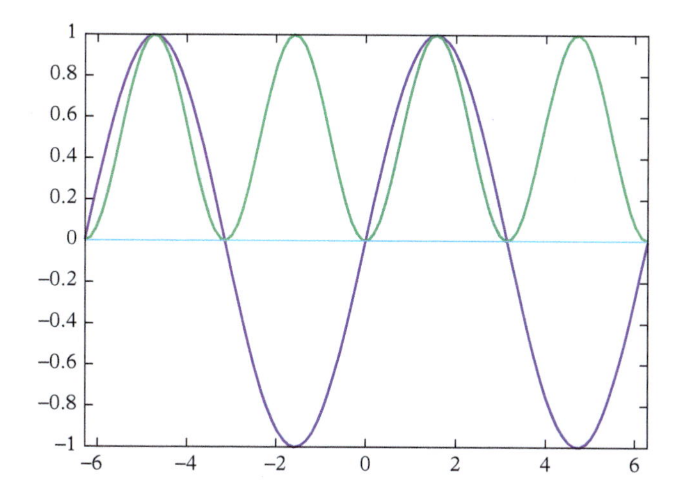

Figure 6.9 *Graph of the function $sin(x)$ (violet) and its square in the interval $(-2\pi, 2\pi)$ (green). The former shows two crests and two dips at $+1$ and -1 respectively, while the latter forms four at $+1$ and 0 respectively. Squaring the sin function generates a twice more compact structure.*

and negative peak, plus or minus one, and vice versa. Now let us square them, i.e. multiply them point by point, to form the product $sin^2(x) = sin(x) \times sin(x)$ or $cos^2(x) = cos(x) \times cos(x)$. In the first place, both functions are positive, because negative by negative gives positive. Most importantly, as one can check by looking at the graph shown below, the sin squared shows a doubled number of peaks, spaced π units apart instead of 2π. This means that by the very act of squaring them, the sine and cosine give rise to a new profile with half periodicity, hence twice smaller scale structure. This makes perfect sense, because the peaks of $sin(x)$ remain peaks of $sin^2(x)$, but the valleys are turned into peaks as well, just because, again, -1 times -1 makes 1. This is how and why squaring a periodic function halves its period, thus literally transferring information from large (period 2π) to half smaller scales (period π).

7

Networks, the Fabric of Complexity

Networks are the fabric of complexity.

(A. Barabasi)

7.1 A must in Complexity

Catchy statements are good at grabbing the attention, but rarely capture the truth in equal proportion. Yet, exceptions exist, and I find Barabasi's previous statement a good example in point. Indeed if there is a broadly encompassing conceptual and computational framework for Complexity, it is the network. Which is why, with thanks to an anonymous referee, I made up my mind to devote a separate chapter to this topic. Besides honouring a must in Complexity, this offers the opportunity to call the readers' attention to a particularly simple instance of networks and associated rule-driven systems, known as Lattice Gas Cellular Automata (LGCA) and its derivative, the Lattice–Boltzmann (LB) method. Both subjects hardly find a mention in the modern books on Complexity, which is a bit surprising as they provide a rare and precious instance of rule-driven discrete systems which solve complex fluid problems in a very quantitative sense, far beyond the level of inspiring analogies.

So, let's begin by discussing the distinction between rules and equations.

7.2 Rules and equations

Einstein is credited for stating that scientists don't think with equations. While in my experience some do, by and large it is true that equations come as the mathematical formalization of qualitative ideas and intuitions about the *rules* which govern the phenomenon we aim to describe. In this respect rules are certainly more general than equations, or, at least the kind of equations Einstein had in mind, known as *differential* equations. Indeed, the most fundamental equations of physics are presented in differential form, meaning that they are derived under the assumption that space and time, if not matter, are indefinitely divisible, the so called *continuum limit*, which we discussed previously in this book. Even though

Sailing the Ocean of Complexity. Sauro Succi, Oxford University Press.
 DOI: 10.1093/oso/9780192897893.003.0007

the spacetime scales of biology set hardly any challenge to the continuum assumption, the fact remains that many complex systems are best described by models in which, not only matter, but also space and time are discrete. And where everything, matter, space and time, is discrete, differential equations no longer hold, while rules and algorithms still do.

This brings us back to Zeno.

7.3 Dismantling Zeno's trick: The lattice world

The reader should appreciate that Zeno's trick would cut little ice in a discrete-time world. To begin with, the procedure simply cannot be iterated *ad libitum*; it comes to an end as soon as the tiny bit of space ahead of the tortoise falls below the minimum allowed space interval. A discrete spacetime is a powerful antidote against paradoxes and infinities, simply because zeros in the denominator of the effect/cause relation are outlawed from the outset. In technical language, such discrete spacetimes are known as, *lattices*, a discrete and finite set of points in space and time (spacetime crystals, in a more evocative language) connected to each other by a regular pattern, such that the minimum distance between two connected points is not and *cannot* be made zero. The simplest lattice in one spatial dimension is a chain of nodes, each connected to its left and right neighbours at the same distance, the lattice spacing, as depicted on the top of Fig. 7.1. The two-dimensional analogue is a grid of nodes, each connected to four or six equi-distant neighbours, left and right, up and down, and so on for higher dimensions (see Fig. 7.1).

The number of connected neighbours is known as *connectivity* and plays a crucial role, together with size, in controlling the properties of the *collective* phenomena which may occur in the lattice world. Amazingly, simple rules on each single node with nearest-neighbour connectivities (very simple networks indeed)

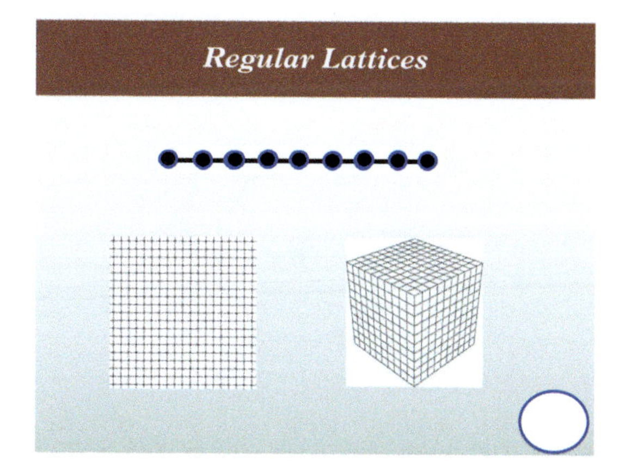

Figure 7.1 *Simple lattices in one, two, and three dimensions.*

prove capable of generating fairly complex behaviour at the scale of the entire lattice. For instance, the whole physics of fluids, a gold mine of Complexity, can be generated this way and some extreme thinkers maintain that not only fluids, but the *entire Universe* works this way [122]! This is an interesting story which played an important role in the historical development of the Science of Complexity, hence worth a few additional words.

7.4 Hamletic fluids: Lattice gas cellular automata

Fluids are a pervasive and paramount presence across virtually all walks of science and industry, not least our own life, as we know it: just think of the powerful triad Air, Blood, and Water. They also stand as a paradigm of Complexity, as witnessed by the breathtaking diversity of shapes they can take, be it the whirls of water in a river flow or the current streams in the sky. Not surprisingly, the wind provides a source of vivid metaphors for the human inability to predict and harness the future, from the Holy Scriptures, 'The wind blows wherever it pleases. You hear its sound, but you cannot tell where it comes from or where it is going. So it is with everyone born of the Spirit' (John 3:8), to the more secular Bob Dylan 'the answer my friend, is blowing in the wind'. Fluids hardly make centre stage in the modern theory of Complexity, which is a pity, given their pervasive role in science at large. It is therefore of some interest to show that simple networks of even simpler entities, known as *automata*, provide eminently interesting and *quantitative* models of fluid flows. The uninspiring name is LGCA, which we are next going to decrypt. Here we go.

Consider a hexagonal lattice in which each node connects to six equispaced neighbours, see Fig. 7.2. At each lattice site, place up to six 'particles', each of which can only move along one of the six links emanating from the site, numbered 1 to 6 counter-clockwise for 1 = East, 2 = North-East, 3 = North-West, 4 = West, 5 = South-West and 6 = South-East. There is a universal clock, an invisible director,

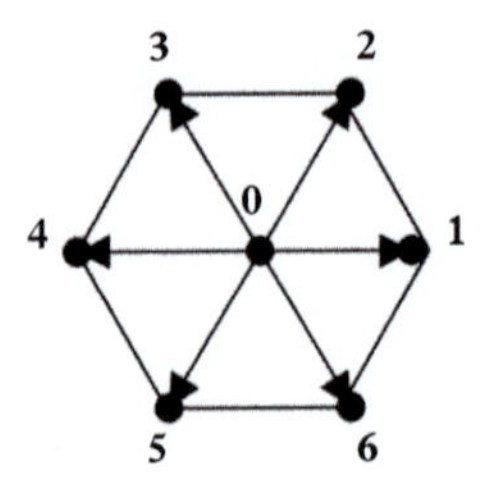

Figure 7.2 *The Frisch, Hasslacher, Pomeau (FHP) automaton, after the official inventors Uriel Frisch, Brosl Hasslacher, and Yves Pomeau. For the record, Stephen Wolfram made important early contributions to the subject.*
Source: reprinted from S. Succi (2001) *The Lattice Boltzmann Equation*, Oxford University Press

that calls the ticks at regular intervals, say one second just to give an idea. At every tick, each particle hops to the neighbouring site along the corresponding link. That is, the east mover hops from the central site 0 to the east site 1, the north-east mover hops from 0 to the north-east 2 node, and so on for the six of them. Most important, the particles are purely ontological entities (automata), in that they have no property other than their presence (to be) or absence (not to be). As a result, the automata are fully defined in terms of a single bit, one per each of the six links, which explains the label 'Boolean'.[23]

Let $n_1 \ldots n_6$, the so-called *occupation numbers* of the links, where $n_1 = 1$ means that there is an automaton moving eastwards, while $n_1 = 0$ means that no such automaton exists. Repeating the idea for each of the six links, it is clear that the state of the automaton at each lattice site is uniquely characterized by a string of six bits $\{n_1, n_2, n_3, n_4, n_5, n_6\}$. Given that there are six bits, the state of the automaton at each node spans $2^6 = 64$ possibilities, ranging from the empty site coded by six zeros $\{0, 0, 0, 0, 0, 0\}$ to the fully-occupied site, coded by six 1s, $\{1, 1, 1, 1, 1, 1\}$. This is a simplified cartoon of a real molecular system, in that, unlike automata, real molecules come with virtually any velocity, not just six, and they can occupy virtually any position in space, not just the hexagonal crystal discussed here (the lattice). Yet, and this is the big thing, if the rules of the game are properly designed (see the following information), this poor cartoon proves nonetheless capable of displaying the full array of highly complex patterns exhibited by real-life fluids! Beware, real-life fluids means that the behaviour of fluids is reproduced in *quantitative* form, not just at the level of 'coloured wall-paper' pictures . . . .

What are these magic rules, then? Here comes the secret.

In a real fluid, molecules interact via rather complicated forces which dictate their erratic motion, a sort of molecular chaos. But Complexity runs deeper than chaos and indeed, notwithstanding their erratic wandering, the motion of the molecules obeys general principles known as *conservation laws*, typically mass, momentum, and energy. What does this mean? It means that in a collision between molecule A and molecule B, the total mass $m_A + m_B$ remains the same as it was before, and so do the momentum $m_A \vec{V}_A + m\vec{V}_B$, and energy $(m_A V_A^2 + m_B V_B^2)/2$. These conservation laws impose stringent constraints on the molecular motion, and it is precisely because of these constraints that the motion of fluids is complex, not random, i.e. a subtle and ever-changing blend of order and disorder. How can the simple lattice automata account for such a subtle and complex blend?

To see how this happens, let us place ourselves on a generic node of the lattice: such nodes receive movers from the six neighbouring sides, i.e. the east

[23] Boolean calculus, from the British mathematician George Boole (1815–1864), refers to the branch of mathematics that deals with variables which can only take values 0 and 1. Boolean algebra is defined in such a way that Boolean variables remain Boolean in the course of the various mathematical transformations they undergo via Boolean calculus. Needless to say, Boolean algebra is key to logic calculus, where 0 means 'true' and 1 'false' (or vice versa) and computer science.

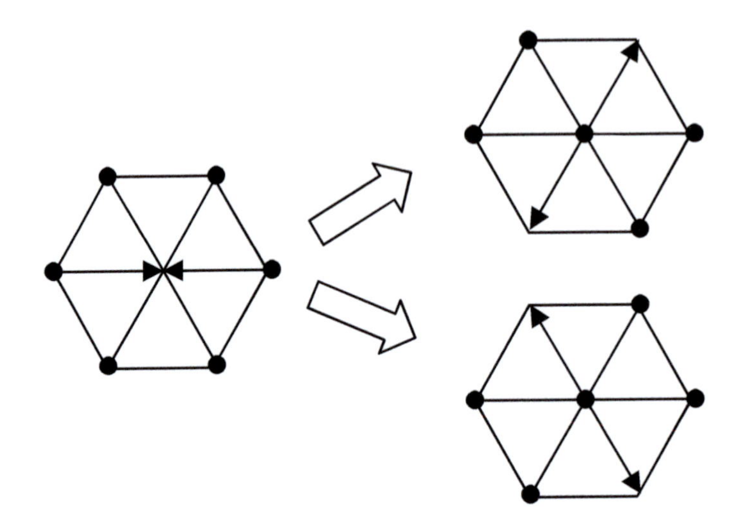

Figure 7.3 *A typical head-on lattice collision, scattering off at 60 degrees clockwise (bottom) or counter-clockwise (top).*
Source: reprinted from S. Succi. The Lattice–Boltzmann Equation, Oxford University Press 2001.

mover from the west node, the west mover from the east node and so on. Once they meet at any given node, the automata interact through 'molecular collisions' which scatter them off along the different directions in the lattice. For instance a head-on collision between East and West movers may results in 60 degrees scattering to a north-east–south-west pair (see Fig. 7.3). Once again, such collisions are only a simplified cartoon of the actual collisions taking place in a real fluid, and yet they can be rigorously shown to fit the hydrodynamic bill. Here is the key: even though the lattice collisions are dramatically simplified cartoons of real molecular collisions, with incommensurably many fewer details, they nonetheless respect the aforementioned *conservation laws*. And, since hydrodynamics is basically the *emergent* behaviour of zillions of underlying molecules, subject to mass-momentum-energy conservations, the two systems, real molecules in a real fluid and Boolean molecules in the lattice, yield *exactly the same* emergent behaviour!

Fig. 7.5 shows one of the first heroic simulations of fluids using Boolean molecules, a subject which captured enormous attention in the mid-80s, as a new revolutionary technique to simulate fluid flows [96, 25]. Even the iconic Richard Feynman (see Fig. 7.4) in the late stage of his career developed a keen interest in this subject.

Unfortunately, LGCA did manage to become the mainstream of fluid simulation for a variety of reasons which have been discussed at depth in the literature. Even though LGCA didn't live up to their (extremely) high expectations, they nevertheless carry a major conceptual value as a working paradigm of 'Emergent Complexity': simple models of complex flows [54]. Incidentally, they also gave

Figure 7.4 *The iconic Richard Feynman, along with Einstein, and Stephen Hawking, one of the very rare scientists who managed to attain worldwide popularity outside the scientific circles.* *Source*: reprinted from it.wikipedia.org.

birth to a pretty successful method, known as LB, which has kept this author happily busy for the last three decades .... [110, 111]. Being part of a small group of people who contributed to the migration from LGCA to LB, I feel like a few words of personal comment are warranted.

7.5 Probabilistic fluids: Lattice Boltzmann

Friends come and go, but enemies accumulate.

(Thomas Jones)

Despite the initial enthusiasm, LGCA didn't make it to a successful tool for practical computational fluid dynamics, and after a few years and very brilliant attempts to make it computationally competitive, it was basically abandoned. Not without fruit though, as it gave birth to the less revolutionary but far more efficient LB method.[24] The reasons for the loss of interest towards LGCA are multifold and pretty technical, but it is nonetheless worth conveying the flavour of the main issues. The first problem was *statistical noise*: since automata are by

[24] Having written two books on this very subject, with the same editor as this one, [110, 111], I probably shouldn't delve much into the subject here.

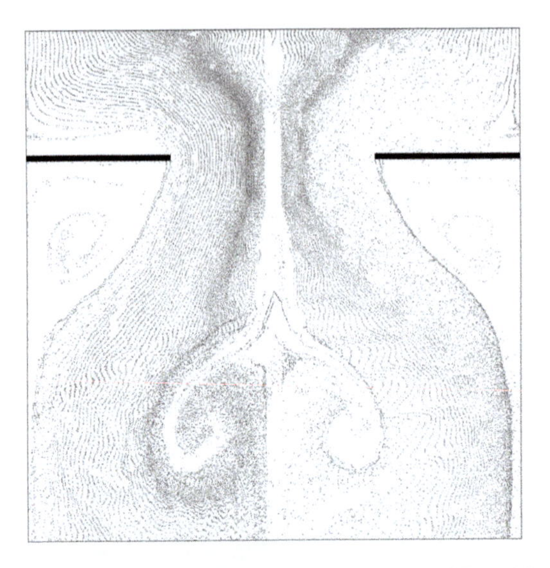

Figure 7.5 *LGCA simulation of a fluid flow through an orifice followed by an aperture. The picture shows evidence of recirculation patterns past the aperture, a typical signature of collective fluid motion. These coherent patterns witness the ability of the lattice gas automata to capture the tendency of real molecules to organize in large groups, all moving coherently. This is the Order emerging from the underlying conservation laws. Beware, Order does not equate to predictability, because ordered motions can prove fairly complex too! This is the wind metaphor discussed in the main text, a literal one, as the wind is a quintessential fluid effect!*
Source: courtesy of Bastien Chopard.

construction Boolean entities 0 or 1, you need a huge multitude of them to give rise to smooth and graded signals, such as those carried by a real fluid. Think of a smooth image on your screen versus a crowd of black and white dots. The second and even more vexing problem was that the *computational complexity* of the collision rules grows exponentially with the connectivity of the lattice. In a two-dimensional hexagonal lattice, with six neighbours per node, there are about $2^6 = 64$ Boolean operations to perform. But the three-dimensional world tells a much more demanding tale; it turns out that the number of neighbours grows from 6 to 24, with the result that the collision rule implies $2^{24} = 16,277,216$, i.e. over 16 million Boolean operations! This proved to be the real show stopper for LGCA.

Lattice Boltzmann was invented on the wake of the first problem, statistical noise. The basic idea is to replace ontology with probability: instead of tracking the automata deterministically, either they exist (1), or they don't (0), the attention is shifted to the *probability* that they occupy a given link of the lattice at any given time. Such probability is a real number which can sit anywhere between 0 and 1, covering the full range of values in between. This automatically eliminates the noise problem because, by construction, the LB probability represents an average over very many Boolean automata realizations. The major and highly unanticipated gift is that the flexibility of the LB representation also permits to overcome the exponential barrier of the Boolean collision rule. Without proof, let me just mention that instead of the 2^{24} LGCA operations, LB gets away with only $24^2 = 576$ (in actual practice less than half of that), i.e. a factor of a hundred thousand savings!

No free lunch, such prodigious speed-up comes at a price: Boolean operations are exact, in that they start with 0 and 1 and always end up with the same 0 and 1, so that there is no truncation error in the process. Probabilities are real numbers, and in principle they come with an infinite number of decimals, say $1/3 = 0.33333333333\ldots$, which cannot be represented exactly within a computer, due to the finite number of bits available for such representation (usually 32 or 64). The result is that real numbers are truncated to so-called *floating-point* numbers, leading to an inevitable approximation error. For instance, on a computer with, say, four-digit representation, the real number 1.25635 would be truncated to 1.256, with an absolute error equal to 0.00035, also known as *round-off error*. In current digital computers the round-off error is much smaller than this, typically about one tenth of a million with 32 bits (single-precision in computer jargon) and one hundredth of a trillion with 64 bits (double-precision). Although these are very small numbers, especially the latter, round-off errors remain a source of major concern for computer simulations of complex systems, for fluids in the first place, which may easily take quadrillions (millions of billions) of floating point operations.

To reconnect with the previous chapters, you may think of round-off errors as very tiny butterflies, whose deteriorating effect accumulates in time to the point of possibly compromising the whole purpose of the simulation. As per Thomas Jones's previous statement, like enemies, round-off errors accumulate.

The important asset of round-off freedom had to be surrendered in order to buy the flexibility of the LB representation and the major ensuing benefits for the simulation of fluid flows. In hindsight, the record shows that this proved to be an excellent bargain. Butterflies didn't hurt after all!

7.6 A new kind of science?

The reasons why LGCA didn't make it as a practical tool for simulating fluids, while LB did, set the ground for a series of considerations which transcend the physics of fluids, to impact on the general theme of simulating physics at large. The first point to notice is that LGCA were not abandoned on account of conceptual issues but practical ones instead, basically computationally efficiency. Indeed, computationally efficiency is strictly to related to technological aspects, most notably the architecture and organization of our current (electronic) computers. It turns out that such organization is highly optimized towards floating-point computing, because of the great generality of the floating-point representation of real numbers. If one would find that the Boolean representation permits to efficiently describe general phenomena in physics, it would certainly be conceivable to design computer architectures specifically targeted to optimized Boolean calculus, i.e. calculus based entirely on binary digits, 0 and 1.[25] This brings us to a conceptual issue, namely the generality of the Boolean representation of the physical laws. In this regard, the LGCA lesson comes down quite neatly: if instead of real molecules fluids were made of tiny Boolean automata, hopping on the sites of a crystal-shaped spacetime with suitable symmetries, the physics of fluids, as we know it, would look *just exactly the same!* We just couldn't tell the difference. Now, the extreme supporters of the automata idea 'simply' (!) maintain that *every* physical phenomenon, no matter how complex, can be traced back to armies of automata, relentlessly and blindly abiding by their rules, literally 'computing out' the Universe in the process. This is the radical stance beneath Wolframs's *A New Kind of Science* [122]. The idea is very appealing: after all, what could be simpler than Hamletic 'to be or not to be' ontological entities, which simply exist or don't? No elementary particle could be more elementary than this! Hence, in a way, Cellular Automata (CA) would achieve the supreme synthesis between reductionism and emergence: macroscopic Complexity from microscopic Simplicity. Within this picture, the emergence of the beautiful whirls of a real fluid cannot be inferred from the simple LGCA rules, no matter how long and sharp you may be willing to stare at them. Emergent phenomena can only be observed upon letting the rules unfold in time, i.e. by actually *running* the simulation! This is another major epistemological leap: we cannot *solve* the automata rules the way we are used to solve the glorious differential equations of physics. No, in order to understand what a complex system does, you have to '*simulate*' it!

7.7 Is the Universe a gigantic self-computing automaton?

As mentioned previously, some extreme thinkers, especially the prominent Stephen Wolfram, take the paradigm all the way, by maintaining that the entire Universe is a gigantic cellular automaton, relentlessly crunching Boolean information underneath this beautiful and mysterious macroscopic world. Wolfram devoted his entire monumental book *A New Kind of Science* to promoting this view, showing examples from virtually all walks of science [122]. A view that he has very recently revamped in another massive single-handed work along similar lines, in which CA's take the more modern form of 'hypergraphs' [123]. For all its undeniable fascination, Wolfram's extreme stance does not seem to strike much of a chord among his peers, no more in 2020 than it did in 2002. The main critique is always the same, namely that many, should I say most, CAs (or hypergraphs) fall short of predicting the complexity of the real world in *quantitative* terms. Put it differently, they provide very inspiring and insightful *analogies* for the

[25] Special purpose LGCA machines were indeed assembled at Ecole Normale Superieure in Paris and seriously considered, but never built, at Los Alamos National Labs.

complexity of the physical world, but rarely pass the litmus test of leaping from the qualitative level of inspiring analogies to the quantitative level of predictive tools.

We have seen this movie before, not necessarily in association with Boolean models. Very elegant and fascinating paradigms, such as self-organized-criticality (SOC) [85, 49] and more recently scale-free networks, have generated landslides of papers and enormous interest across many walks of science and society. As the very title suggests, in his book *How Nature Works*, Per Bak states a strong case for SOC as a fairly general mechanism for Complexity, [87]. Self-organized criticality is a property of nonlinear dynamical systems far from equilibrium to evolve towards a critical state, with no need of external tuning of the control parameters. The SOC paradigm is the apparently humble sandpile: imagine you are on the beach, killing time with a handful of dry sand on your hand, slowly trickling through your fingers. A small cone of sand forms below your hand, hosting a a sequence of micro-avalanches of various sizes, through which the sandpile grows both in width and height yet keeping the same (critical) slope angle. SOC means that the humble heap of sand attains this precise critical angle without any specific tuning; all you have to do is let sand flow through your fingers, as easy as that (see Fig. 7.6). The fascination of the idea cannot be escaped.

However, on closer scrutiny, the question of whether nature likes to sit right on the spot of these paradigms, or just close-by instead, is indeed still open to debate. Heaps of real sand seldom break and generate avalanches according to simple SOC algorithm devised by Bak, Tang, and Wiesenfeld in the mid 80s

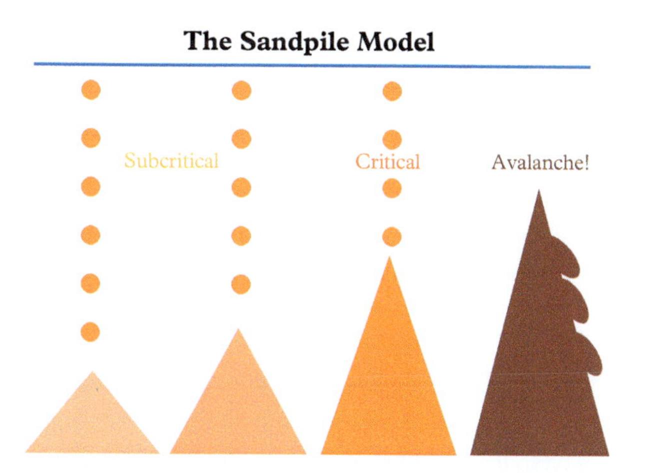

Figure 7.6 *The sandpile paradigm underneath self-organized criticality. By letting sand grains fall to the ground, a sandpile forms which grows until it reaches a critical slope above which avalanches are triggered. The sandpile adjust to its critical slope without any external fine-tuning, just lets sand grains fall down.*

[85]: some do, for instance rice grains, but it is an exception rather than the rule. Scale-free networks, one of the leading avenues of modern Science of Complexity are very inspiring, but again, it looks like real-life systems rarely abide in full to this paradigm [16]. This is not to detract from the fascination and intellectual import of these paradigms, but simply to caution against the temptation of accepting them as quantitative descriptors of complex systems based on their inspirational power alone.

7.7.1 The take-home lesson of lattice fluids

To conclude with lattice fluids, we shall retain just a single, yet far-reaching lesson, namely that *simple regular and uniform networks, equipped with simple (Boolean) rules, prove capable of supporting complex emergent behaviour.*

Emergence is one of the most defining features of complex systems, one that straddles across vastly different disciplines within and beyond the natural ones, social sciences, and psychology scoring high in the list. This paradigm may or may not be grand enough to encompass the entire Universe, but it covers nevertheless a large variety of complex systems, hence providing a major conceptual (and sometimes) practical contribution to the Science of Complexity to modern science and society at large. The litmus test, though, always rest on a quantitative match to mother nature. As far as I can tell, this test still filters out many paradigms of Complexity, no matter how mathematically elegant and conceptually inspiring. As mentioned previously, even though they receive little or no attention at all in the Complexity literature, LGCA and LB stand out as one of the few precious examples which pass the aforementioned litmus test.

7.8 From lattices to general networks

In the previous sections, we emphasized the power of simple and uniform networks known as lattices, showing that, notwithstanding their simplicity, they support fairly complex emergent behaviour, such as fluid turbulence.

The natural question is: how about irregular ones? By definition, irregular lattices do not display any regular pattern, the nodes are laid down irregularly in space and their connectivity may change from node to node, the latter case usually going by the label of 'unstructured' lattices (see, for instance, Fig. 7.7). Such irregular lattices are conveniently described as networks, made of nodes and connecting links, sometimes also known as vertices and edges, respectively. The nodes carry the variables that specify the state of the system, and the links host the interactions which govern the evolution of the variables placed on the nodes. Due to its great generality, this 'Node' and 'Link' representation embraces a broad variety of systems, whence the power of networks to describe Complexity.

Given their paramount role of networks, a few additional words are warranted, pointing the detail-avid reader to the large number of excellent books on this specific subject [77, 19, 14].

Figure 7.7 *A non-uniform and irregular lattice, typically used for solving the equations of fluid and solid mechanics on digital computers. The flexibility of triangles permits to accurately represent nontrivial geometrical shapes, such as the dolphin in the picture.*

7.9 Network basics

Let us lay down a set of N *nodes* in the plane, numbered $1, 2 \dots N$ and stipulate that the generic nodes i and j are *'connected'* if there exists a link ij between the two. It is standard practice to formalize such connection by introducing a *connectivity matrix* C_{ij}, whose entries take the value 1 if nodes i and j are connected, and 0 if they are not. Note that 'connected' must be understood in a broad sense, not just physical, social networks being a chief example in point.

Of major importance is the connectivity of each given node i, say C_i, which is the sum of C_{ij} over all other nodes $j = 1, N, j = i$ excluded. For instance, in the Fig. 7.8, the node 0 has connectivity 4 as it connects to nodes 2,3,5, and 8. Simply said: the connectivity of a given node, also known as *rank* in the network literature, is the number of its connections, or *active links*. Highly connected nodes are called *hubs*, and they represent the crucial nodes of the network, as we shall clarify shortly. The link may be either one-way, say i to j, or two-ways both i to j and j to i, like in a two-lane highway. In the former case, the link is said to be *directed*, and in the latter is called *undirected*, i.e. it goes both ways. A global property of major interest is the total connectivity of the network, defined as the sum of the single-node connectivities, i.e.

$$C = \sum_{i=1}^{N} C_i = \sum_{i=1}^{N} \sum_{j=1; j \neq i}^{N} C_{ij} \tag{7.1}$$

Hence, each network can be characterized in terms of two global numbers, its size N, the total number of nodes, and its connectivity C, the total number of active links. A fully disconnected network (is this really a network?), features $C = 0$, a fully connected one, in which each node is connected to each other,

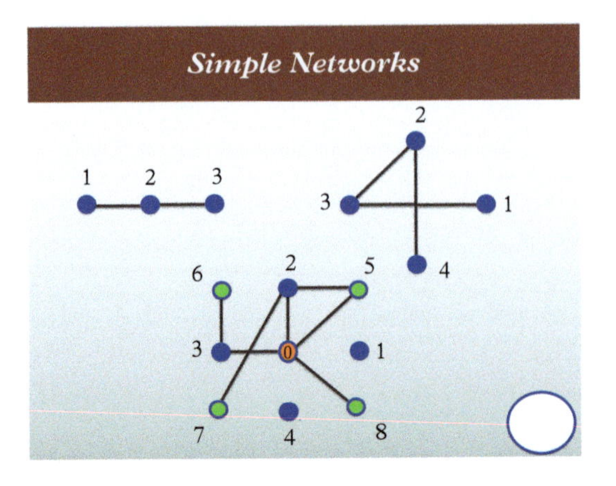

Figure 7.8 *Networks in one and two dimensions. The first is a one-dimensional regular lattice with $N = 3$ and $C = 2$, the second is a two-dimensional lattice with $N = 5$ and $C = 3$ and the third is another two-dimensional with $N = 9$ and $C = 7$.*

scores $C = N(N-1)/2$ if the links are directed and $C = N(N-1)$ if they are undirected. Note that we have excluded self-connections, i.e. $C_{ii} = 0$. These are the global numbers which characterize the network as a whole.

7.9.1 Statistical properties

Of great relevance is the distribution of the single-node connectivities, namely the statistics of the number of nodes $N(c)$ which feature a connectivity c. For instance, in the second example of Fig. 7.8, there are two nodes (1 and 4) with connectivity 1 and another two (2 and 3) with connectivity 2. Finally, there is no node with zero connectivity. Hence, the statistical distribution of the connectivity is $N(0) = 0$, $N(1) = 2$, and $N(2) = 2$. The social interpretation is self-explanatory: the nodes with highest connectivities are the most visible ones: the 'stars' of society, a huge and highly dynamic network indeed. No joke, this is how social networks work, for good or for ill, and network visibility can turn into a very effective money-spinning machines, as any successful 'influencer' would be happy to attest. We shall return to this shortly. In a regular network the distribution of the connectivity has no statistical spread: for instance in the hexagonal lattice of the Frisch–Hasslacher–Pomeau automaton, every internal node has the same connectivity $c = 6$ (we are making abstraction of boundary nodes, which feature $c = 3$). In a general network, however, there is a statistical spread of the connectivity, often identified with its degree of disorder [9].

7.10 Metric versus topological networks

Eventually, each link ij can be given a weight W_{ij}, which measures the strength of the connection between the two nodes. In this case, the connectivity matrix can take any value between 0 (disconnected) and 1 (fully connected). In general, the weight matrix needs not be symmetric, that is W_{ij} does not need to be equal to W_{ji}. This is readily understood in terms of sentimental relations, Bob (node B) may be desperately in love with Alice (node A), with a much milder return from Alice, if any. This means $W_{BA} = 1$ and $W_{AB} < 1$ for mild return and $W_{AB} = 0$ for no return at all: a rather dry way of saying 'unreciprocated love'.... The value of the weight may or may not depend on the physical distance between the nodes. In physical systems it usually does, because the links carry interactions, which usually depend strongly on the distance. For instance, in a network of gravitational bodies, each body interacts with every other with a force proportional to the inverse of their distance squared. These are called *metric networks*, since the spatial location of the nodes has a huge impact on their behaviour. The same goes for traffic networks: the airline link taking from Rome to Los Angeles carries a different weight (defined here as the time duration of the flight) than the link between Rome and New York. Neither link is symmetric, since it is well known that flying eastwards is faster thanks to the favourable tailwind.

In many other complex systems, especially social ones, such as the internet and derivatives, distance is no player, and what matters is only the number of active links, a direct index of popularity, or *social visibility*. Such networks are often called *topological*, because physical distance plays no role. More precisely, in a topological network, the distance between two nodes is defined as the number of links that have to be traversed to move from one to the other. For example if node i is connected to node j via node k, i.e. from i to k and from k to j, the topological distance from i to j is 2, regardless of their geometric distance. But why doesn't the geometric distance matter? Basically, because light travels fast and signals on modern telecommunication devices, be they electrons on wires, waves on optical fibres or satellites, travel at light-like speed (as a simple reminder, light goes around the whole globe almost eight times in a single second ...). This is fantastic, as you can Skype or Whatsapp your friends in real time anywhere in the world, but it comes with dire flip sides as well. The same speed that enables your Skype-call also powers the financial transaction that dissolves your lifelong savings at a keystroke. And if you think that this is being overly dramatic, the 2008 financial bubble shows that we are not talking disaster movies, but real (modern) life instead. And, the current COVID-19 pandemy tells an even more dismal tale. I hope this conveys the idea of how crucial networks are to modern society.

7.11 Random networks

Let us now spend a few words on a class of networks which have played a paramount role in the whole development of network and graph theory to this day: random networks. Random networks were developed by the legendary Hungarian mathematician Paul Erdös (1913–1996) (*The Man who only Loved Numbers*) and less legendary but still very prominent Alfred Renyi (1921–1970), also a

Hungarian mathematician (no surprise). The recipe to build a random network is utterly simple: 1) for each node i, interrogate all subsequent other nodes $j = i+1, N$, that is; 2) choose a number q between 0 and 1 and draw a random number r also between 0 and 1; 3) if r is smaller than q a link is established, otherwise the two nodes remain disconnected. That's it! The randomness is self-explanatory, as it is built into the very generation process described by the previous three steps. Despite the deceptive simplicity of the rule, random networks come with a bag of far reaching properties.

Here, we just content ourselves by noting that the statistical distribution of connectivities in a random network obeys the bell-shaped Gaussian statistics, (see Fig. 7.9):

$$N(c) = Ae^{-(c-m)^2/2\sigma^2} \tag{7.2}$$

where m is the mean connectivity, σ is the statistical deviation and A is a numerical a prefactor fixed by the total number of nodes on the network. In the previous calculation, $N(c)$ is number of nodes with connectivity c, and the sum over c returns N, the total number of nodes in the network. In a Gaussian distribution, all nodes share more or less the same number of links, m in our case, up to a statistical fluctuation, $\sigma = 1$ (Disorder).

As an example, if the average connectivity is, say, $m = 4$, and the statistical deviation is $\sigma = 1$, this means that the vast majority of nodes features a connectivity between $4-1=3$ and $4+1=5$. The probability of finding nodes with much higher or much smaller connectivities decreases exponentially fast, according to the relation 7.2. For instance, the probability of finding one node with more than 7 links, hence 3σ above average, is about one in thousands, and if instead of 7 we take 9, the probability collapses to just about one in a 3.5 million! Incidentally, this is the famous five-sigma rule which marks the all-important border between accident and discovery in science, most famously in high-energy physics. In other words, *random networks are dominated by the average and outliers are heavily suppressed.* In Nassim Taleb's colourful terminology, this is Mediocristan, the kingdom of Joe

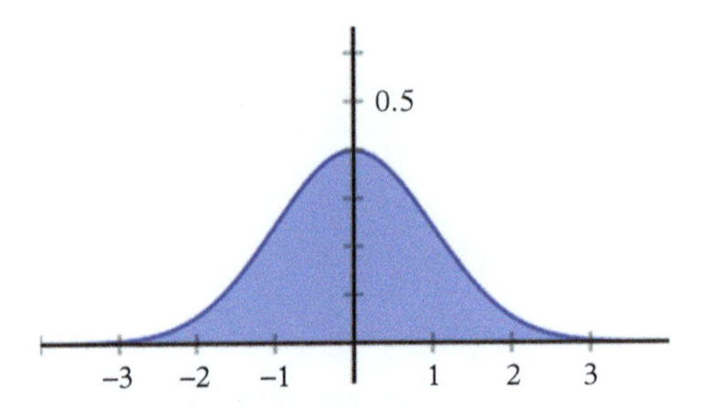

Figure 7.9 *The characteristically bell-shaped Gaussian distribution for the case of zero mean, $m = 0$ and unit statistical deviation, $\sigma = 1$. Note that the tails are heavily suppressed.*
Source: reprinted from en.wikipedia.org.

Average and Plain Jane [112]. This does not sound very exciting, agreed, but don't be fooled: random networks are rich and fascinating objects on their own right, independently of their conformity to the real world.

Which, by the way, is not nil.[26] But do complex systems abide by the random network model? The answer is that most of the time they don't, which also means that connections in complex systems do not arise at random. So, if not random, what then? This takes us to the next item, scale-free networks, one of the major highlights of the modern theory of Complexity [8].

7.12 Scale-free networks

The first question we tackle is this: what does *scale-free* mean at all? Well, simply that if you rank the single-node connectivities in descending order, from the most connected (popular-visible in social terms) to the least one (the least visible), the statistics of the connectivity does not follow a bell-shaped Gaussian curve, but a power-law instead. In math terms:

$$N(c) = A/c^{\alpha} \tag{7.3}$$

where A is again a numerical prefactor and α is the so-called *scaling exponent*, which characterizes the specific power-law decay of the number of nodes with connectivity c.

A few comments are in order.

First, the name scale-free. The Gaussian distribution depends on two numbers, the mean m and the statistical fluctuation σ. The latter, in particular, fixes the natural scale of the distribution, i.e. the natural units in which c should be measured. In a power-law distribution, there is just one free parameter, the exponent α, which means that there are no longer any natural units to measure c, the distribution is, in technical terms, *scale-invariant*. What this means is that if you *rescale* c, i.e. multiply it by any given number, say K, the corresponding distribution remains the same, a power law with the same exponent α, only multiplied by A/K^{α} instead of A. The curve is shifted up or down, depending on whether the rescaling factor is larger or smaller than one, but its shape is left unchanged.

[26] Erdös (simple instead of two acute accents is a typographical necessity) was quite a character. He spent most of his life wandering from conference to conference mostly hosted by friends and collaborators, which he visited for as long as needed to write one or more joint papers. Legend has it that he would show up at the door with his travel bag and the statement 'my brain is open'. This lifestyle led to over a thousand published papers, a world record for mathematicians. Given his genius and productivity, it is to be assumed that his colleagues were generally happy with this habit. With a few exceptions though. I don't remember where, but I seem to have read somewhere that he once came to the point of asking a friend for a pedicure. With all due respect to his supreme math, enough is enough, and instead of pedicure, he just got kicked out of the door! Kathleen Fearn, the gentle controller of my English style, informed me that a similar story applies to the visit of Hans-Christian Andersen to Charles Dickens house, with pedicure replaced by shaving. They didn't publish any book together, though.

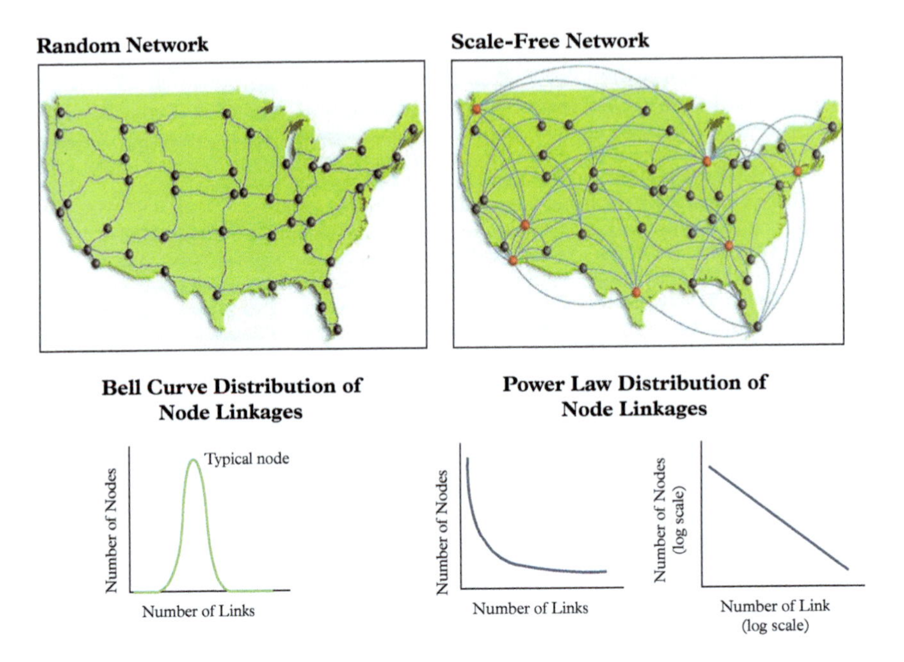

Figure 7.10 *A random network (left) and a scale-free one (right), with the corresponding distributions of the links. The random network follows a gaussian statistics while the scale-free obey a power-law statistics. The very few nodes with a large number of links are the hubs of the network. They are the strength and the weakness of the network at a time.*
Source: redrawn from https://www.google.com.

That's what scale-invariant means. The upshot is that power-law must arise in connections with rules which are indifferent to a rescaling of the number of nodes. An example of random and scale-free networks is reported in Fig. 7.10.

A defining point is that vastly different systems, say athletes and academics, often share similar exponent because they both operate under similar competing conditions and they tick at the same tock, the famous organizing principles that we discussed in Chapter 2.[27] This, for one, shows the great unifying power of scale-free networks. Let us comment further on another major defining point: tolerance to outliers.

7.12.1 Tolerance to outliers

A minute thought's reveals that in a power-law distribution rare events (outliers) are not so rare after all, and certainly way more frequent than they would under a Gaussian statistics, Joe Average and Plain Jane reign no more. To get a sense of what this exactly means, take the case $\alpha = 1$ (highly cited academics feature

[27] In case the reader may doubt it, I can assure that fierce competition is pretty usual in academia.

$\alpha \sim 0.9$). Under such an exponent, the second-best performer gets half of the links of the top score, the third-best performer one third and so on. It takes the tenth best to get a factor ten times fewer links than the best; a very slow decay. Compare this now with *Gaussian statistics* (with unit variance), in which the third performer gets about 13/100 (thirteen in hundred) less links, the fifth just about 3/10000 (three in ten thousands) and the tenth a ludicrous two in a billion of billions. To quote a James Bond movie, 'The World is Not Enough', seven billions is not enough for the tenth performer to stand a chance of showing up at all! Borrowing from Taleb once more [112], this is Extremistan, the land where Joe Average and Plain Jane must abdicate.

Is this good or is it bad? As usual, the answer goes with the Sibyl: it depends.[28] For good or for ill: dreams may come true, like in the Land of Oz, and so do nightmares, like Nassim Taleb's popular black swans [113]. But why is it so? What is the informing principle which grants outliers with chances so heavily negated in Mediocristan? I am not sure that a consolidated theory really exists, but a few popular hunches stand from the crowd. A most popular mechanism is the so called *preferential attachment*, namely the tendency of famous people to attract more followers than anonymous or less known ones (the snowball effect). If you wish, the long-known power of success, the *rich get richer and poor get poorer* (apparently, first coined by Percy Bisshe Shelley in *A Defence of Poetry*, 1821), a total opposite of the Mediocristan rule in which everyone is supposed to get more or less the same.

It so happens that systems ruled by the *preferential attachment* mechanism, social ones decidedly taking the place of pride here, tend to develop scale-free structures. This is a very strong correlation mechanism, which flies in the face of the randomness which governs the Gaussian world. And the correlation driver is that 'famous' people are more 'attractive' than unfamous ones (note unfamous instead of infamous, since the latter are also 'attractive' in the web sense of the world, as they definitely attract attention). Good or bad, that's apparently how humans tend to function: people strive to connect with the rock star, athletes, or Hollywood superstars, who cares about Joe Average or Plain Jane? If you think it's ugly, I would subscribe, but that's how the story goes, at least when success is the overwhelming and only driver in town. Before I move to another subject, let me say a few possibly unconventional words in this respect.

7.12.2 The Matthew effect

Preferential attachment is sometimes quoted as 'Matthew effect', with reference to the famous Parable of Talents from the Gospel of Matthew (25:14–30). The parable tells of a master who entrusts his property to his servants before leaving for

[28] The word Sybil comes, via Latin, from the Ancient Greek *sybilla*, for prophetess. The point is that their prophecies were less than renowned for their clarity, hence, in Italy at least, the Sybil is an epitome of ambiguity.

travel, according to their abilities (talents). More precisely, one servant received five talents, the second two, and the third only one, for a total of eight, a substantial amount of money for the time. Upon returning home, after a long absence, the master asks his three servants for an account of the talents he entrusted to them. The first and the second servants explain that they each put their talents to work, and have doubled the value of the property with which they were entrusted.[29] Both were congratulated by the master and awarded even more talents: 'His lord said unto him, Well done, good and faithful servant; thou hast been faithful over a few things, I will make thee ruler over many things: enter thou into the joy of thy Lord' (Matthew 25:23).

The third servant, however, the one who received just a single talent, couldn't think of anything better than burying it in the ground, for fear of losing it. He receives a very different reaction: his only talent is given to the servant who already had more, and he is sent 'into outer darkness: there shall be weeping and gnashing of teeth' (Matthew 25:24–30). The Matthew effect is usually identified with the 'rich get richer, poor get poorer' 'Wall-street' paradigm, but in my view, it goes way beyond and, most importantly, in a completely different direction. And the direction is that of a call for responsibility on the talents entrusted to us, an exhortation not to let them be wasted. Seems to me that this is just the opposite than the pursuit of success for the mere sake of one's own benefit. But this another book, so let's leave this ground and go back to another important consequence of power laws: *small worlds*.

7.13 Small worlds

A major feature of hubs is their double-edged nature, the superconnectors, the superstars of the network, but also the main carriers of its liabilities. The prime point is that hubs make the network a so-called small world, meaning that the nodes are never too distant from each other. But what does distance mean here? For one, by all means not the physical distance. As explained before, complex networks are mostly topological, which means that the distance is measured by the number of links it takes to go from one to another, not by their geometrical or geographical distance. Here comes Erdös again. Indeed, the typical example of topological distance is the so-called *Erdös number*, also known as 'collaborative distance', which counts the number of mathematicians you work with before you get to Erdös. The meaning is plain: if you work with Renyi, who certainly worked with Erdös, but you didn't work with Erdös himself, your Erdös number is 2, while Renyi's Erdös number is 1. Now take it to the social: what is your Erdös number versus, say the President of the USA? At first, you would expect a very

[29] For the record, at the time of the New Testament, the talent was a pretty substantial amount of money. In currency units, a talent was worth about 6,000 denarii, a denarius being the usual payment for a day's labour. At one denarius per day, a single talent was therefore worth 16 years of labour!

large number, and so would I. But, no, it isn't. I don't know mine with Donald Trump, but with Barak Obama it was just 2. Not because I'm any familiar with the White House, but simply because I happen to know a person, in fact more than one, who knows him first hand. They are my hub to the White House! (For the record my Erdös number with new President-elect Joe Biden is again 2). As I said, I don't know my Erdös number vis-a-vis of Trump, but I do know that it can't be much larger than a mere 6. How do I know? Just because it has been shown that many social networks are 'small worlds', [109], i.e. they contain hubs which boost the connectivity of the network and make it 'small' in a topological sense (see Fig. 7.11).

At first, this is surprising, but in light of power laws, less than it seems, after all. Let us see why.

Taking the world population at one billion for simplicity (it's actually seven) you may notice that 6 is not far from 9, the logarithm of one billion in base 10. This is precisely what happens in small-world networks, the number of links connecting, on average, any node to any other one, is approximately given by the logarithm of the number of nodes in the network. This is again the result of the presence of hubs in the net. Small worlds are generally perceived as good news, as they make us feel close to each other, and hubs deserve much of the credit for this sense of commonality. At this point, we have ascertained that hubs are the big guys in town, they do most of the work in sustaining the network activity and functions, be they physical, biological, or social. But big does not mean invincible, and since they are so preciously rare, losing them for whatever reason, natural or due to malicious attacks, inevitably exposes the whole network to dire consequences. That's how fragility kicks in.

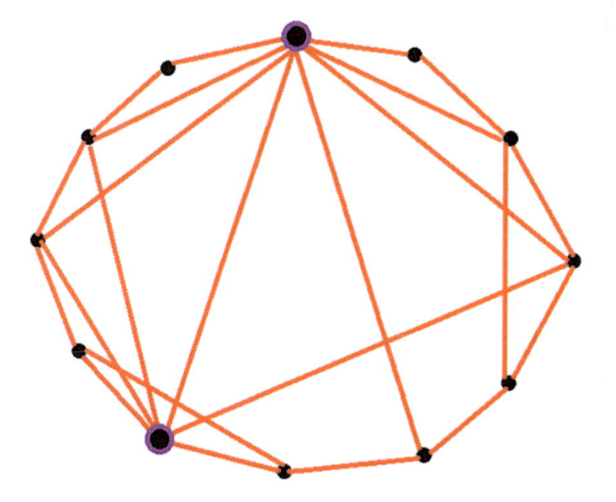

Figure 7.11 *Sketch of a simple small-world network with two hubs.*
Source: reprinted from en.wikipedia.org.

7.13.1 Entry fragility

If you somehow manage to impair or disactivate a hub node, which is a rare event since hubs are rare themselves in the first place, the whole connectivity is compromised and since connectivity is in control of the correct functioning of the network, the whole enterprise may crash. This is something we have seen time and again in modern times; power outages, financial bubbles, cyber-attacks and, lately, the devastating COVID-19 outbreak as well. The message is pretty plain: open and highly connected societies work wonders in terms of efficiency, but the more sophisticated and efficient they get, the more fragile they become too.

Although I am no network specialist, I sense it is becoming increasingly apparent among the experts that a desperate focus on efficiency and performance, relying heavily upon fewer and fewer super-hubs, comes at a very dangerous detriment of robustness. If the hub is protected, super-performance, super-optimization, will result, beer is cool and life is good, this is how turbo-economies fly sky high. But if by some accident or attack, a rare event, less rare than we thought, the infamous black swan hits the hub, the super-system turns suddenly into crumbles.

At some point of this writing, I was locked down in my native city for three weeks by the COVID-19 outbreak: someone in far-away Wuhan ate the wrong animal, as they might have done at any point over the centuries, and all of a sudden, the entire modern hyper-connected world is down on its knees. With no clue on when our pre-COVID lifestyle will be back, if at all. Had it happened just twenty years ago, the maybe cynical reaction would have been, well that's China, we wish the best to them, but no big deal here. Nowadays, if was just matter of a few days before the virus was knocking at the European doors. This, too, is a small-world effect.

Wise people say that there'll be deep lessons to be learned from this disaster, and they are probably right. I guess everybody agrees that modern society is in a very serious need of reconsidering the crucial tradeoff between Efficiency and Fragility, and the best Science of Complexity surely has a major role to play in designing this new and hopefully better landscape. For my part, I miss traveling a lot, a point to which shall return in the Epilogue. For a source of insightful thoughts in this direction, I warmly recommend the inspiring book by Helga Nowotny, *The Cunning of Uncertainty* [80].

7.14 Complex adaptive networks

In the previous section, we have discussed the basic rules which inform the growth of both random and scale-free networks, random selection, and preferential attachment. These are just two out of the many possible mechanisms which govern the dynamic evolution of the network, i.e. how it forms and grows in time (see Fig. 7.12). This is one of the leading themes of modern research in Complexity, and again the interested reader is best directed to the excellent books in the literature [14]. Here, I just wish to conclude with a mention to so-called *adaptive networks*, namely networks in which nodes and links come and go in response to the external stimulus from the environment as it changes in time. Or to the traffic they are meant to support in order to deliver their functions. More often than not, the nodes of adaptive networks are 'thinking units', called *agents*, to indicate that they scheme and take decisions, sometimes to the point of gaming the rules 'on the fly', as typically occurs in trading and financial systems, which often change faster than most single agents can make sense of. To reconnect to the matters discussed in the preface, they change the very *laws* which rule them! Adaptive networks are central to a large variety of complex systems, to begin with biological ones, metabolic, neural, cardiovascular, respiratory, you name it. But they vastly transcend biology, into the social and economic, complex adaptive systems (CAS) being the general label for systems organized according to such adaptive networks. It is fair to say that CAS represent one of the leading paradigms of the modern Science of Complexity.

We conclude by displaying two networks par-excellence, the Internet and the human brain (see Fig. 7.12). The latter counts about a hundred billion nodes (neurons), the latter is expected to exceed ten billions soon, thus outnumbering the world population. This pair of outstanding CAS offer an exemplar case of what we called chicken-egg causality: the internet, surely the result of our brain activity, most definitely affects back the brain that originated and fuels it, as social media stand tall to witness. The analogy and inter-relation between these two

Figure 7.12 *Two paramount complex adaptive networks: the internet and the human brain. The brain generates the internet and the internet shapes back the way brains act. If only in loose terms, this fundamental chicken-egg loop might be responsible for much of the Complexity of modern society. The emergence of a 'collective' brain is a scary prospect.*
Source: reprinted from commons.wikimedia.org and en.wikipedia.org

paramount complex adaptive networks is shallow at the best from a strict scientific viewpoint, yet intriguing and somewhat scary.

7.15 Summary

Networks provide a very general, elegant, and powerful framework to develop quantitative models of Complexity in both natural and social systems. Networks need not be complex to support complex emergent behaviour, as clearly proved by Lattice Gas Cellular Automata for the case of fluids. However, complex networks, with large numbers of nodes, high connectivity, and Disorder, are capable of displaying an enormous span of complex behaviours. This is particularly true for complex adaptive works which change in time along with the data traffic they support and sustain, as well as with the environment the live in. The internet, the brain, epidemic networks stand out as major examples in point. A quantitative grip on the mechanisms that control the flow of information across a dynamically adaptive network, holds the key to a deeper understanding of Complexity in a huge variety of systems. The question remains, though, as to how much *system-specific* information has to be injected within this beautiful framework before mother nature responds in the affirmative on quantitative grounds. This is a most fascinating and, as far as I can tell, still widely open question.

Part II

The Science of Change

8

Old but Gold: Thermodynamics

Il calore non può essere separato da fuoco, o la bellezza dall'Eterno.

(Dante Alighieri)

8.1 The science of (slow) change

In the previous chapters, we have seen that Complexity has much to do with the way systems change in time, particularly when such change is erratic and hard to predict. But we have also seen that, besides hard-to-predict changes, Complexity has a lot to do also with subtly orchestrated ones, usually associated with the delivery of specific functions.

Our own life is quintessential Complexity, and it should be appreciated that the orchestra of complex processes that keep us going, takes not only energy in the first place, but also and above all its ability to *morph* from one form to another. For instance, the chemical energy contained in the food transforms into mechanical energy which sustains the vital functions of our body, like pumping blood in our veins or breathing air in our lungs. The study of these changes forms a fundamental pillar of science known as thermodynamics, literally the science of 'heat motion' [6, 125, 78, 83]. Why heat, is not hard to grasp, as it has been known since the dawn of mankind that heat, say fire for clarity, changes things: '*Ignis mutat res*', as the ancient Romans said. Even if the first control of fire (by *Homo erectus*) dates to no less than a million years ago, we had to wait one million minus just a tiny two hundred years, before we could make rational sense of the way heat transforms things around us. Indeed, the foundations of thermodynamics were laid down in the nineteenth century by a comparatively small group of inspired scientists and engineers in France, Germany, and England, and they consist basically of two fundamental laws, known as the First and Second Principles of Thermodynamics.[29] As we shall see, such principles bring on stage two paramount characters, energy and entropy, acting as the true prima donnas

[29] Peter Atkins mentions four laws: zeroth, the definition of temperature, first and second energy, and entropy, the same as in this book, and third, also mentioned in this book, the impossibility to achieve zero absolute temperature. The classification is different but the bottom line is pretty much the same.

Sailing the Ocean of Complexity. Sauro Succi, Oxford University Press.
 DOI: 10.1093/oso/9780192897893.003.0008

of the cosmic scene. Indeed, virtually any change we see happening around us, is motored by energy and entropy and the way they lead their cosmic dance in space and time.

So, let us make the acquaintance of the two cosmic ladies.

8.2 The cosmic prima donnas

Energy is definitely the more popular of the two, owing to its direct impact on daily life: virtually all of our activities depend on the availability of some form of energy, be it the fire of the domestic gas cooker or the fuel in our car, the electricity in our bulbs, you name it. An interruption of the energy supply means immediate 'shut-down' of modern society, as geopolitics shows all too well. The other prima donna, entropy, on the other hand, is much less known outside the circle of specialists, typically physicists, chemists and information scientists, even though it has no less of an impact on our lives. The two prima donnas are at no risk of being mistaken for each other, as their distinctive traits are completely different. Energy is a byword for motion, hence life and proactivity in general, a pretty positive press image indeed. Entropy, on the other hand, comes with almost the opposite flavour, as it speaks for ageing, decay, and ultimately, (thermal) death. No doubt, not the most popular guest for a dinner party .... In a burst of political incorrectness, the image comes to my mind of a colourful classy lady for the former and a grim and kind of scary, but still fascinating one, for the latter.

In tragic terms, life and death. But, as usual, this black/white picture bares little resemblance to the truth: nature is invariably much subtler and more inventive than our metaphors. True, the cosmic primadonnas often compete with each other, but they are also capable of enchanting cooperation: entropy can be constructive and at the end the day, the beauty of our world comes precisely from a continued duel turning into a very fertile duet instead. It's a beautiful story indeed.

So, let's start by digging a bit deeper into the shiny part of the pair: energy.

8.3 Energy: Transformation and conservation

Energy is the source of motion: to move from here to there, we need energy. If we are to trust a certain Aristotle (384–322 BC), 'The energy of the mind is the essence of life', our thoughts need energy too, which is literally true. On possibly a more mundane note, Yoko Ono agrees: 'Energy is so important. If you don't have it, forget about rock and roll'. With Aristotle and Yoko Ono aligned, we can safely move along. In the first place, energy is two fairly different but complementary sisters: kinetic and potential. As mentioned previously, kinetic energy, underscores the idea of motion ('*kinesis*' for the Greeks). We have met it before in his book, and its mathematical formula goes like this: mass (m) times velocity (V) squared, divided by two. In equations

$$E_{kin} = \frac{1}{2}mV^2 \tag{8.1}$$

Thus, kinetic energy is the energy associated with motion, standing objects have none. And, as noted previously, it grows quadratically with velocity: doubling the velocity makes kinetic energy four times higher, a fact that we are justly reminded of on highway displays, showing the distance to come to a halt (braking distance) at a given speed. Kinetic energy goes up with the square of the velocity and so does the breaking distance (just in case, a handy rule to compute the latter is to divide the speed, in km/h, by ten and square the result: that is, at 50 km/h, the distance is $5 \times 5 = 25$ metres, and at 100 km/h, it's 100 metres). Potential energy, on the other hand, has no association with motion, still material bodies are filled with it.

But, why potential? Because, although it does not go with motion, it has potential to cause it, which is actually what it does all the time. How? Simply by morphing into its kinetic sister and vice versa.

Possibly, the most popular form of potential energy is the one which goes with gravity, as the following example should clarify. Take a ball of mass m and keep it standing at an elevation h above the floor. If g denotes the gravitational acceleration (9.81 metres per second squared), its potential energy is

$$E_{pot} = mgh \tag{8.2}$$

Now let the ball go fall from your hand: it starts falling towards the floor, acquiring an increasing downward velocity in the process (see Fig. 8.1). This is potential energy morphing into kinetic energy. By the time the ball hits the ground floor ($h = 0$), all of its potential energy is converted into kinetic one and the ball reaches its maximum velocity. Since all the potential energy has been converted to kinetic energy as the ball hits the floor, the maximum velocity is readily computed from the condition

$$\frac{1}{2}mV_{max}^2 = mgh \tag{8.3}$$

which delivers $V_{max} = \sqrt{2gh}$. This shows that the velocity of the ball as it hits the ground scales less than linearly with the height: to double the impact velocity you must let the ball from a four times higher altitude.

Note that the maximum velocity does not depend on the mass, a tennis ball and a lead cannon ball hit the ground with the same speed! This flies in the face of our intuition: we all expect the heavy cannon ball to move faster than the light tennis ball, don't we? And indeed this is exactly what happens in ordinary life, in which objects are subject to the friction exerted by air, i.e. dissipation. But if the cannon ball and the tennis ball were left falling in perfect vacuum, no surrounding air, they would definitely move at the same speed! It took the genius of Galileo (1564–1642) to sort out this apparent paradox, giving rise to the modern science of mechanics.

Energy Conservation

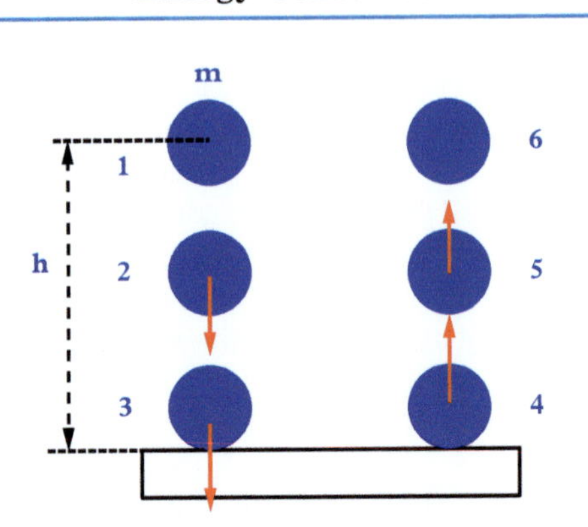

Figure 8.1 *The conversion of potential energy into kinetic one. Left, top to bottom: 1) the ball at an elevation h starts falling down under the effect of gravity, 2) the ball acquires speed by converting potential energy into kinetic one, 3) the ball hits ground with maximum speed, all the potential energy being converted into kinetic one. Right, bottom to top: 4) the ball bounces back on the elastic floor with no energy loss, 5) the ball starts raising and loses speed by converting potential energy into kinetic one, 6) the ball reaches the elevation it started from with zero velocity, as all the kinetic energy is converted back to potential energy. The sum of the two energies is exactly conserved throughout the process.*

Back to our ball. After hitting the floor, the ball bounces back and starts moving up at a now decreasing speed in time, because kinetic energy gives what it earned on the way to the floor back to its potential sister. If no energy were lost on the rebound (mission impossible on planet Earth) and air were not to take its toll (mission impossible number two), the ball would reach exactly the same elevation it started from: the initial condition is restored, ball in your hand, kinetic energy gone, all potential energy again. Back to square one, as if nothing had happened. This simple example illustrates the magic *flexibility* of energy: it moves back and forth between kinetic and potential and if dissipation could be neglected, nothing would get lost in this mutual exchange. It is fair to say that this flexibility is essential to sustain the Complexity of the natural world, living organisms in the first place (see Fig. 8.2).

As a matter of fact, no floor would rebound the ball for free, and same for the surrounding air. Some energy toll is inescapable, which means that the ball cannot regain exactly the same elevation, but it reaches a bit lower than that, the deficit being proportional to the energy dissipated on the floor and in the air met by the ball along the way. The same exercise also conveys a far-reaching message: barring out dissipation, energy can change suits, from kinetic to potential and back, but the sum of the two does not change in time. In other words, *energy is conserved*, and the importance of this conservation law cannot be overstated.

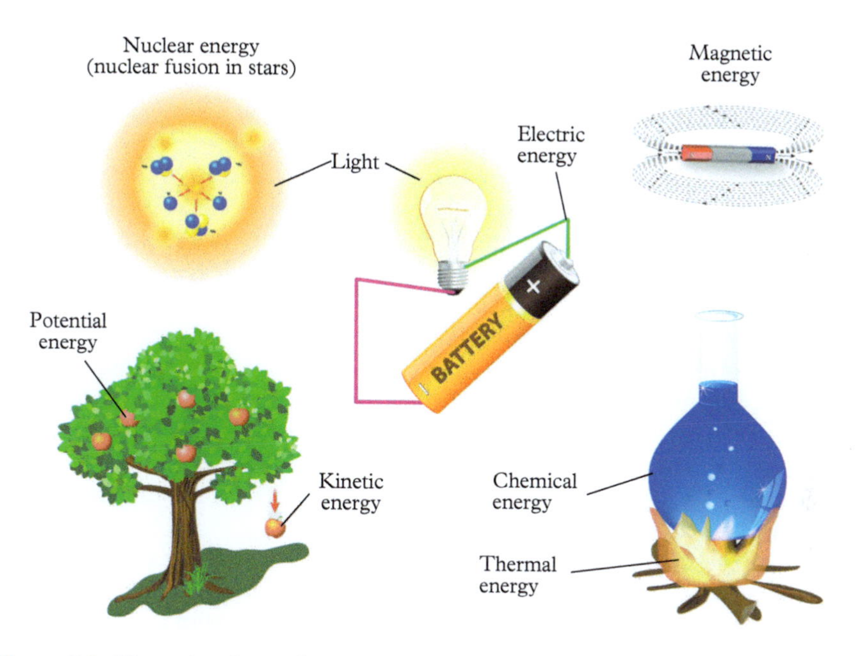

Figure 8.2 *The various forms of energy.*
Source: redrawn from fanak-theetowahgroup.com.

Indeed, such conservation is usually promoted to the status of First Principle of Thermodynamics [6].[30]

8.3.1 The many faces of potential energy

Earlier on, we have presented energy in the form of two sisters, but the energy sisters are many because, while kinetic energy is one, potential energy comes in multiple versions. The previous example deals with a specific and very familiar form of potential energy, the one associated with gravity, but there are many others potential energies around: chemical, nuclear, electromagnetic, you name it, (see Fig. 8.2). They describe very different phenomena, but they are all equally respectful of the first principle.

[30] It would be interesting to spend a few comments on the ultimate reason *why* energy is conserved. Directing the interested readers to Atkins's book, let me just mention the main point. And the point is that the laws of physics do not change in time. If they did, energy would no longer be conserved. Some branches of modern physics speculate indeed that the laws of physics, and particularly the values of the fundamental constants of nature, such as the speed of light or the gravitational constant, might have been different in the past. For the moment, this is however still in the wilds of speculation.

For instance, chemical energy, the one stored in the chemical bonds which form atoms and molecules, is typically released upon breaking them. That's how we get calories from food. Nuclear energy follows the same path, except that the bonds are between nuclei (protons and neutrons), which makes a whole world difference in terms of how much you can get out of it, in fact millions of times more! Chemical energy is usually measured in electronvolts (eV), which is the energy acquired by a single electron moving across a voltage of 1 Volt, named after the Italian physicist Alessandro Volta (1745–1827). It's really a very tiny amount: to convince yourself, compare it with the amount of energy it takes to increase the temperature of one litre of water by just one degree, an amount of energy called a *calorie*, from the latin *calor* for heat. The number is 2.6 followed by nineteen zeroes, basically twenty-six billions. The 'culprit' for such a disparity is the tiny electron, which is really a small object as compared to the human body .... Be that as it may, chemistry speaks in electronvolts, but since there are many many molecules in any macroscopic object, those many many molecules make for the amount of energy that we need to sustain our body (or lose weight, depending on personal wishes). Just for the record, it takes about 300 calories a day to lose one pound of weight, OK?

Nuclear energy is like chemical energy, except that it applies to much smaller particles: protons and neutrons in nuclei, instead of electronic bonds. Nuclei are approximately five orders of magnitude smaller than atoms, with a corresponding five orders of magnitude increase in energy.[31] Here the typical scale is millions electronvolts, MeV for short. This is a lot: with a MeV available per nuclear reaction, just a single gram of uranium delivers enough energy to power about three thousand American households for an entire year! Unfortunately, the public image of nuclear energy is mostly associated with the atomic bomb, unquestionably one of the darkest pages of modern history. However, one should not forget about its many peaceful and actually highly beneficial outcomes, nuclear medicine at the forefront. Mind-boggling as these figures are, modern physics has taken a major leap beyond nuclear; the Large-Hadron-Collider at CERN is currently operating in the range of 10 TeV, for Tera electron-volts, i.e. ten millions times larger than the MeV!

Another key form of energy is the electromagnetic energy, the one that powers our household appliances, cell phones, computers, trains, you name it. Light sits under the same umbrella and its carrier, the photon, can be thought of as the 'elementary particle of light' or equally well as an elementary oscillation, just like a wave. This wave-particle duality is a central feature of matter and energy at the atomic scale and below. The photon is a weird character in many respects,

[31] There is a famous principle in physics, named after the German physicist Werner Heisenberg (1901–1976), according to which the amount of energy required to 'see' a given object is inversely proportional to its size. If you want to see small things you need powerful microscopes It is the very same reason why to dig deeper and deeper in the ultimate constituents of matter, we need larger and larger accelerators.

in the first place because it carries kinetic energy without mass, in other words it's pure motion! Indeed, its energy is proportional to its frequency, i.e. the number of oscillations per unit time (space), times a universal constant named after the German physicist Max Planck (1858–1947), the founding father of quantum physics. In equations

$$E = hf \tag{8.4}$$

where h (not to be confused with the altitude in the formula of potential energy) is the Planck constant (see also Appendix 19.3 on quantum mechanics). Note that mass does not appear at all in the previous expression: energy does not need mass. Planck's constant takes an astronomically small value in macroscopic units $h \sim 6.6 \times 10^{-24}$, 0 followed by 23 zeros before you meet the first nonzero digit, 6. This does not mean though that electromagnetic waves cannot carry sizeable energies, which is exactly what they do at very high frequency. The fastest photons ever observed in the Universe, super-high energy gamma rays from cosmic explosions, can reach up to a few hundred trillions of electronvolts! Being massless (as far as we can tell) the photon moves really fast, in fact as fast as one can get in our known Universe, namely the speed of light, about three hundred thousand kilometres per second. The speed of light is a universal constant, but the frequency covers an impressively broad range of values. For instance, visible blue light has a spatial period (the speed of light divided by the frequency) of about four hundred nanometres, which means that blue light waves oscillate about twenty-five millions times in a metre. On the other hand, a standard radio wave oscillates just once every 10,000 kilometres!

The previous considerations convey a sense of how broad and flexible energy is: it pervades the entire Universe from the tiniest elementary particles, all the way up to the grand Cosmos. A great prima donna by all means. How about the other one, entropy? Before discussing it, we need to spend some time in illustrating another chief thermodynamic quantity, in fact the one derives its very name from: Heat ($\theta\varepsilon\rho\mu$ ὅτητα for the Greeks).

8.4 What is heat?

Heat is a pervasive and familiar presence in our life and indeed we all have a very distinct sense of what it does for us: to say it quickly, heat is what makes things hot! A bit more precisely, heat changes the temperature of material bodies, which get hotter upon absorbing heat and colder upon giving it away, (see Fig. 8.3). What we perhaps do not immediately realise is that such temperature-changing ability may be put at our service, not only to get hot water but also to get our car going. In other words, heat can do useful work for us. Indeed, at the dawn of the industrial revolution, engineers, mostly French, British, and German, realized that, although they did not know what heat really was, for sure they knew that it could set 'things'

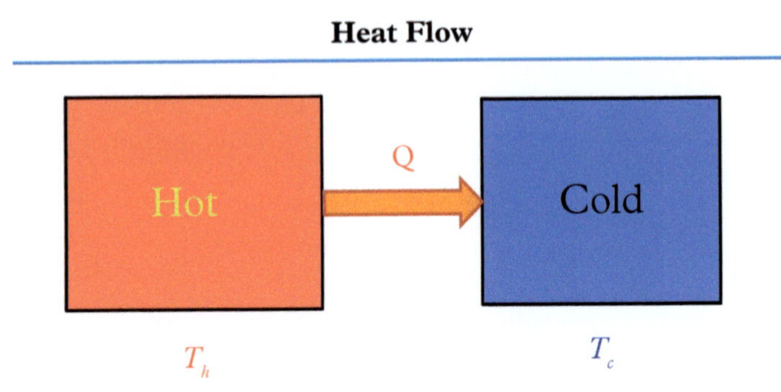

Figure 8.3 *Heat (Q) flows from a hot body at temperature T_H to a cold body at temperature T_C. As a result, the hot body gets colder and the cold one gets hotter. Heat acts as a temperature equalizer.*

in motion. The prototypical example is the humble piston-cylinder system. Fill a cylinder chamber with gas and cap the top with an actionable piston, then heat up the cylinder and the gas will start expanding, thus lifting the piston upwards in the process. Heat moves the piston head, it actually converts into motion (kinetic energy), and a pretty useful one at that! No surprise that engineers and scientists of the time started thinking very hard about ways to extract useful work from thermal machines and do it as efficiently as possible. The first quantum leap was made by the young French engineer Nicolas sadi Carnot (1796–1832), who devised a procedure now known by his name, the so-called Carnot cycle, (see Figs. 8.3 and 8.4).

8.4.1 Work and zero temperature

Before we delve into details, let's face the most immediate question upfront, namely whether it is possible to convert *all* the available heat into mechanical work. Before we make it clear that the answer is a plain NO, let us inform the reader that, in thermodynamics, work has a very precise definition: it is the product of pressure times the change in volume produced by pressure itself. In math terms

$$\delta W = P\delta V$$

where δW is the amount of work made by a gas expanding at pressure P and pushing the piston up so as to increase the volume of the gas by an amount δV.

It is natural to argue that the impossibility of a full conversion of heat into work is expected, since devices with pistons, solid walls, wheels and all this kind of moving machinery, must necessarily experience a lot of friction and energy losses. This is true, but the reason why heat cannot be turned entirely into

Figure 8.4 *Nicolas Sadi Carnot: his eponymous cycle unravels the fundamental limitations on the amount of heat which can be converted into useful mechanical work.*
Source: reprinted from en.wikipedia.org.

mechanical work has a little precious nothing to do with friction! Instead, it is a *fundamental limit* of the natural world which prevents material bodies from attaining zero temperature, also known as Third Principle of Thermodynamics. This statement may sound a bit arcane for arcane it is indeed: what do we mean by zero temperature? If you think of ice, which I bet at this moment you actually do, I'm sorry, but you have to think again, for water freezes indeed at zero temperature, but this is not the zero temperature the third principle refers to! For (possibly) historical reasons, which have little to do with making life simpler, temperature is in fact measured in three (yes, three!) major scale of units.

The first and most popular (in Europe) is called Celsius (symbol ^oC), from the Swede Anders Celsius (1701–1744), who proposed it back in 1742. Zero Celsius is precisely the temperature at which water turns into ice (and vice versa). The second scale is named after Lord Kelvin (born William Thomson, 1824–1907), and it is simply Celsius plus 273.15, which means that in Kelvin degrees, ice forms at $T = 273.15^oK$. Kelvin is the relevant temperature when it comes to thermodynamics, hence the fundamental barrier is zero Kelvin, also known as *absolute* zero, equal to -273.15 Celsius.

Where does such fundamental limitation come from? The explanation had to await about a century, with the advent of quantum mechanics, which made it clear that matter cannot be placed completely at rest; some residual motion is always left, and it is precisely this motion that is responsible for non-zero temperature. For the record, the third temperature is named after the German scientist Daniel Fahrenheit (1686–1736), and the conversion from Fahrenheit (symbol ^oF) to Celsius goes like this: subtract 32, then multiply by 5 and divide by 9. When you are in the USA, remember that 32 Fahrenheit is the name of ice on the road.

Back to Carnot. Having realized that by heating cylinders one can set pistons in motion, the next practical question is how to do it on a continued basis, and not just once. The immediate solution is to proceed in cycles: once the piston has reached its upmost position, at the end of expansion, one can push it back to the original one by exerting pressure on it from the exterior. To be noted, this implies that this time you *spend* energy to bring the piston back where it started from. This is the price of realizing a cycle than can be restarted again. So, the name of the game is to extract mechanical work from the expansion and spend it back in the compression step, making sure that the former exceeds the latter. Happily enough, this is indeed possible by a clever sequence of transformations, now known as Carnot cycle.

And since it is a cycle, you can in principle repeat it indefinitely, which is precisely how thermal machines work to this day: current cars go through a few thousands such cycles per minute, as indicated on their dashboard. And if this is not enough to impress you, dear reader, let me kindly remind you that besides driving our cars around, this mechanism is also responsible for all the metabolic functions that keep us alive .... Not that we should search for pistons in our bodies, but cyclic nano-machines converting heat into work and vice versa are a commonplace in the human body. Not coincidentally, they are called molecular motors. The Carnot cycle is so fundamental to warrant general acquaintance beyond the scientific realm, which is why we next proceed to illustrate it in some more detail. For those willing to dig a bit deeper, further information is provided in the Appendix 8.1.

8.5 Much ado about something: The Carnot cycle

Consider a gas in a piston in a contact with two 'heat reservoirs' at different temperatures, which we label as 'hot' (T_H) and 'cold' (T_C) for convenience. Reservoir here means a large system, capable of absorbing or releasing heat without any sensible effect on its temperature, a sort of 'thermal elephant'. The Carnot cycle is composed of the following sequence of four thermodynamic transformation (see Figs 8.5, 8.6, and 8.7):

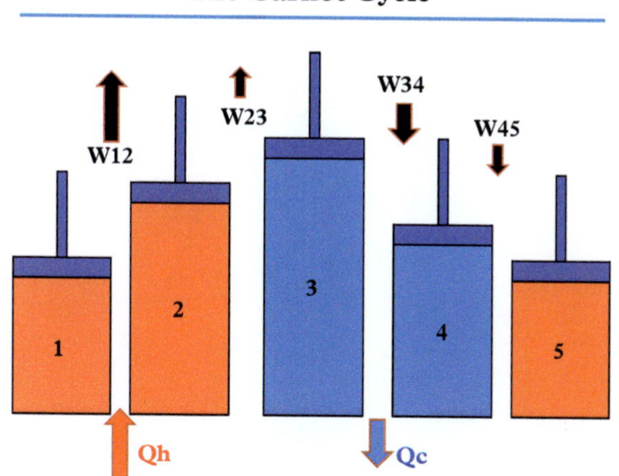

Figure 8.5 *The Carnot cycle with gas and pistons. From left to right: hot isothermal expansion ($1 \to 2$), the gas absorbs an amount of heat Q_H at high temperature T_H. Adiabatic expansion ($2 \to 3$), the gas still expands without absorbing any heat, thereby lowering its temperature from T_H to T_C. Cold isothermal compression ($3 \to 4$), upon being compressed at constant temperature T_C the gas delivers an amount of heat Q_C to the environment. Adiabatic compression ($4 \to 5$), the gas is compressed to the initial volume without releasing any heat to the environment, and consequently it heats up to the initial temperature T_H, thus completing the cycle. The black arrows on top of the diagram represent the work performed in each of the four transformations, arrow up means that the piston performs work (expansion) on the environment, while arrow down (compression) means that the environment performs work on the system. The bottom arrows represent the heat exchange, arrow up for heat absorbed by the gas and arrow down for the heat released. The final balance is $Q_H - Q_C = W_{12} + W_{23} - W_{34} - W_{45}$, i.e. $Q = W$ and $\Delta E = 0$.*

1. *Hot isothermal expansion at* $T = T_H$,
2. *Hot to Cold adiabatic expansion from* T_H down to T_C,
3. *Cold isothermal compression at* $T = T_C$,
4. *Cold to Hot adiabatic compression from* T_C up to T_H.

In the previous list, isothermal means that the temperature remains constant in the transformation, while adiabatic means that no heat is exchanged between the system and the reservoirs.

The key point is that *the heat absorbed by the system in the isothermal expansion is larger than the heat released in the isothermal compression.*

This thermal delta is exactly the amount of work that the system delivers to its environment, or, in equivalent terms, the mechanical work that we can extract from the system to perform useful tasks, such as setting the piston in motion. In

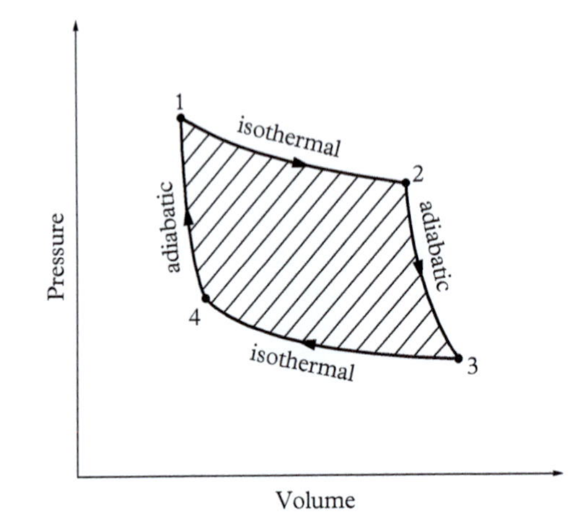

Figure 8.6 *The Carnot cycle in geometrical representation: the pressure-volume thermodynamic plane. The dashed area within the cycle corresponds to the work done in the cycle, which is the difference between the area subtended by the upper and lower isothermal curves. This is the geometrical analogue of the formula $W = Q_H - Q_C$.*
Source: redrawn from https://www.google.com.

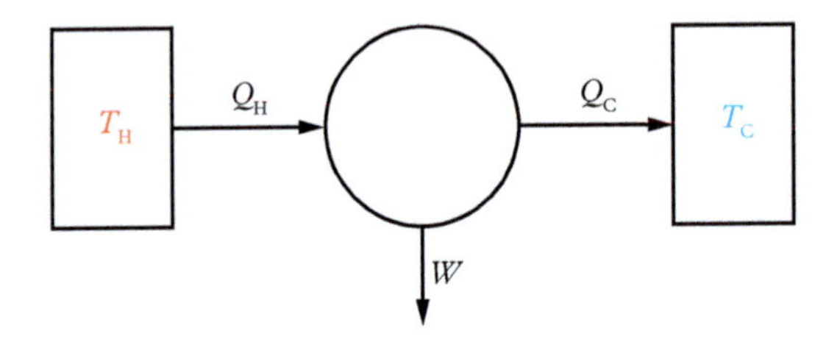

Figure 8.7 *Schematics of the Carnot cycle. Upon absorbing an amount of heat Q_H at 'hot' temperature T_H, the system performs an amount W of mechanical work, and as a consequence it must release an amount of heat $Q_C = Q_H - W$ to a 'cold' reservoir at temperature $T_C < T_H$. That is how heat sets things in motion.*
Source: reprinted from en.wikipedia.org.

simple math,

$$W = Q$$

where W is the work performed by a system absorbing a quantity of heat Q, which, in turn is given by the difference between Q_H, the net heat absorbed at high temperature T_H and Q_C, the heat released at low temperature T_C. But why would the work match exactly the net amount of heat absorbed Q?

The reason rests on energy conservation, which reads simply as:

$$\Delta E = Q - W \tag{8.5}$$

where ΔE is the energy change of the thermodynamic body in the course of the transformation. This is the essence of the First Principle of Thermodynamics: upon absorbing heat, the body increases its energy, and upon performing work to the external world, the body loses it to the surrounding environment. Hence, since according to the first principle energy is conserved, the change in energy in any given thermodynamic transformation must match the net heat absorbed minus the work done in the transformation. As simple as that.

Next comes a point which may sound trivial but it's not. Given that the initial and final states of the cycle are the same by definition, their energy is also the same, which means that in the Carnot cycle energy cannot change, $\Delta E = 0$. From this it follows that the work performed by the system is precisely equal to the net amount of absorbed heat, $W = Q_H - Q_C$. The reason why this is true is that the energy is a so-called 'thermodynamic function', i.e. it depends only on the thermodynamic state of the system, i.e. pressure, volume and temperature, not on the specific transformation through which such state is attained (see Fig. 8.6). The subtle point is while neither heat nor work are thermodynamic functions, their difference, energy, *is*! In other words, the amount of heat and work spent in taking the system from state A to state B does depend on the specific thermodynamic transformation, but their difference doesn't! This is why, while they are literally responsible for every thermodynamic change, heat and work don't share the same 'noblesse' status as energy (and entropy as well, as we shall see shortly).

8.6 The mirage of perfect efficiency

At this point, it is important to appreciate that the heat released at low temperature is lost forever, i.e. there is no spontaneous process that can send it back into the system. The reason is heat does not flow *spontaneously* from cold to hot bodies (it actually can, but only upon spending work in the process, which is what our fridge does). It is then natural to define the thermodynamic efficiency of the Carnot cycle as the work gained versus the heat invested in the process, namely:

$$\eta = \frac{Q_H - Q_C}{Q_H}$$

This shows that perfect efficiency, $\eta = 1$, implies $Q_C = 0$, no heat should be released in the cold isothermal compression. Is this possible at all? Unsurprisingly, the answer is a square NO. To appreciate why, the next point to realise is that the amount of heat exchanged at constant temperature is proportional to temperature itself (see Figs. 8.6 and 8.7). This means that Q_H is larger than Q_C, implying a positive amount of work. Since the heat is proportional to temperature, the efficiency can also be recast in terms of temperature, namely:

$$\eta = \frac{T_H - T_C}{T_H} = 1 - \frac{T_C}{T_H} \tag{8.6}$$

This expression informs us that efficiency benefits from operating the Carnot cycle between largely different temperatures, very hot versus very cold (that's why our car consumes less fuel in a cold winter night than in a hot summer day). The same expression also suggests two routes towards the mirage of perfect efficiency, $\eta = 1$. The first is to set the hot reservoir at infinite temperature, $T_H \to \infty$, the second is to set the cold one at temperature zero $T_C = 0$. Of course, both spell chimaeras, but while the former is immediately obvious (infinity is a no-no in physics), the latter is not. Indeed, Carnot et al. were definitely aware that some heat should be released in the course of the cycle, although they did not know why.

As mentioned earlier on, it took the microscopic insight of quantum physics, a century later, to realize the fundamental reason why no material body can be brought to literally zero (Kelvin) temperature. The actual world record is around the nanokelvin, namely one billionth of Kelvin degree, which is really a chilly number, but still not zero. Thus, the Carnot cycle is 'ideal', as it sets the upper bound for the efficiency of any real-life thermal cycle, an upper bound that cannot reach the perfect value $\eta = 1$, because no material body can be cooled down to zero Kelvin. *This is a fundamental limitation, long before it is a practical one*!

And don't be fooled, this limitation has *nothing* to do with dissipation! This is the reason why the Carnot cycle, besides providing the steppingstone for the industrial revolution, also represents a cornerstone of modern science at large. For the sake of the argument, a typical car would feature $T_C = 300$ and $T_H \sim 500$, so the efficiency of the Carnot cycle would be $200/500$, about forty per cent. For any given pair of hot and cold temperatures, this sets an ideal limit that cannot be exceeded by any real-life engine. Albeit unrealizable, chimaeras are mostly useful in science, for they tell us where the fence lies between the possible and impossible. As noted earlier, the previous expression shows why our car does better at night, when heat is released at a lower temperature, than during the day. In general, the higher the contrast between hot and cold, the better the efficiency. It is possibly of some interest to remind where the thermal fence lies at the present day: the hottest temperature ever achieved on Earth is about 10^{13} Kelvin degrees (ten thousand billions), in high-energy colliders. The lowest one is about 10^{-9} (one billionth), obtained in ultracold quantum gases. If one could operate the Carnot cycle between these extreme states of matter, one would achieve an efficiency which misses the ideal value of just 1 one part in 10^{-22}, i.e. ten thousand billionth of billionths. The day engineers can find a smart enough super material sustaining a Carnot cycle between high-energy colliders and the magnets that keep them going (at about 1 Kelvin), efficiency 1 will no longer be a dream.

Alas, technology can't go that far ....

8.7 Heat and dissipation

Men do not die from overwork, they die from dissipation and worry.

(Charles Evans Hughes)

We commented previously that one must resist the inclination of attributing the impossibility of reaching perfect efficiency to dissipation. The point is that Carnot's cycle is ideal because it implicitly assumes that *all transformations proceed at an infinitely slow pace*, precisely to avoid dissipative losses due to mechanical friction. But even then, efficiency cannot attain the ideal value, 1. We are facing a brain-twister again: what does 'infinitely slow pace' mean? Let's be serious, a transformation proceeding at infinitely slow pace, that's no transformation at all, isn't it?

The point is that when we wrote that we heat the cylinder and the gas starts expanding, we tacitly assumed that the heat should be supplied in very tiny amounts, step by step, so that the resulting motion of the piston is so slow that friction on the wall can be neglected to any practical purpose. More precisely, the piston should move slowly enough to allow internal relaxation of the gas molecules, before moving again (remember the Deborah number of Chapter 2). The attentive reader may sense the scent of Zeno's paradox again: the assumption is that energy and heat are continuum quantities which can be broken down in an indefinitely large number of indefinitely small pieces, just like Zeno did for time. It is therefore no coincidence that the zero temperature issue was solved by quantum mechanics, namely the science that originated precisely from the observation that energy cannot be broken into indefinitely small pieces, but comes in tiny discrete packets instead, known as quanta (see Appendix 19.3 on quantum mechanics at the end of the book).

We can be even more specific with the help of little math. We wrote earlier on that a volume change δV at constant pressure P releases an amount of work $\delta W = P\delta V$. This expression says nothing about the *time*, call it δt, it takes to expand (compress) the volume from V to $+\delta V$, i.e. the *speed* of the transformation.

Let us expose this speed by writing the volume change as follows:

$$\delta V = \frac{\delta V}{\delta t}\delta t \tag{8.7}$$

where the speed is precisely the first term on the right-hand side, namely the change of volume per unit time, while the second term is the duration of the process, (see Fig. 8.8). Despite its name, classical thermodynamics is the science of *infinitely slow change*, where 'infinitely slow' means that the processes involved proceed in the limit of zero speed, i.e. $\delta V/\delta t \to 0$. Here we run face-up into a serious brain twister: the left-hand side of (8.7) is nonzero by definition. So, how can this be, given that the speed at the right-hand side is zero?

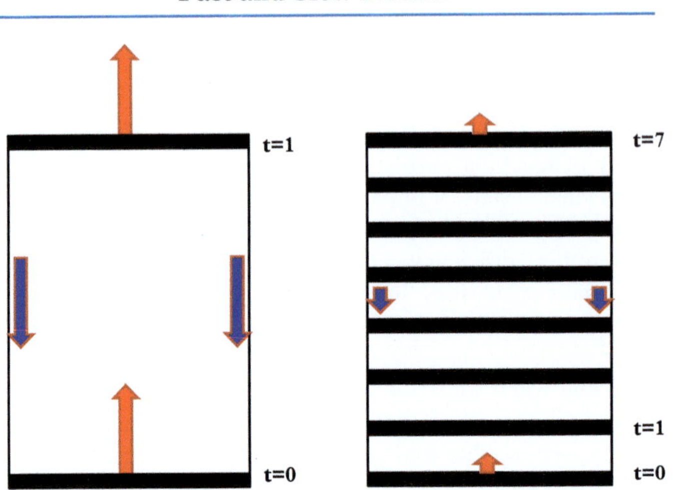

Figure 8.8 *A fast (left) and slow (right) pistons driving the expansion of the gas in the cylinder chamber. The left piston moves seven times faster (upward arrow) and dissipates* $7^2 = 49$ *more energy per unit time (the energy dissipation rate is proportional to energy, hence the square of the piston speed). By making the piston slower and slower, dissipation can be made increasingly small, to the point of being negligible. The downward arrows indicate the friction force acting at the solid wall of the chamber, in direct proportion with the piston speed.*

The way out of this paradox is to consider the limit of infinite duration, i.e. $\delta t \to \infty$, so that the product of zero times infinity returns a finite number! If this sounds like a math hocus-pocus, it is because hocus-pocus it is indeed, but an honest one: it is rigorously true that zero and infinite can be multiplied together and the result can be ..., anything: it can be zero, it can be infinity and, finally, it can be any finite number you wish! Before you get totally mad at me and stop reading, dear reader, let me hasten back to common sense. And the common sense is that, in actual practice, 'infinitely slow' means that the piston speed driving the volume change, must be slow enough to make dissipation on the walls completely negligible as compared to the energy of expanding gas. That's all it means.

In summary, although the Carnot cycle is an idealized chimaera, in fact precisely because of this, its monumental importance cannot be overstated. Carnot's vision of the thermodynamic cycle as the conceptual heart of *any* thermal machine, stands as tall as ever to this day. With all these preparations, we are now ready to introduce the second prima donna: stay tuned, it's time for entropy!

8.8 Entropy

Life is understood backwards, but must be lived forwards.

(S. Kierkegaard)

Trying to capture the ultimate nature of Entropy can get to pretty rarefied peaks of abstraction, with a multitude of connections with the notions of probability, information, uncertainty, irreversibility and the passage of time. Yet, the way entropy entered the scene of modern science was a far cry from abstraction: instead, it had to do with heat, steam, and all sort of very down-to-earth machinery we have been discussing in the previous sections. We talked about heat and work as the drivers of energy changes, the first principle, hence one may be led to identify them as independent forms of energy, even though, as noted earlier on, neither is. Here is the point again in some more detail.

Energy, by definition, is a function of the state of the system, namely its thermodynamic variables, typically density, pressure, temperature, denoted by the triplet NPT (N stands for number, from which density derives by simply dividing by the volume) or density, volume, temperature, denoted as NVT. This means that the change of energy in any given thermodynamic transformation taking the system from state A to state B depends only on the values of the thermodynamic triplet in these states, and not on the specific *path* taken from one to the other (path, here, means thermodynamic transformation, like the ones we met in the Carnot cycle).

It turns out that both heat and work -do- depend on the specific thermodynamic transformation, hence they are -not- functions of the thermodynamic state, but their difference, the energy gained or lost in the process, does not! The change in energy depends only on the initial and final state, not on the path in between, hence energy is a function of state. As noted previously, this is the reason why, in a cycle, where A is equal to B by construction, the energy is left unchanged (see Fig. 8.6).

In a similar vein, it was noted that another such path-invariant (state function in technical language) can be formed by dividing the heat exchanged in a given transformation by the temperature at which the transformation occurs. In mathematical terms,

$$dS = \frac{\delta Q}{T} \tag{8.8}$$

The left-hand side defines the change of entropy in the process by which the system exchanges the quantity of heat δQ at *costant* temperature T. The notation dS, instead of δS, stands precisely to indicate the path-independence of entropy change. Here again some magic occurs, heat is not a state function, but upon dividing by the temperature, it acquires that status!

How come?

Here it goes: since the heat exchanged during an isothermal transformation is proportional to the temperature at which the heat exchange takes place, by dividing the heat absorbed during the expansion at high temperature by the high temperature itself, we obtain exactly the same number we obtain by dividing the heat released during the low-temperature compression by the low temperature itself.

Figure 8.9 *A rather intimidating Rudolf Clausius.*
Source: reprinted from en.wikipedia.org.

The formula says it better:

$$\frac{Q_H}{T_H} = \frac{Q_C}{T_C} \tag{8.9}$$

We have already used it, without mentioning entropy, when we defined the efficiency of the Carnot cycle in terms of temperature. Thus, even though the heat balance is non zero, the heat balance divided by the respective temperatures is. This means that heat over temperature is a function of state, and that function of state is precisely what we call entropy.

The mastermind of entropy, and the one who coined the term, was the German Rudolf Clausius (1822–1888) (see Fig. 8.9), with the following rationale, Greek words, 'en-tropie' [intrinsic direction]. In his own words I have deliberately chosen the word entropy to be as similar as possible to the word energy: the two quantities to be named by these words are so closely related in physical significance that a certain similarity in their names appears to be appropriate.

So similar in many ways and so distant in many others, especially when it comes to talk about time, as we shall see in the final chapters. Given that entropy is on a par with energy as a state function, it follows that it does not change in the Carnot cycle either, which is precisely what we read from the expression (8.9)

$$\Delta S = \frac{Q_H}{T_H} - \frac{Q_C}{T_C} = 0 \tag{8.10}$$

The condition of no entropy change defines *reversible transformations*, i.e. transformations that can be seamlessly run back and forth in time, like in a movie. But real life is not a movie, which points us straight to the Second Principle of Thermodynamics.

8.9 The Second Principle of Thermodynamics

As we have seen, in a loop transformation, (initial and final states being the same) a function of state cannot change by definition. But in a generic transformation from equilibrium state A to equilibrium state B, it generally does. Take energy: if the heat absorbed in going from A to B exceeds the work done, the change is positive and vice versa. So, energy can change both ways, and if the system is closed (neither heat nor work exchanged), it does not change at all.

At variance with energy, though, *entropy can only increase or stay the same.* In formulas

$$dS \geq 0 \tag{8.11}$$

Transformations for which $dS = 0$ are called *reversible*, while all the others, for which $dS > 0$, are called *irreversible.* As the very name indicates, reversible transformations can occur spontaneously either ways, from A to B and viceversa. Irreversible ones spontaneously go one-way only. This embarrassingly simple-looking inequality encodes the Second Principle of Thermodynamics, one of the most fundamental, in fact possibly *the* most fundamental and inescapable limitations of all in the natural world!

Before we delve into these key matters, let's unfold the idea with a simple example. You're back home late on a winter night and your room temperature says zero Celsius degrees (by now you know it can't be zero Kelvin, do you?). No good, so you turn the heating on, to set it to a more palatable +20; barring malfunctioning of the heating system (always lurking behind the corner ...), heat starts to flow from the heater to the room. What does entropy do in the process? The heater delivers some amount of calories, say Q, at temperature $T = 273 + 20 = 293$ (in fact slightly above, but let us gloss over that for the moment) so that it loses $Q/293$ units of entropy to the room. The room, on its side, absorbs the same Q calories at temperature $T = 273$, so that it gains $Q/273$ entropy units from the heater. The total entropy balance is then

$$dS = -Q/293 + Q/273$$

which is a net positive, simply because 293 is larger than 273, as simple as that!

Somehow, we have just stated the obvious: heat spontaneously flows from hot bodies to cold ones, and, by its very definition, entropy can only grow in the process of bringing both to a common temperature because the entropy given away by the hot body (the heater) must necessarily be smaller than the one absorbed by the cold one (the room). The two only come to a balance when they attain the same temperature, namely at thermal equilibrium, in which case entropy stops growing. That's what thermal equilibrium is. A minute's thought exposes a deep message behind the obvious: if heat did the opposite, flow from cold to hot, cold bodies would become colder and hot ones hotter, no way one could reach the equitable compromise known as thermal equilibrium. In other words, *entropy growth is a byword for thermal stability*. Likewise, entropy saturation in time is a signature of thermal equilibrium. Note that the same reasoning, although left implicit, applies to the Carnot cycle as well. When we said that the gas absorbs heat at high temperature T_H from the hot reservoir, what we meant was that the reservoir was at a slightly higher temperature than T_H, otherwise no heat could flow to the cylinder. The point is that, by definition, the reservoir is so large that this surplus is in practice so small that we can actually neglect it for most purposes. But not if we take a closer look to the entropy budget.

The entropy change in the isothermal expansion is $dS_H = -Q_H/T_{HR} + Q_H/T_H$, where the temperature T_{HR} of the hot reservoir is just very slightly above T_H. And precisely because of this, dS_H is slightly positive. One may naively expect that this entropy increase would be balanced out by a corresponding entropy decrease in the isothermal compression. Another minute's thought shows that, no, entropy still goes up. The computation is plain: $dS_C = -Q_C/T_C + Q_C/T_{CR}$, where the temperature T_{CR} of the cold reservoir is slightly below T_C, whence dS_C positive again. Hence, the total change of entropy in the process is the sum of two positive contributions $dS = dS_C + dS_H > 0$. This brings up another basic aspect of the second principle: the entropy increase applies to the overall system, the piston plus the two reservoirs, usually called the *Universe*, to indicate that nothing is left out of it. By definition, the Universe is a closed system, simply because there is nothing 'outside' it to interact with! Hence, *the second principle applies to closed systems*.

This is worth being borne clearly in mind, because it is fundamental for our own existence, as we shall discuss in more detail in Chapter 9. The reader may argue that this is maybe purely coincidental, other processes could behave differently. Actually, the answer is a solid no: if this were the case, entropy would have no place in science. The fact is that any process taking *spontaneously* from a given equilibrium to another implies no decrease of entropy. Here, any means literally any, no exception known in the Universe (see previous discussion,) as we know it. Entropy growth is one of the most inescapable laws of nature, for many, the most inescapable of all: as per Atkins's book, these are the basic rules of the game [6].

8.10 The arrow of time

I'm older than I once was and younger than I will be. But that's not unusual, no, it isn't strange.

('The Boxer', Paul Simon)

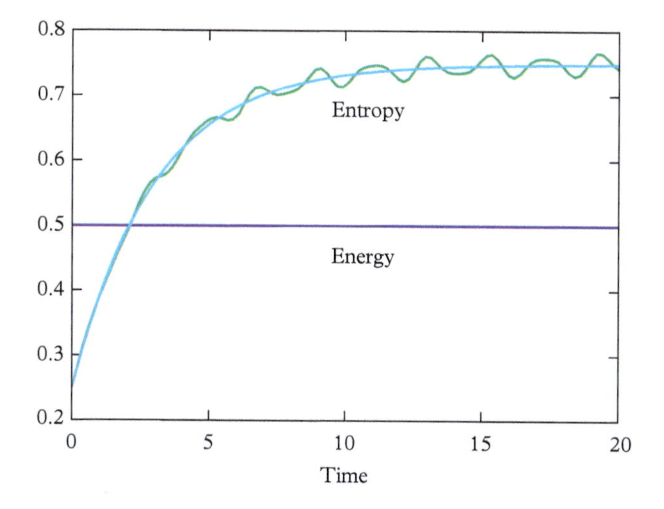

Figure 8.10 *The energy and entropy of a closed system as a function of time. The former is conserved, while the latter can only grow until saturation, namely thermal equilibrium. Note that second principle does not rule out occasional oscillations in time, but the average trend is always for growth until saturation.*

The inequality (8.11) expressing the second principle unravels the one-sided nature of entropy, which can only grow or stay the same (see Fig. 8.10). This immediately clicks with another key quantity which goes inexorably one-way only: Time. We can move right and left, up and down in space, as it best pleases us (of course I am referring to free space, definitely not an urban environment!), but we cannot move backward in time, no matter how hard we try. We are helpless against the one-sided nature of time: we inexorably 'run' towards our future, incapable of stopping the flow of time, let along revisiting our past (modern physics speculates of so called wormholes which would allow to connect to the past, but this is not the story here).

That's why entropy is also known as the 'arrow of time', to borrow the very successful metaphor of the British cosmologist Sir Arthur Eddington (1882–1994) (best known for his famous experimental verification of Einstein's prediction that light is bent by the Sun). Thus, we are faced with a key distinction between the two cosmic primadonnas, the first moves the world around by 'changing suits', i.e. clever transformations from one form to another. Yet, none of such transformations says anything about time because energy keeps the same value all along. Entropy, on the other hand, is all about the one-sided nature of time, from past to future: in a word, ageing.

And when future equals past, the game is over: no change, thermal equilibrium.

Indeed, entropy growth spells stability, which at first glance sounds comfortable, until one realizes, as Clausius et al. quickly did, that the ultimate form of stability, the grandest equilibrium of all, is death itself, 'thermal death' in a more scientific, yet no less depressing, wording. This the grim side of entropy, the dark lady of the cosmic duo, evoking Macbeth's famous 'walking shadow ...' (Macbeth, Act 5, scene 5, lines 16–27).

We shall have more to say about these matters in the course of this book, but before doing so, it is of some interest to take a look at the microscopic foundations of thermodynamics, for they provide a precious clue to the microscopic physics underlying the principles of thermodynamics. The towering hero here is the great Austrian physicist Ludwig Boltzmann (1844–1906), to whom we shall devote Chapter 9.

8.11 The rules of the game

To conclude, thermodynamics, the science of change, lays down the basic rules of the game which govern the functioning of the natural world on a macroscopic scale, namely the Three Principles of Thermodynamics:[32]

1. *First Principle*: the energy change in a thermodynamic process equals the heat absorbed minus the work performed in the course of the transformation. For a closed system (no heat, no work) the change is zero and energy is conserved. Yet, it can change from one form to another, and these transformations move the world around. In compact math form

$$dE = \delta Q - \delta W$$

2. *Second Principle*: the entropy change of the Universe (system plus environment) in a thermodynamic process is always in positive or zero, the latter condition denoting thermal equilibrium. Entropy growth marks the natural tendency of stable systems towards thermodynamic equilibrium, a condition of no-change also known as thermal death. Out of equilibrium, zero entropy growth (reversibility) can only be attained by infinitely slow transformations. In math form

$$dS \geq 0$$

3. *Third Principle*: no material body can attain strictly zero temperature in Kelvin degrees. The implication is that no thermodynamic transformation, not even one proceeding infinitely slow, can ever convert heat entirely into mechanical work (perfect efficiency). In math form

$$T > 0$$

[32] As mentioned earlier on in this chapter, Atkins adds a fourth principle, which he gives number zero, the concept of temperature. I defer the discussion of temperature to Chapter 9.

All of the previous calculations come from purely empirical observations, with no microscopic underpinning from the basic physics underneath, which was developed in the subsequent years, with the advent of statistical and quantum mechanics. Yet, their monumental importance rests precisely with their generality and universality, i.e. independence of the underlying micro-physics. That's why they genuinely deserve the status of 'Rules of the Game'. Complex systems take full advantage of the great freedom left within the bounds of the three principles, but, to the best of my knowledge, none of them has ever managed to cross over them.

8.12 Summary

Energy and Entropy rule the physical world. They compete and cooperate and their duel/duet gives rise to virtually all we observe around and within us: the rules of the game. At a first glance, this is yet another Ying-Yang story, Energy the Constructor, Entropy the Destroyer, ultimately Life versus Death. Reality is much subtler than that, because energy can destroy as well and, conversely, entropy comes with its own constructive power. Yet, this tragic Life versus Death metaphor is vigorously fuelled by the second principle, which informs us that, at the end of the day, the last word is always for (thermal) death, simply because this is the ultimate equilibrium, the pinnacle of stability. Depressing as it is, this picture says nothing about the microscopic why's underneath thermodynamics, a crucial matter which we proceed to discuss in Chapter 9.

9

The Man Who Trusted Atoms

O! Immodest mortal! Your destiny is the joy of watching the evershifting battle.
(Ludwig Boltzmann)

9.1 The microscopic roots of thermodynamics

In Chapter 8 we covered the fundamentals of thermodynamics, the science of change, and made the acquaintance to the basic rules which govern such changes, namely the three thermodynamic principles. Such principles have been derived from empirical observations, dealing with very practical and down-to-earth machinery, such as gases and moving pistons. Despite its empirical nature, this enterprise leads to deep and far-reaching conclusions, such as the notion that any natural system left in isolation runs inexorably towards its thermal death. For all the depth of such conclusions, thermodynamics remains completely silent as to the *microscopic* underpinning of its principles: where do they come from, and what is the basic physics beneath them?

At this point, it is worth placing thermodynamics in its proper historical perspective: in those days, the pinnacle of theoretical science was mechanics, which, in sharp contrast with thermodynamics, spoke a language of idealized perfection.

This is well expressed in the famous sentence by the French mathematician Pierre Simon de Laplace (1749–1827), who envisaged a 'intellect' capable of computing everything. The sentence goes more or less like this: 'Such an intellect would embrace in the same formula the movements of the greatest bodies of the universe and those of the lightest atom; for it, nothing would be uncertain and the future, as the past, would be present to its eyes'. This is the pinnacle of Newtonian science and a literal manifesto for determinism. Not that Laplace would not concede that there would be uncertainties attached to the previous calculations, but they would be of experimental nature, not computational one. This is the idealized perfection of mechanics, which makes no distinction between past and future, both equally present to its eyes. This shows a blatant clash with thermodynamics, and most notably with the second principle, which emphasizes the one-sided nature of time (irreversibility) (see Fig. 9.1).

It took the genius of the Austrian physicist Ludwig Boltzmann (1844–1906) to build a far-reaching bridge between the two, a bridge called Statistical Mechanics,

Sailing the Ocean of Complexity. Sauro Succi, Oxford University Press.
 DOI: 10.1093/oso/9780192897893.003.0009

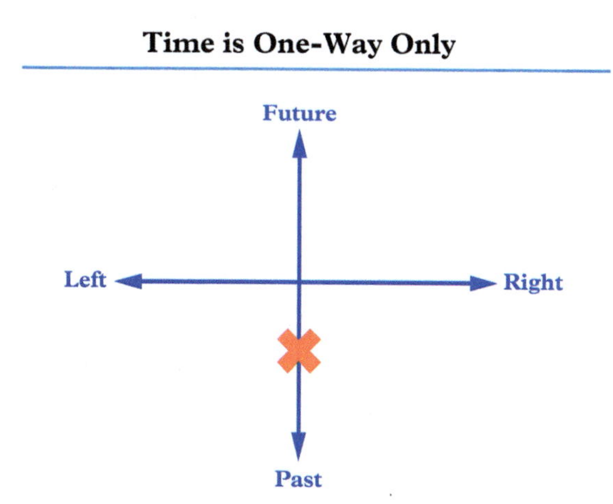

Figure 9.1 *Unlike space, time runs one-way only: you can move left or right, but you can only move towards the future, no return leg to the past.*

the *ante-litteram* example of unification in physics. The conceptual and practical import of such unification cannot be overestimated, as we proceed to discuss in the present chapter.

9.2 Ludwig Boltzmann

Duino, Italy (then Austria), September 5 1906: the lifeless body of Ludwig Boltzmann hangs from a noose, bringing to a tragic conclusion the mortal trajectory of one of the towering figures of the science of all times, just a little step (maybe) below the immortal trio of Galileo, Newton, and Einstein. Ludwig Boltzmann was born in Vienna in 1844 and, despite its tragic end, his lifeline offers little scope for the clichè of the misunderstood genius, as he collected glory and honours galore in his own lifetime. And yet, his life was a tormented one nonetheless, ending up in suicide in the morning of 5 September, 1906, during a brief holiday in the Italian city of Duino, near Trieste (the same town where the Bohemian-Austrian poet Rainer Maria Rilke (1875–1926) would compose his *Elegies* a mere 6 years later).

9.2.1 Kinetic theory of gases

Boltzmann is credited for being the founder of the kinetic theory of gases, a less than inspiring name for a theory that served as the portal to one of the deepest and broadest areas of theoretical physics, known as statistical mechanics. As its very name betrays, statistical mechanics occupies itself with the problem of connecting the properties of the macroscopic world, the one we perceive with our senses, to the underlying laws of the microscopic world, the 'things that we can't see'. Since the microscopic world is populated by crowds of atoms and molecules, a statistical description is mandatory, whence the name statistical mechanics. It should be realized that in Boltzmann's time, the microscopic world was a big mystery, to the point that even the very physical existence of atoms was fiercely contended. True, back in ancient times, the Greek philosopher Democritus (460–370 BC) speculated about the existence of tiny indivisible units of matter (that's what the word atom means in Greek, indivisible) and built up a picture of the entire Cosmos out of it. But it is one thing is to speculate about the material world, and a totally different one to provide scientific evidence of the existence of atoms.

Boltzmann fought his entire life for this cause, (see Carlo Cercignani's book *The Man who Trusted Atoms* [21]), facing the fierce opposition of the so-called energetic school, whose tenet was that matter is a sort of continuum fluid, like energy. Legend has it that this harsh fight eventually led Boltzmann to take his life, but the story is much less straightforward than that: he was known to suffer bipolar attacks, from very high spirits all the way down to deep bouts of depression [51]. In addition, in his final years he suffered from increasing problems of vision, which surely didn't help his feelgood factor, as it separated him more and more from his passion for piano playing. And, it is somehow sadly ironic that the dispute about atoms was definitely put to a rest just a few years after his death, basically by Einstein's 1905 model of Brownian motion as the result of collision between particles, as experimentally confirmed only 3 years later by the french physicist Jean Perrin (1870–1942, physics, Nobel 1926).[33] The fact remains that Boltzmann is, and will always be, a monumental figure of the science of all times.

So, what did Boltzmann contribute to modern science? In a nutshell, he built the first quantitative bridge between the microscopic world of atoms and the macroscopic laws of thermodynamics. More precisely, he wrote down an equation, now known by his name, from which he was able to derive the existence of a physical quantity that can only increase in time and eventually come to rest only at attainment of equilibrium. Precisely what entropy does in the macroscopic world. In passing, his equation also provided a neat physical interpretation of all other thermodynamic quantities, such as pressure, temperature, heat, and work. In brief, he laid down the bridge between the microscopic mechanics of atoms and molecules and the law of thermodynamics. In more lyrical terms, the bridge between what we can and cannot see.

Here goes the story.

[33] Brownian motion, after the Scottish botanist Robert Brown (1773–1858), refers to the erratic motion of microscopic particles, pollen in Brown's case, due to the bombardment of surrounding molecules. However, the record shows that it was first Jan Ingenhousz who described the irregular motion of coal dust particles on the surface of alcohol in 1785. This is yet another example of the so-called Stigler's law of eponymy (1980), according to which no scientific discovery is ever named after the true discoverer. Stephen Stigler, a professor of statistics in Chicago, probably escapes his own law ....

9.3 Entry probability

Boltzmann took a very original and bold step: instead of insisting on the molecular trajectory of each individual atom or molecule, he focussed his attention on the *probability* of finding an atom or molecule at a given position in space, with a given velocity at a given instant of time. This is called the Boltzmann probability distribution function, mathematically denoted as $f(x, v, t)$.

More precisely, the product

$$f(x, v; t)\Delta x \Delta v \tag{9.1}$$

denotes the *average* number of molecules (atoms) with position x within a spatial region of size Δx and velocity v within a velocity region of size Δv, at time t.

The innovative content of this approach should not go underappreciated. To begin with, atoms/molecules were not accepted as real at the time, which costed him harsh opposition by a group of colleague scientists, mostly in the circle of the highly influential Ernst Mach (1838–1916). In addition, the very idea of using probabilities at a time when the queen science of mechanics was utterly deterministic, was itself regarded with more than little suspicion. But Boltzmann proceeded undeterred along his way, and good for us that he did! He then wrote down an equation describing how this probability changes in space, velocity, and time.

The Boltzmann equation is no walk in the park even for professionals, for a number of good reasons. First, it lives in *seven* dimensions, that is three coordinates for the position x in space, another three to prescribe the molecular velocity v at each given position in space, plus time. This alone makes it solution a very hard task to this day. Second, the change of the probability in space, velocity and time is dictated by two basic and competing processes: molecules move from place to place (streaming) and when they meet in the same place, they interact according to rather complicated collision rules (collision). Collisions steer the system towards its ultimate fate, thermal equilibrium, while streaming constantly works in the opposite direction: non-equilibrium. This is the famous 'evershifting battle' Boltzmann referred to in his lyrical quotation at the beginning of this chapter. Once again, a competition/cooperation story. For the curious, a few more details on the Boltzmann equation are provided in Appendix 9.1.

9.4 Local and global equilibria

Like with any other competing scenario, the natural question is: does the evershifting battle ever settle?

The expectation is that it does whenever streaming and collisions come to a balance, hence no longer have any effect on the statistical distribution of the molecular velocities. The story is however a bit subtler than this because while collisions take place when molecules sit basically on the same location in space, streaming reflects the motion of molecules from one place to another. Hence, streaming and collision can come to a balance only when *both* have no effect on the statistical distribution of the molecular velocity. Since streaming moves molecules around in space, its effects are cancelled only when the system is uniform in space, same density, same velocity and same temperature everywhere. This is called *global equilibrium* .

Collisions, however, are less demanding: they can leave the molecular distribution unaffected place by place even though different places may exhibit density, velocity or temperature. Hence, *local equilibrium* does not require uniformity. The reason is as follows: suppose we focus our attention on the number of molecules with a given speed v. If one such molecule collides with another molecule at speed, say w, and change their respective velocities from v to v' and w to w', such *direct* collision decreases the number of molecules with velocity v. In other words, a direct collision leads to a *loss* of molecules with velocity v. Conversely, *inverse* collisions taking a pair of molecules with velocities v' and w' before the collision into v and w after, result in a *gain* of molecules with velocity v. Hence, collisions are in balance whenever direct and inverse collisions match exactly one another (see Fig. 9.2).

Since this can happen at different density and temperature from place to place, this is called a *local equilibrium*. Please note that local equilibrium is all but a boring state of dull idleness, molecules come and go all the time, but they keep a *dynamic* balance between direct ad inverse collisions. Using the financial analogy we brought up in Chapter 3, local equilibrium is like having have several

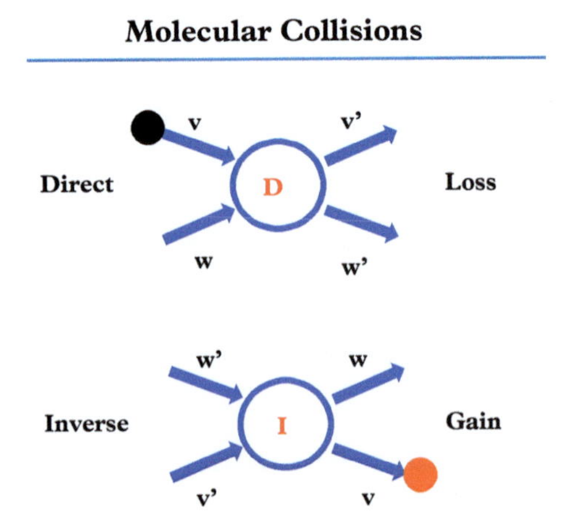

Figure 9.2 *Sketch of direct and molecular collisions. Direct collisions (top) between molecules with velocities v and w, turn their velocities into v' and w', hence remove one molecule with velocity v, a loss term. Inverse collisions do just the opposite, hence they increase by one unit the number of molecules with velocity v, a gain term. Local equilibrium results from the balance between the two Gain = Loss. Yet, neither is zero, for this is a dynamic equilibrium.*

individuals in different places, whose spending equals their earning every single day, although both earning and spending are different for each individual, say 100 Euros for the first, 50 for the second, 75 for the third and so on. At global equilibrium, all individuals earn and spend the same amounts, say 75 for everyone. This is in fact the state relevant to thermodynamics.

9.4.1 Maxwell–Boltzmann local equilibrium distribution

Having clarified the distinction between global and local equilibrium, the next question is: what does the local equilibrium distribution look like in velocity space? Boltzmann's detailed analysis showed that at local equilibrium, the probability distribution is a typical bell-shaped curve named after the great German mathematician Friedrich Gauss (1777–1855). In the specific framework of the kinetic theory of gases, this is best known as Maxwell–Boltzmann (MB) distribution, after Boltzmann himself and James Clerk Maxwell (1831–1879), universally known for his paramount theory of electromagnetic interactions. For the record, Maxwell derived the MB distribution a few years before, on independent grounds, although with no relation to the second principle of thermodynamics, a major leap that we owe to Boltzmann.

In equations

$$f^{eq}(x, v) = An(x)e^{-\frac{m(v-U(x))^2}{2k_BT}} \tag{9.2}$$

where $n(x)$, $U(x)$ and $T(x)$ are the local density, velocity and the temperature position x in space, and A is a temperature dependent constant. The symbol k_B is also number, now known as, guess what, Boltzmann constant.

The MB distribution informs us that the probability of finding a molecule with relative speed $v - u$ decays exponentially with the ratio of its kinetic energy $m(v - U)^2/2$ to the average thermal energy k_BT. It is indeed possible to show that k_BT is precisely the average kinetic energy taken across the entire population of molecules (the precise value is $3k_BT/2$, but this is immaterial to our purposes). As a result, the total kinetic energy of an ideal gas of N non-interacting molecules is given by

$$E = N\frac{mU^2}{2} + \frac{3}{2}Nk_BT$$

The first term on the right hand side is the macroscopic kinetic energy of the ensemble of atoms with mean velocity u, i.e. the energy associated with the ordered macroscopic motion of the gas The second term, on the other hand, is the kinetic energy associated witty the erratic motion of the molecules around the average speed U. Note that this latter quantity is linearly proportional to the number of molecules and to their temperature.

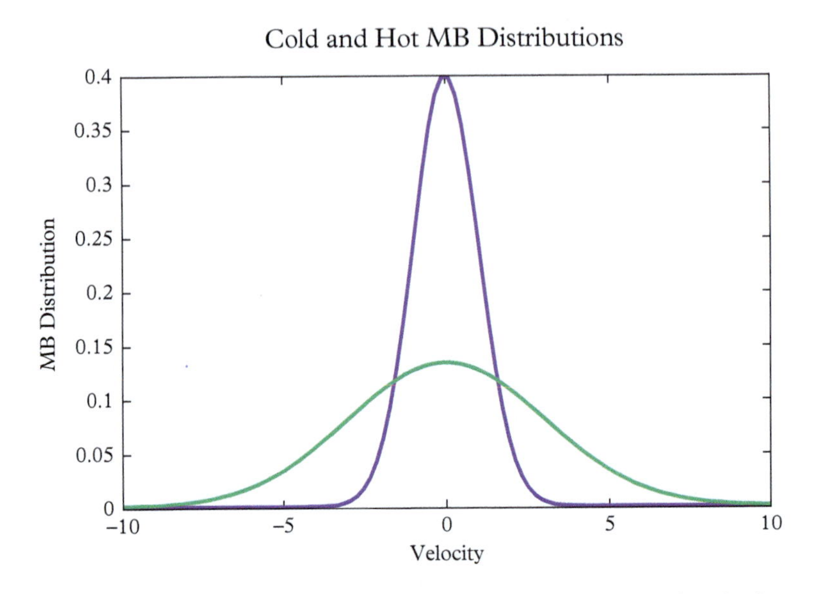

Figure 9.3 *A 'cold' MB distribution at temperature $T = 1$ (narrow curve) and a hot one at temperature $T = 9$ (broad curve). As one can see, at higher temperature, the high-velocity region is exponentially more populated. Note that the distribution is symmetric around the state of no motion, $v = 0$, which is also the most probable one. This is because we assumed that the macroscopic gas of molecules is at rest, i.e. zero net motion. If there was a net macroscopic motion at a mean velocity $v_{mean} = u$, the MB distribution would stay exactly the same, except for being centered around such a mean velocity. Hence, the mean velocity measures organized macroscopic motion (mechanical energy in Chapter 8) and temperature measures the statistical scatter due to molecular fluctuations. Order and Disorder take now a very precise connotation, and so does their coexistence.*

This means that molecules whose energy significantly exceeds the thermal value are exponentially rare (see Fig. 9.3). The import of the exponential Boltzmann factor cannot be overstated across virtually all walks of science, not least biology.

Summarizing, Boltzmann showed that there's a battle in the battle, first direct versus inverse collisions and then collisions versus streaming. The first battle settles down when the molecular velocities arrange according to the MB distribution (local equilibrium), the second when the macroscopic quantities, density and temperature, do not change anymore in space and time (global equilibrium). The two are typically well separated in time, first comes the direct-inverse local battle, which is fought independently place by place within the gas of molecules. Then the global streaming-collision battle, which implies communication between the different local equilibria across the entire gas. It is only at the end of the two that both sources of change, streaming and collisions, become zero and 'the fat lady sings', the unwelcome song of thermal death. Interestingly, however, it is in the first battle that the entropy kingdom starts to build up.

Let us see how.

9.4.2 The fatal attraction of conservation laws

Regardless of the initial statistical distribution of the molecular velocity, the competition between direct and inverse collisions inexorably drives the molecular distribution towards the MB shape. It is a sort of 'fatal attraction' which cannot be escaped as long as the total mass, momentum (mass times velocity) and energy of the molecules involved in the collisions is left uneffacted by the collision itself. These are the paramount *conservation laws*, which feed order within the erratic world of molecules. And this order shows up at the level of the organized, mechanical motion we have discussed in Chapter 8. Such order coexists with molecular disorder, as witnessed by the statistical spread of the MB distribution, which is precisely measured by the temperature. Since *any* initial distribution must converge to MB, the information contained in such initial condition gets literally lost in the process. The only way to reclaim it back would be to run time in reverse, but this a no-go for macroscopic creatures, such as we are. Hence, information is destroyed, entropy grows, and time moves forwards only. This is the content of the famous *Boltzmann H-theorem* (see Appendix 9.1), the microscopic underpinning of the second principle of thermodynamics. Not bad for a plain gas of molecules, right? I really hope that by now the reader appreciates why Boltzmann largely deserves a place among the greatest scientists of all times, [21, 20, 44] (see Fig. 9.4).

Figure 9.4 *Ludwig Boltzmann's image in the monumental cemetery of Vienna.*
Source: reprinted from commons.wikimedia.org.

9.5 From molecules to entropy

From his celebrated H-theorem, Boltzmann was then able to infer that, once the evershifting battle is settled, the entropy of a given macroscopic system, say a gas of N molecules in a box of volume V at temperature T, is proportional to the number of microscopic arrangements of the molecules which give the same three values of (N, V, T), i.e. the same *macrostate*. The formula, which is engraved on his tomb in the monumental cemetery of Vienna, reads as follows:

$$S = k_B log\ W \tag{9.3}$$

where W is precisely the number of microstates we alluded to previously, (see Fig. 9.5).

This is one of the most important equations in the history of science and will always be. Incidentally, it provides the operational key to navigate Ocean Complexity, for it shows that macroscopic creatures do not depend on the number of microstates W, but on its logarithm instead, which is the precisely the entropy S of the macrostate. And as we know (see Chapter 19, Appendix on Numbers), the logarithm is a formidable compressor of numbers: for instance it turns the astronomical Avogadro number $\sim 10^{23}$ into a very modest and manageable 23!

It is curious and kind of ironic that Boltzmann never wrote down this equation himself! Although he most definitely laid down all the conceptual background behind it, the formula was first written down by Max Planck, the founding father of quantum mechanics.

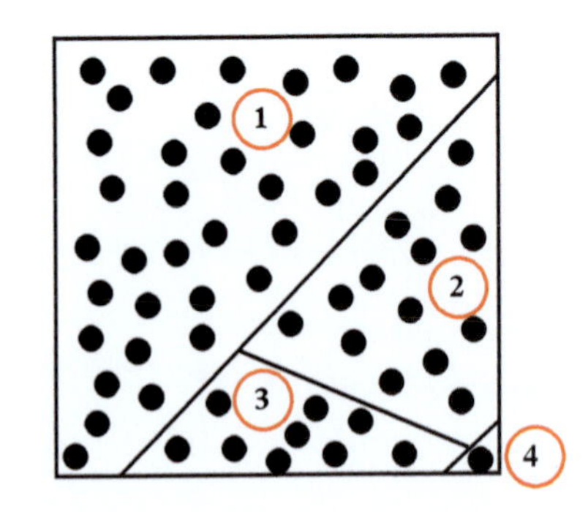

Figure 9.5 *Sketch of the statistical meaning of the Boltzmann's formula. The full space of micro configurations, each represented by a black dots, is partitioned in four domains, each corresponding to four different macrostates. In other words all microstates in the same domain share the same value of the macroscopic variable, for instance energy. The entropy of each macrostate is proportional to the logarithm of the number of microstates which fall within the corresponding domain. On the assumption that microstates fill up the domains uniformly, this number is proportional to the area of the domain. In this case,* $S_1 > S_2 > S_3 > S_4 = k_B log(1) = 0$.

We have just observed that the equation (9.3) reflects the astronomical separation between the combinatorial Complexity of the microscopic world (W) and the information we respond to in the macroscopic world (S). Failing such separation, we would be lost in Ocean Complexity. The story is less obscure than it seems, let us see why.

9.6 Entropic forces

The room where I am sitting in this very moment contains millions of billions of billions of molecules. Suppose I had the chance of cherry-picking one just next to me and swap it with one on the corner of the room. Would I experience any effect of such a swap? In other words, would I perceive a different temperature or pressure in the room? Most certainly not. Now think of how many such swaps I could make without ever noticing any difference: the logarithm of this monster number is precisely the entropy of the air in my room. Next, suppose I could amass every single molecule into a heap in a corner, nothing left elsewhere in the room: then, I would definitely experience a big change in pressure and temperature, because I wouldn't be able to breathe in the first place, given that there would be no molecule left around me, hence, no pressure nor temperature and ... nothing to breath! In other words, I would have changed the macroscopic state: the number of molecules in the room stays the same, but they occupy a much lesser volume and exert no pressure on me, since none of them ever hits my body any longer. Which of the two macrostates is more likely to occur on a spontaneous basis?

This is a patently rhetorical question: who has ever seen all the molecules in a room spontaneously migrate all together towards the corner, instead of spreading across the entire room and fill it up more or less uniformly? In thermodynamic terms, the all-in-the-corner macrostate has a much lower entropy than the ordinary macrostate, simply because there are incommensurably many fewer molecular micro-arrangements which can realize it.

We can make the example even more explicit, if a bit artificial.

Consider a room with ten light bulbs, each of which can either be 'on' or 'off'. If they are all on, we have full brightness, top score 10, whereas if they are all off, we have total darkness, bottom score 0. The score, just one number, is the analogue of the macrostates, the list of the ten numbers, 0 or 1, is the analogue of the microstate. And, as we shall see, there are many more than just ten possible microstates. In fact, their number is 1024, since each bulb can take two values and there are then of them, so the total number is 2 multiplied ten times by itself, which is $2^{10} = 1024$. Thus, total darkness is realized by a list of ten 0-s and full brightness is a list of ten 1-s. Note that the two are perfect mirror images; by turning zeros into ones and vice versa, darkness and brightness turn into each other. These are the most ordered states, since every bulb takes the same value, no fluctuations, no disorder in the system.

How many microstates are compatible with these top-ordered states?

The answer is plain: just one, since any change in one of the ten zeros (ones) is enough to compromise total darkness (brightness), bringing some disorder into the system. If you wish, you could describe these top-ordered states as very fragile, for any bulb-flip would spoil it. Using Boltzmann's equation, their entropy is $S = log\ 1 = 0$, where we have taken $k_B = 1$ for convenience. Now let us ask ourselves about the minimal brightness, score 1, just one bulb on out of ten. How many microstates are compatible with it? Since any of the ten bulbs 'on' would do, there are precisely ten possibilities, hence the state of minimal brightness can be realized in ten possible microways. The Boltzmann equation gives $S = log\ 10 \sim 2.3$. Now move on to the next state of brigthness, score 2. The number of lists compatible with it is the number of pairs of bulbs 'on', which can be formed out of ten bulbs. Simple combinatorics or plain enumeration shows that this number is $(10 \times 9)/2 = 45$, nearly five times larger than the macrostate 1. In fact, there are ten ways to select the first bulb 'on', and once this is done, there are nine ways to select the second bulb 'on', their product being divided by two because swapping the two 1s gives the same configuration. The story repeats number by number, macrostate 3 can be realized in $(10 \times 9 \times 8)/(3 \times 2 \times 1) = 120$ ways, giving $S = log\ 120 \sim 4.78$). It is not hard to see that macrostate 5 is the one with most microscopic realizations, their number being $(10{\times}9{\times}8{\times}7{\times}6)/(5{\times}4{\times}3{\times}2 = 252)$, with an entropy $S = log\ 252 \sim 5.52$. Going beyond 5 takes the show downhill, because the number of realizations giving 6 for brightness is the same as the one giving 4 for darkness, so it is clear that the top value is realized when darkness and brightness are in exact balance, which gives total brightness 5. If one plots the number of microstates compatible with a given macrostate, the result is the staircase bell-shaped curve I will now give, known as binomial distribution, (see Fig. 9.6).

This curve shows that the most disordered state, the one scoring 5 in total brightness, is the most populated, 252 microstates, hence the most probable one.

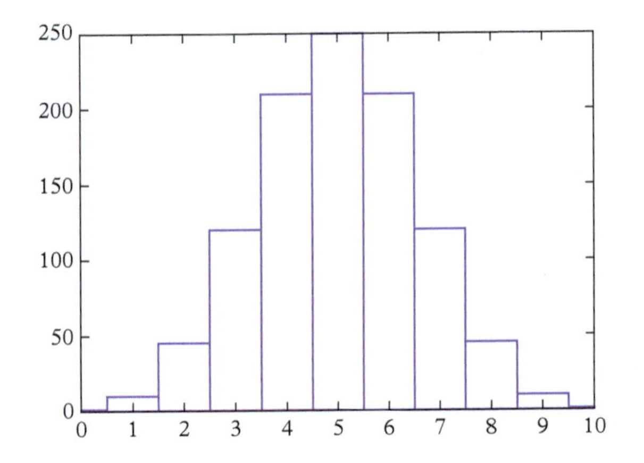

Figure 9.6 *Binomial distribution for N = 10, a staircase version of the Gaussian distribution, to which it converges in the limit of infinite N.*

And if instead of ten light bulbs you have Avogadro numbers of molecules, the contrast is immensely larger. But the entropy grows much more slowly than this because of the logarithm in Boltzmann's formula. These are the molecules filling up the room versus the heap in the corner state!

9.6.1 Entropy and happiness

Now we see why entropy is a measure of the disorder of the microscopic world. The system evolves towards maximum entropy states just because these are the easiest (most probable) states to realize on a purely combinatorial basis: the widest door. Tolstoy's opening of *Anna Karenina*, which we invoked before in this book, says it best again: happiness is much harder to achieve than unhappiness because there is only one (or very few) ways of being happy and vastly more to be not. Happiness demands all 'good' conditions to be met, one failing being often one too many, hence, enough to spoil the show. Happiness is a very *fragile* condition, unhappiness, on the contrary, is very robust, there are many ways of changing things and still remain as unhappy as before, or even more .... In brief, Unhappiness is the wide door, Happiness is the narrow one.[34] The effect is very real, to the point that scientists introduced the notion of *entropic force*, precisely to denote the probabilistic drive towards high entropy states. Entropic forces are the ones promoting stampeding for the widest door and they are always in action, even in the absence of any other interaction. In fact, *especially* in the absence of interactions, i.e. when the system is completely at the mercy of the vagaries of pure combinatorics (*molecular chaos*). A freely expanding gas is the prototypical example: the gas does not expand under the effect of any interaction, since the molecules don't exert any force upon each other, the force driving the expansion is purely statistical, this is what an entropic force is! And this is precisely the force pushing up the pistons in the expansion stage of Carnot's cycle. In this chapter we have been dealing with gases, but similar considerations (although less straightforward) apply to solid and liquids as well (see Fig. 9.7).

9.7 Temperature: The entropic trigger

Before Boltzmann, temperature was a bit like time, if asked, no doubt we know what it is, but once asked to explain, we really don't. Sure enough, temperature is a close relative of energy, since bodies get hotter upon absorbing heat, but what form of energy exactly? Boltzmann provided a very cogent picture of the microscopic roots of temperature: it is the kinetic energy associated with *disordered* motion of the molecules. What do we mean by disordered molecular motion? A daily life analogy might help.

[34] This is only a pedagogical metaphor, as one should definitely refrain from identifying perfect order with the most desirable condition, as we shall discuss in the Chapter 11.

Consider cars on a highway, and say that they proceed at an average speed of 100 km/h. Does this mean that each and every car proceeds exactly at a speed of 100 km/h? Most certainly not; one moves perhaps at 95 another at 105, yet another at 90 and yet another one at 110. The average over four cars is $(95 + 105 + 90 + 110)/4 = 100$, but, somehow ironically, none of them moves exactly at their average representative speed! This is because there are individual departures from the average, called *fluctuations*, reflecting the individual freedom of the molecular world: this is aptly called *molecular chaos*. Temperature is precisely the kinetic energy associated with these individual fluctuations, namely the statistical deviation from the average. This is a quintessential emergent property, namely one that only makes sense as applied to a statistical collection of cars. Each individual car has energy, not temperature, since temperature is a property of the entire collection of cars. Now imagine shrinking cars down to the size of molecules, and what we perceive as the net motion of this 'fluid of cars' is 100 km/h, the macroscopic speed of the fluid, whereas temperature is the kinetic energy contained in the fluctuating velocities. This shows that temperature is the actual manifestation of molecular disorder whereas the average speed is the manifestation of order, namely the net motion we perceive in the macroscopic world. In the example of the room, the average speed is basically zero, but this by no means implies that the molecules of air in the room are just sitting idle: quite on the contrary, they move pretty fast, approximately three times faster than a Ferrari! The reason why I don't perceive any net motion in the room is that for each molecule moving along one given direction in space, there is another molecule moving along the opposite direction, so that the two balance each other and no net motion results. Net macroscopic motion only results when this statistical balance breaks down.

With these microscopic insights, we are now in a much better position to revisit the operational definition of entropy given in Chapter 8, namely the heat absorbed divided by temperature. Heat raises temperature, hence disorder, but since temperature is itself a measure of disorder, the increase of entropy is smaller than it would be at lower temperature, because disorder on top of disorder (heat transferred at high temperature) is less effective than disorder on top of order (heat transferred at low temperature). Incidentally, this also shows why $T = 0$ is a chimaera, at zero absolute temperature, even the tiniest amount of heat would generate an infinite amount of entropy! This is the microscopic foundation of the definition of entropy $dS = \delta Q/T$, given in Chapter 8.

Beautiful, isn't it?

9.8 Under pressure

We have just seen that temperature is basically the kinetic energy stored in the erratic motion of molecules. What about pressure? Once the atomistic picture is endorsed, the microscopic face of pressure emerges seamlessly: pressure is the force per unit area, exchanged between the gas molecules and the walls of the

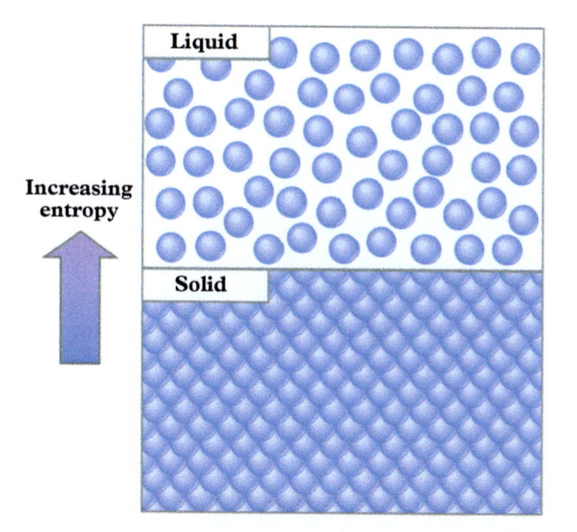

Figure 9.7 *Entropy as a measure of disorder. The ordered solid (bottom) has less entropy than the disordered liquid (top). Beware though, there exist solids which have more entropy than liquids, ice being the most famous example in point. This means that the widely held identification of entropy with disorder does not capture the full picture, as we shall discuss in more detail in chapter 16.*
Source: reprinted from commons.wikimedia.org.

container. In other words, each time a gas molecule hits the wall and gets bounced back, it exerts a force to the wall in the process, and the pressure is just that force per unit area of the wall. This is how the gas molecules push the piston up or down upon impinging on its surface and reflected back.

Consider the isothermal expansion of the Carnot's cycle: the volume increases, temperature stays, hence pressure must go down because the same number of molecules occupies a larger volume, resulting in fewer rebounds (per unit time) on the walls. This is why in an ideal gas the product of pressure times the volume is proportional to the temperature. If the temperature stays constant, it means that the disordered motion of the gas molecules has the same intensity throughout, hence the force transferred in a single rebound is the same. What changes, though, is the frequency of the molecular rebounds on the solid wall, which decreases due to the expansion (say, the piston moving upwards, see Fig. 9.8). If, on the other hand, the temperature goes down at constant volume, pressure goes down too, because, due to the cooling effect, each single molecule is left with less fluctuating kinetic energy.

Incidentally, we note that pressure is the principal character when it comes to produce useful work, like pushing a piston up or down. In Chapter 8, we have seen that the amount of work produced by a piston expanding the volume of the gas by an amount δV at constant pressure P, is given by $\delta W = P\delta V$. As usual, the symbol δW stands to indicate that, just like heat, the amount of work performed

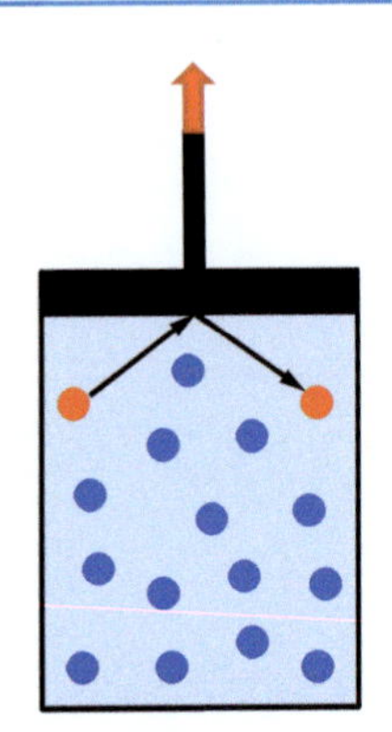

Figure 9.8 *The microscopic origin of pressure. Gas molecules impinge on the piston head and are elastically bounce back into the gas, thereby exerting a force on the piston. Molecules are tiny, but they are many, hence their cumulative effect lifts the piston up. This is a quintessential entropic force in action.*

in a given transformation does not depend only on the initial and final states, but also on the specific thermodynamic transformation. Since pressure is defined as the force exerted by the impinging molecules per unit area of the piston, the amount of work can also be rewritten as $\delta W = F\delta h$ where $\delta h = \delta V/A$ is the change in elevation of the piston, A being its area. This reconnects to the basic definition of mechanical work as the product of the force times the displacement of the body it acts upon. Summarizing, the molecular hypothesis gives full microscopic account of the empirical observations which laid down the foundations of thermodynamics.

9.9 The microscopic face of dissipation

In technical terms, energy is defined through its changes, namely the work done by a given force to bring the system from state A to state B. This work, in turn, is the product of the force times the displacement produced by the force itself, as just illustrated in Section 9.8. As discussed in chapter 8, energy is a state function, i.e. it depends only on the initial and final states, not on the specific path taken in between. If you climb a mountain by taking a comfortable path through the woods, or a steep rocky wall, the change in potential energy is exactly the same: your mass, times gravitational acceleration, times the height of the mountain, $E = mgh$ (we assume you start at zero altitude, $h = 0$).

This may sound nonsensical in the first place, because you have to put much more work (and skill) to go via the rocky path, but this is none of the energy's business. The huge difference between the tourist's versus the professional paths goes entirely on the account of dissipation, the energy lost in friction, never to return to the macroscopic world. Let's go back to the molecular world. Take a fluid flowing across a solid wall; the layer of fluid molecules next to the wall experiences an attraction from the molecules of the solid wall and gets slowed down in the process. From the macroscopic viewpoint, the fluid molecules experience a net friction, which is proportional to their speed relative to the wall and to the frequency of the collisions with the solid molecules. In the absence of external forces, such friction ultimately puts the fluid molecules at rest, meaning that the energy of the fluid molecules is entirely lost to the wall in the form of heat. This heat flows from the fluid to the solid, which transforms the ordered motion of the fluid molecules into disordered oscillations of the solid molecules. If the 'fluid' molecules did not interact with the 'solid' ones, they would flow freely over the wall without any dissipation. But 'no molecule is an island' and some interaction always takes place (in so-called superfluids, such interactions are very low but still not zero) which is why friction and dissipation are a trademark of real life. Likewise, it would be great if one could return the heat from the solid back to the fluid just by reversing the fluid velocity, but we know all too well that this is not the way things work in the actual world: whether it moves rightwards or leftwards, the fluid molecules always give energy away to the walls.

As mentioned before, such dissipation is *not* part of Carnot's picture, which is an idealized zero-speed process in which all transformations take a virtually infinite amount of time, so that friction never has a part in the scene. But in the real world, things need be accomplished within a given finite-time schedule, lest they are rendered useless or even damaging. Timing, in life like in music, is everything! Hence, dissipation cannot be escaped, and it adds a further energy toll on any real life (finite-time) thermodynamic process. Beware: if we could measure the energy of every single molecule, nothing would be lost, the net energy of the fluid would be found intact in the vibrations of the solid molecules. But this energy is 'disordered' and we cannot retrieve and convert it back for useful purposes, like pushing the piston up or down. For macroscale creatures, such as we are, this energy is lost forever, (see Fig. 9.9).

Summarizing, mechanical work is the organized brother of disorganized heat. Even though they are the true actuators of thermodynamic change, neither enjoys the pedigree of a 'thermodynamic function', because they both depend on the specific process and not on the initial and final states only. Remarkably, their difference, energy, does not and this is what buys the 'noblesse' status of thermodynamic function. A similar statement goes for entropy versus heat, not without the crucial intermediate of temperature. Cosmos and chaos again, now in specific micro-thermodynamic hats. And as we have learned well by now, both are needed: there is no life without net motion, there is no life without temperature and dissipation either. Life can't be pure chaos but not a perfect-order game either ....

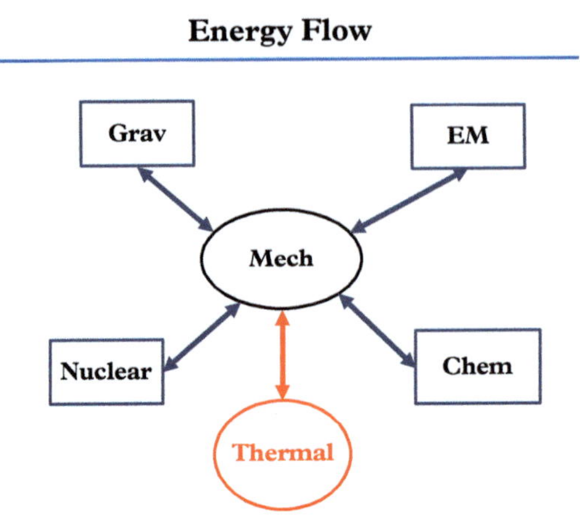

Figure 9.9 *The energy conversion between its manifold forms. Ovals: kinetic energy in its ordered (mechanical) and disordered (thermal) forms. Boxes: potential energy in its gravitational, electromagnetic, chemical, and nuclear forms (chemical energy is basically electromagnetic). The various forms of potential energy convert to mechanical energy, which in turn converts to thermal (and back) via heat flow. If such transformations could proceed at an infinitely slow pace, no dissipation would occur, and thermal energy could be converted back into mechanical energy. Even so, a complete conversion would demand zero absolute temperature, which is impossible. Entropy measures the amount of information lost in the thermal energy that cannot be converted back to mechanical energy.*

9.10 Entropy

Only entropy comes easy.

(Anton Chekhov)

We conclude this chapter with a brief discussion on the connections of entropy with three major themes of Complexity: time, information, and disorder.

9.10.1 Entropy and time

In Chapter 8 we saw that heat spontaneously flows from hot to cold bodies, and this necessarily entails an entropy increase till saturation occurs (thermal equilibrium). Since entropy increases in time, the two notions are strictly connected to each other, and we could even turn the statement upside down by saying that time is the direction of entropy growth, the arrow of time, time grows in entropy. The contrast with the other prima donna is striking: energy morphs into various forms, but in an isolated system without friction, it does not change in time, it is conserved. Which means that, at variance with entropy, energy is timeless and 'forever young'. Thus, if life was purely energy-driven, it would be timeless as well: for good or for ill, everybody knows that this is not how the story goes.

Here comes the 'dark side' of entropy: to us, humans, at least past a certain age, the passage of time is perceived mostly as ageing and deterioration, the unstoppable trend towards the end of the line, an end known as thermal equilibrium, the condition where nothing changes anymore. This bleak interpretation surely did not escape the attention of the founding fathers, particularly Clausius, who coined the less than joyful expression of 'thermal death'. And the universal character of the second law added an inexorable sense of depression conveyed by the idea of entropy as the carrier of ageing, deterioration and ultimately thermal death, no exception to the rule, including the Universe itself.

A true skyfall scenario.

Fortunately, as we shall see in the next chapter, modern thermodynamics offers a less apocalyptic vision, but for the moment, let us stay with the depressing picture as it emerged in the early days of thermodynamics. Depression notwithstanding, the scientists tried to understand where the second principle was coming from, i.e. its microscopic foundations. The operational definition of entropy did not shed any light in this respect, and even in hindsight, there was no way it possibly could because, at variance with energy, entropy is an *emergent* property, which only makes sense as applied to groups of objects/individuals. It is an inherently statistical quantity. A qualitatively new level of understanding, a literally new vision of the microscopic world was needed to this purpose, and this precisely Ludwig Boltzmann's monumental legacy. This brings us to the next entropic connection: entropy and information.

9.10.2 Entropy, information, and scrambled eggs

As we have seen, entropy is a measure of disorder: there is way less entropy in a heap of molecules amassed in the corner than in the same number of molecules spread around the entire room. Molecules don't spontaneously amass in the corner. Likewise, there's much less entropy in the intact eggs as you buy them in your grocery store than in the omelette you get by scrambling them in your pan. Scrambled eggs from intact ones for a tasty breakfast is a commonplace, intact eggs from unscrambled omelettes is a no-show in this world, (see Fig. 9.10). This provides an additional deep clue, namely that disorder also means loss of information: once the eggs are scrambled, we have no way to tell which egg each tiny bit of the scrambled omelette came from. This information is lost forever. Why? Simply because we are macroscopic objects and have no way of tracking the motion of each and every molecule in the course of the process of turning intact eggs into scrambled ones. And not because we don't know the equations of motion, which we do, but because we are not able to solve them, at least, not to the level of accuracy required to answer the previous question (something Pierre-Simon de Laplace did not anticipate). Hence, entropy is a measure of our increasing *Ignorance* of the microscopic world. Please, note that Ignorance, here with a capital I, has all but a dismissive connotation. In fact, the opposite is true, it is a

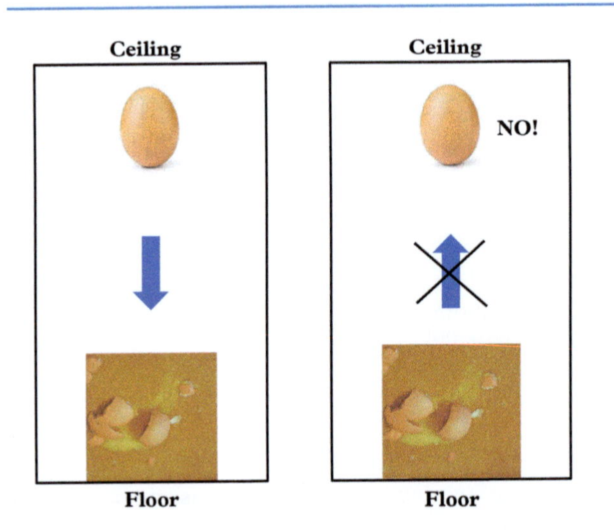

Figure 9.10 *An egg falling from the ceiling breaks down and scrambles on the floor (left). Inverting gravity (right panel) does not cut any ice to the cause of restoring the unbroken egg .... The act of scrambling destroys the information encoded within the structure of the unbroken egg, thereby killing reversibility. The transition unbroken-broken egg cannot be undone.*

distinctive noble trait of human condition. We may want to call it unknowledge for the sake of a better wording, but the meaning is the same.

The surprising, I'd say breathtaking, fact is that we have the means of surviving in this ocean of mounting unknowledge, a gift we neither understand nor deserve, to quote Eugene Wigner again. To put it bluntly, we can happily enjoy our omelette with zero need of knowing which egg each piece came from. This Ignorance does zero harm to our wellbeing .... And this is only a trivial example out of a constellation which fills up our daily life: *insensitivity to unknowledge* The detailed reasons are unclear, but a sure point is the following: being much larger than the atoms/molecules we are made of, we don't respond to the individual fate of each and every one of them, but only to their collective behaviour. We experience density, pressure, temperature, which form out of zillions molecules, not the individual molecular trajectories leading from unscrambled to scrambled eggs, which are astronomically more numerous than those making the reverse journey. This is the basic reason why we can survive in the ocean of microscopic Unknowledge which surrounds us in all directions.

It is probably no exaggeration to say that life, as we know it, depends crucially on this divide between Order and Disorder, namely the information which we need for our survival versus the one we can do without. And again, it is an amazing gift that life can thrive in the face of the relentless increase of the latter versus the former. In light of the previous information, another question naturally is: why are we so much larger than the atoms and molecules we are made of? Again, a solid scientific answer is still open, but the fact remains that in order to develop the kind of Complexity required to perform vital functions, large groups of molecules are required to work in sync. Otherwise, we could be made of just one single big molecule, after all, which would be dangerous indeed because if by chance that single molecule would stop functioning, the whole game of life would be over, no second chance. With many cooperating and competing microscopic agents we are protected against microscopic vagaries, which is a key to sustain life as we know it. This said, if these macroscopic vagaries keep accumulating in a coherent fashion over time, then the aforementioned statistical protections may go in jeopardy, which is how disease develop in the first place. This is an extremely important area of application for the Science of Complexity, but one which lies beyond scope of the present book.

9.10.3 Caveat: Entropy and Disorder

We wish to conclude this chapter with a cautionary remark, and the remark is that we should not hasten to identify entropy with disorder. Having noted myself in this chapter that entropy is a measure of microscopic disorder, the reader may be baffled at the previous caveat. Please, don't be, here comes the point. It is intuitively clear that solids are more ordered states of matter than liquids, hence one would expect that they always carry less entropy. Yet, this is actually not true: there are solids which feature more entropy than liquids! And the reason is that there are regular crystal arrangements in which molecules have *more freedom to move* than in the disordered arrangement typical of liquids. This has been first shown to be the case for (apparently) simple systems like a collection of hard spheres: beyond a given concentration, a disordered collection of spheres (liquid) *spontaneously* orders into a crystal! And if you think that this is an artificial nicety, you might be interessted to know that this exactly what water molecules do when liquid water freezes to ice. We shall more to say on these matters in chapter 14. The only point we wish to make here is that possibly the deepest meaning of entropy is not disorder but *freedom to move*, or even more generally, *propensity to change*. And while it is often the case that spontaneous change is accompanied by increasing disorder, this should not mistaken for a necessity. What is always true, instead, is that as long as there is room for an entropy increase, there is room for change as well. Thus, in the end, what entropy really measures is the potential for change.

9.11 Summary

Ludwig Boltzmann is responsible for a timeless achievement: by means of a statistical formulation of classical mechanics, he laid down the microscopic foundations of thermodynamics, thus unifying two apparently vastly disparate fields. Although still largely unknown to the wider public, I believe his place in the history of science is right next to the greatest ever, Galileo, Newton, Maxwell, and Einstein. The reasons are best discussed by professional historians of science, my personal speculation being merely bad chronological luck. Boltzmann died in 1906, shortly ahead of the period which saw the explosion of the two superstar theories of physics of all time, to this day, namely relativity and quantum mechanics. How could an old-fashioned world of classical particles possibly survive competition with the mind-boggling scenarios opened by the two superstars? The same is true to a large extent today, statistical physics (statphys) still receives less limelight than headlines-making big science, as per in the opening chapter of this book. However, as time went on, Boltzmann's ideas proved far more general than the physics they were born from and for, namely classical mechanics, and they continue to hold a central stage in most forefronts of modern science. We shall return to this in the final chapters of the book.

9.12 Appendix 9.1: Boltzmann's kinetic theory

Perhaps the most far-reaching result of Ludwig Boltzmann was the identification of a quantitative link between the world of atoms and the second principle governing the macroscopic world. The math is dense, but we can cut it to the bones.

Boltzmann took a very imaginative and bold conceptual move: he focussed on the *probability* of finding an atom or molecule at position r in space with a velocity v at time t, call it $f(r, v, t)$. More precisely, the number of atoms in small volume of space Δr, with a velocity within an interval Δv, is given by $\Delta N = f(r, v; t)\Delta r \Delta v$. He wrote down his eponymous equation which governs the change of such probability in time, space and velocity. The Boltzmann equation takes the following general form:

$$\frac{\partial f}{\partial t} + v\frac{\partial f}{\partial x} = G - L \tag{9.4}$$

The first term on the left-hand side represents the change in time of the probability distribution function at a given position x in space and velocity v. The second term on the right hand side is the change in space due to the molecular streaming. The right hand side reflects the competition between inverse collisions (Gain) and direct ones (Loss). Local equilibrium is defined by the condition $G = L$, indicating that direct and inverse collisions dynamically balance each other, causing no change in the local probability distribution. Such local equilibrium can change from place to place and become global, i.e. the same everywhere, when the system attains uniformity in space, so that the streaming term is also nil. This is the state of global thermodynamic equilibrium. One more observation: the symbols G and L hide a great deal of mathematical complexity which in turn reflects the microscopic Complexity of molecular collisions. As a result of such complexities, both G and L depend on product of $f(x, v, t) \times f(x, w, t)$ of the Boltzmann distribution at location x and time t bit with two different velocities v and w. This shows that the Boltzmann equation exhibits a quadratic nonlinearity. Moreover such product needs to be summed upon all possible partner velocities w, indicating that, besides being quadratically nonlinear, the Bolztmann equation is also non-local in velocity space. These two features, along with the fact of inhabiting a six-dimensional space plus time, explain why the solution of Boltzmann equation still presents a very hard mathematical and computational task.

Based on his equation, he was able to show that there exists a quantity, called H-function for rather unknown reasons, which can only increase or stay constant in time, precisely like entropy! Just for the sake of the argument, Boltzmann's H-function reads as follows

$$H(t) = -\sum_{r,v} f(r, v; t) log f(r, v; t) \Delta v \Delta r$$

The initiated reader would recognize the microscopic version of the far more famous formula of entropy:

$$S = k_b log\ W$$

For the sake of completeness, we also report the expression of the main thermodynamic quantities.

Number of molecules:

$$N = \sum_{r,v} f(r, v; t) \Delta r \Delta v$$

Kinetic energy:

$$E_{kin} = \sum_{r,v} \frac{mv^2}{2} f(r, v; t) \Delta r \Delta v$$

Macroscopic velocity:

$$U = \sum_{r,v} v f(r, v; t) \Delta r \Delta v \ /N$$

Thermal energy:

$$\frac{3}{2}Nk_BT = \sum_{r,v} \frac{m(v-U)^2}{2} f(r,v;t)\Delta r \Delta v$$

Pressure:

$$P = Nk_BT/V$$

For a professional discussion of Boltzmann's kinetic theory, see [58, 60].

10

Biological Escapes

Entropy is the price of structure.
(I. Prigogine)

10.1 The new alliance

The two previous chapters chronicled the basic traits of the two cosmic ladies, energy and entropy, along with their microscopic interpretation. By now, the reader should be ready to appreciate how similar and different they are at the same time. They both stem from the very same common sweaty ground, steam engines, but energy speaks for organized motion and ordered structure, while entropy seems to stand for basically the opposite, all about limitations, including the most inescapable and conclusive one, thermal death, the triumph of disorder.

The picture sounds pretty bleak indeed, doesn't it?

Fortunately, as we shall discuss in this chapter, developments in modern non-equilibrium thermodynamics have revealed a richer and appreciably less dismal picture than this. In particular, they have unveiled a sunny side of entropy, namely its constructive role in shaping Order out of Disorder in the (apparent) face of the second principle. A role which proves particularly prominent and decisive for the formation of structures at all scales in our own Universe, including biological ones. So, let us thus take a look at this much needed smile from lady entropy, from Bios and Cosmos. Let's begin with the first, leaving the second to Chapter 11.

10.2 Biological escapes

Clausius's bleak picture of entropy speaks of an inexorable downhill run of the Universe towards its thermal death. Yet, biology runs a very different show, one centered on the formation and growth of organized structures: we see flowers blossoming in spring time, caterpillars turning into butterflies, babies coming into the world. In other words, we see Life in action in multiple and magnificent forms. How come? How does one reconcile the blossoming of life with thermal death? As we shall see, a closer inspection of the second principle unveils the key. Beware,

Sailing the Ocean of Complexity. Sauro Succi, Oxford University Press.
 DOI: 10.1093/oso/9780192897893.003.0010

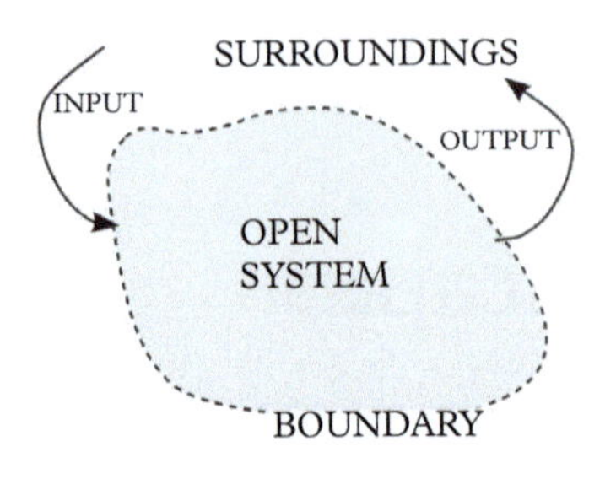

Figure 10.1 *Sketch of an open system. The system communicates with the surroundings (environment) through the boundary, across which mass, momentum and energy are exchanged. For isolated systems, such an exchange is zero (closed systems).*
Source: reprinted from en.wikipedia.org.

the key is not forever, it comes with a deadline attached, we cannot outwit entropy growth for good, but for a while we can, and this 'while' is precisely our own precious lifetime. There are two basic keys to the biological escape: 1) living systems are generally *open* and 2) function *far from equilibrium* (see Fig. 10.1).

Let's dig a bit deeper into the far-reaching implications of the previous statements.

10.3 Open systems

As mentioned earlier, the second principle applies to *closed* systems, i.e. systems which do not communicate with their environment. Remember the adiabatic step of the Carnot cycle: the system expands and then contracts with no heat exchange with the two reservoirs: it is thermally isolated. Not so for the isothermal steps, in which the system heats up and cools down by absorbing and releasing heat from/to the thermal reservoirs. Obviously, most systems we see around us, ourselves in the first place, are not isolated at all: we don't last long without food, less without water and just a few minutes airless.[35] The point is so obvious that one wonders whether any closed system would exist at all, at least in the living world. It turns out that there is definitely one, and certainly not in a footnote, as we are talking nothing short of our own Universe! Technically speaking, the Universe is defined as the union of any given system plus, literally, everything else, the name of this everything else being 'Environment'.

Symbolically:

$$\mathcal{U} = \mathcal{S} + \mathcal{E}$$

[35] Incidentally, the record is a rather incredible 22 minutes and 22 seconds by the German Tom Sietas ....

where $\mathcal{U}$ stands for Universe, $\mathcal{S}$ for the system in point and $\mathcal{E}$ for its environment.

Both $\mathcal{S}$ and $\mathcal{E}$ are open and exchange mass, energy and entropy across their interface. They sum up to the entire Universe, which is the ultimate closed system by definition. Hence, the second principle states that the entropy of the Universe cannot decrease in time, it either grows or stays. Granted that as a 'noble' human species, we should all be interested in the ultimate fate of the Universe, a less noble version of ourselves would hardly place such eschatological worries ahead of our own 'little universe', perhaps just the weather forecast for next weekend.

Briefly, we are chiefly interested in the actual system, or better said, *subsystem*, we live in. Leonard Cohen (1934–2016) says it best, 'I am mostly interested in what contributes to my own survival' [64]. And such subsystems, most notably the biological ones, are patently *open*, their very survival is entirely dependent on their ability to exchange mass, energy and information with their surrounding environment: that's how animal and vegetable life both work. The plain but key conclusion is that while the entropy of the Universe is bound to grow in time or stay, the same does not necessarily apply to each of its subsystems. In particular, a given subsystem can develop Order (i.e. structure) from Disorder by producing entropy and dumping it down on its environment, reducing its own entropic content at the expense of the environment.

This might seem selfish and selfish it is indeed, but that's exactly what biological systems do for a living: they lower their entropy content by increasing that of the environment, the final entropic bill is always sent to the Universe .... inglorious as this is, that's what we are: entropy polluters, to the point that a cosmic censor would probably be entitled to establish an entropy tax (please don't mention it to the politicians ...). For the sake of clarity, take our cherished planet Earth as the system $\mathcal{S}$, our equally cherished star, the Sun, as the environment $\mathcal{E}$, and close the Universe to just the sum of the two: Earth plus Sun. Of course, this is only an approximation, since there is way more to the Universe than our beloved planet and its burning star, but for the sake of our argument, let us ignore the rest.

The Sun supplies energy to Earth in the form of photons (light) over a band of wavelengths between 100 nanometres (UV) up to millimetres (Far IR). This energy is then exploited to sustain all sort of chemical processes which keep life going on our planet. At the end of the day, the remaining energy is radiated back to the outer atmosphere, with a substantial shift of the spectrum towards long wavelengths. More precisely, per each high-energy photon absorbed on Earth's surface, there are about 20 less energetic photons radiated away (see Fig. 10.2). Since the entropy of the photons is approximately equal to their number, Earth delivers to the outer space about 20 times more entropy than it receives, and such entropy 'deficit' is precisely the price to sustain the ordered and living structures on our planet (See Fig. 10.3). Being much more numerous, the long photons happen to be in a state of higher disorder, hence higher entropy, and that's precisely how planet Earth arranges its entropic bargain. In Prigogine's words, this is the price of growing structures on planet Earth, including ourselves .... Simple as this is, it identifies nonetheless a remarkable escape route from Clausius's bleak

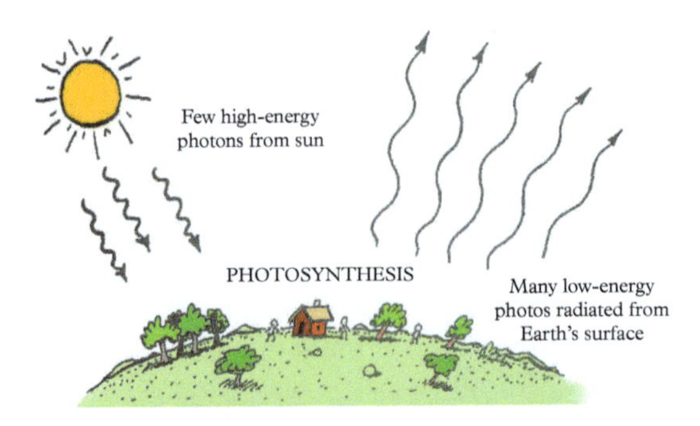

Figure 10.2 *The entropy supply from the Sun. Planet Earth absorbs heat from the Sun in the form of ordered high-energy photons (temperature about 6,000 Kelvin) and releases it to the Universe (temperature 3 K) in the form of disordered low-energy photons (temperature 300 K).*
Source: redrawn from https://www.google.com.

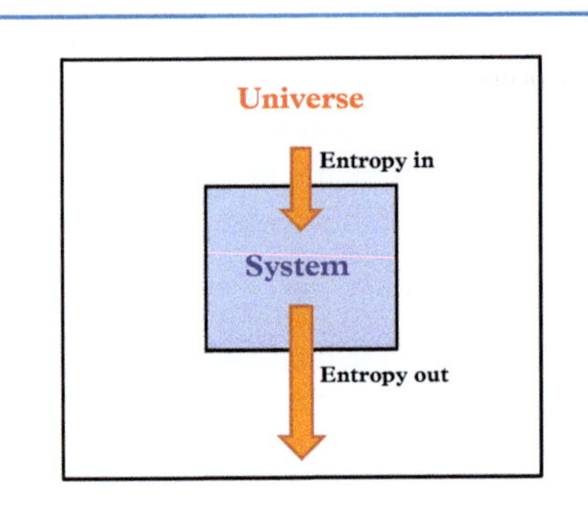

Figure 10.3 *Sketch of the entropy dumping mechanism which allows subsystems of the Universe to build up coherent structures by dumping more entropy to the Universe than they receive from it. A job at which biological organisms seem to excel, whence the title of this chapter.*

picture: the second principle cannot be escaped *globally*, but it can be sidestepped *locally*, in finite regions of space and time, at the expense of others. And since we are definitely finite in both space and time, as long as we don't aim for immortality, that is good enough.

This is what Prigogine referred to as to the New Alliance, a new deal between nature and humans: more to follow.

10.4 Out of equilibrium

The next key point about Clausius's formulation is that it refers to systems at equilibrium. But again, this by no means the usual state of affairs with biological

Figure 10.4 *Erwin Schroedinger's epoch-making book What is Life?, along with the author.*
Source: reprinted from en.wikipedia.org.

systems, which are typically kept out of equilibrium precisely by the aforementioned mass and energy exchanges with the environment. *Metabolism* is the word for this non-equilibrium process. The great Erwin Schroedinger (1887–1961), one of the founding fathers of quantum mechanics, in an exceedingly influential foray into biology, his epoch-making book *What is Life?* [100] (see Fig. 10.4), famously stated that 'It is by avoiding the decay into the inert state of equilibrium that an organism appears so enigmatic. What an organism feeds upon is negative entropy.' He came to the point of giving a name to negative entropy, 'negentropy', which, although it does not appear to have cut much ice among scientists, conveys well the idea. Leaving aside names, the idea that thermodynamic systems can develop structure by borrowing entropy from the environment is sound and far-reaching, once proper quantification is put in place.

A more technical and less than inviting word is NESS, for 'Non-Equilibrium Steady-State', meaning by this states that don't change in time (steady), not because they are at the thermodynamic equilibrium, but because they are kept away from it by external supplies. Once these external supplies go, NESS follow suit. We know it all too well: stop eating and after a few days you die, stop drinking and you die even faster

Perhaps, the most celebrated post-Clausius formulation of the second principle is the theory of *Dissipative Structures*, by the Russian-Belgian chemist Iliya Prigogine (1917–2003), who coined the term to describe the Nobel Prize in chemistry for his insights into non-equilibrium thermodynamics [91]. In a burst of poetical philosophy Prigogine called this a New Alliance between mankind and nature, meaning by this, I suppose, that nature is more benevolent than Clausius's picture would have [92, 93].

This does not take the last word away from thermal death, no, but it may suspend and defer it long enough for interesting things to happen on the space and time scales that interest us. Our own life is all about such a suspension, literally, *time borrowed away from the second principle*. This sentence is perhaps a bit too high in philosophical romanticism, but not devoid of scientific meaning. To elicit the point, let us describe a concrete example of a prototypical dissipative structure: Bénard cells, from the French physicist Henri Bénard (1874–1939).

10.5 Dissipative structures

Consider a simple household experiment, a flat layer of liquid is sandwiched between two solid plates and heated from below. The bottom plate is 'hot', at temperature T_H, and the upper plate is 'cold', at temperature T_C, smaller than T_H. The thermal delta between the two plates, $\Delta T = T_H - T_C$, triggers heat flow across the liquid, from bottom to top, in order to match the temperature constraint imposed by the two plates (see Fig. 10.5).

Now the liquid has a problem to solve, namely, how to comply with the thermal constraints imposed by the plates (the environment). As long as ΔT is sufficiently small, this can be achieved by individual molecular collisions; the hot molecules near the bottom hit the cold molecules just above and transfer part of their energy to them in the process. Thus, the hot molecules below cool down and the cold molecules above heat up. The molecules above do the same with the molecules in

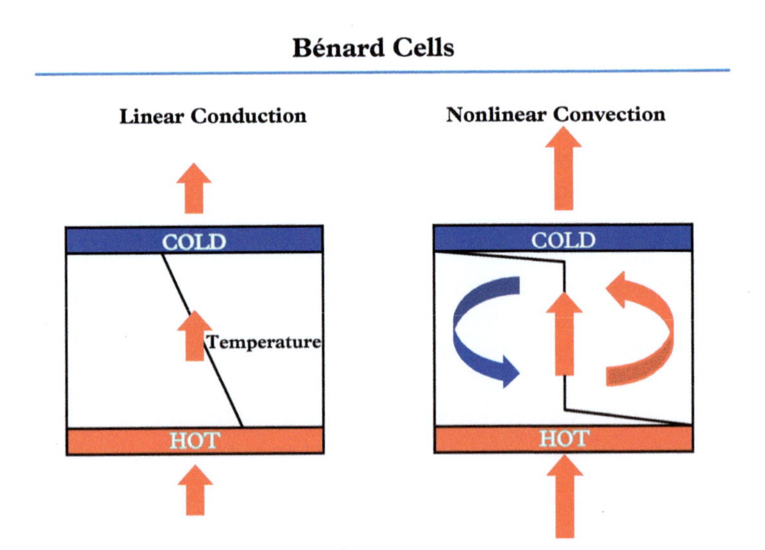

Figure 10.5 *The mechanism of Bénard cells (right): fluid heated from below (red) reaches up to the cold wall, where it deposits its energy, getting cool (blue) and then falling down to the bottom again. Note the drastic change of the temperature profile with respect to the case of linear conduction (left). The overwhelming share of temperature change is localized in the near-wall layers.*

the layer just above them, and so on, layer-by-layer across the liquid, all the way up to the top wall.

If the top wall were adiabatic, i.e. did not allow any heat out, in the course of time it would heat up and equilibrate with the bottom wall, leading to a uniform temperature profile, $T_U = (T_H + T_C)/2$, across the entire liquid. But in the Bénard setup the top wall is kept at a constant temperature T_C, which means that the heat received from the bottom wall and transferred across the fluid, is given away to the environment. The result is a temperature profile which decreases in linear proportion to the elevation across the fluid, starting from T_H at the bottom decreasing to T_C at the top. This is the linear regime physicist call 'conduction', one which is entirely sustained by molecular collisions. In this regime the heat transfer across the fluid is simply proportional to the thermal delta $T_H - T_C$: double the former, the latter doubles too. Linear science reigns here.

10.6 Conduction and convection

Now make the thermal delta a bit larger and see what happens.

Up to a point, nothing changes, the temperature profile remains linear across the fluid, and all is well in Linearland.

But at some point, beyond a critical temperature, a completely new movie premieres. The featureless fluid is replaced by a regular pattern of *coherent structures*, known as Bénard cells, whereby parcels of hot fluid from the bottom lift up directly to the top wall, where they deposit their energy, without much interaction at all with the intermediate molecules in between. Upon releasing energy to the cold wall, the hot parcels cool down, get denser, and consequently, they start to travel in the opposite direction, top to bottom, until they hit the hot wall, where they get heated back again, ready to restart a new cycle; remember Carnot?

A literal *thermal highway* opens up in the fluid, one that simply didn't exist before in the molecular world of Linearland! Besides serving immediate purposes in our daily cooking, Bénard cells are of utmost importance in most atmospheric and geophysical phenomena, such as the formation of tornadoes or magma transport in the planetary crust. What we wish to emphasize here, though, is its conceptual import as a paradigm of *structure formation via dynamic instability*.

This sounds a bit obscure, but it's actually not. Structure formation indicates the emergence of organized patterns, the rolls. Dynamic instability means that the linear temperature profile is not capable to sustain temperature differences above a given threshold, which means that above such threshold a new transport mechanism must arise, and the instability of the linear profile is precisely the gateway to this new regime (see Fig. 10.6). This is the New Alliance Prigogine's referred to. But let's take a closer look.

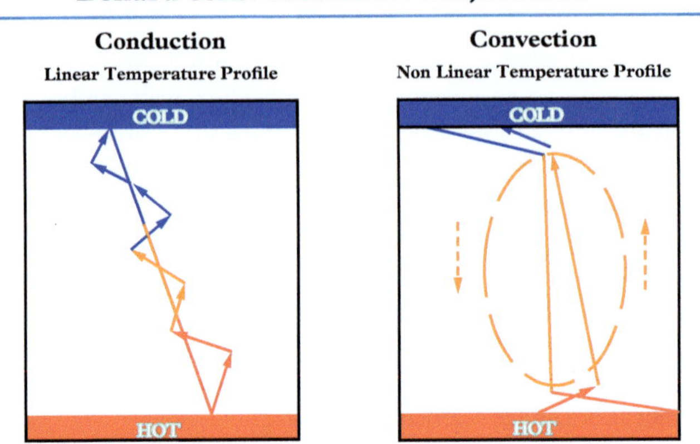

Figure 10.6 *The mechanism of Bénard cells (right) again with a cartoon of the underlying molecular trajectories (short arrows). In the conduction case, molecules lose their energy gradually and uniformly bottom-up (red-orange-light blue). In the convection case they lose most of their energy in the bottom layer (red to orange), then basically keep the same energy until the reach the vicinity of the top layer in which they suddenly lose another major amount of energy (orange to dark blue), so as to match the thermal constraint* $T = T_C$ *on the cold wall. The long and thin orange lines indicates that molecules move much faster because they are carried out by the coherent structure they form in the first place.*

10.7 A new thermal deal

The linear conduction regime is structureless, no organized patterns in sight. Beyond a given threshold however, such regime is no longer capable of coping with the thermal constraints imposed by the boundaries: individual molecular collisions are simply incapable of transferring the amount of heat required to sustain the imposed thermal delta. This presents the system with a make-or-break bifurcation: either it enters a new functioning regime, or it is doomed to burn out from failure to release heat at the imposed rate.

Happily enough, a new regime is indeed possible, and one which *emerges* from the ashes of the first, via a dynamical instability. What this means, in concrete terms, is that the linear profile becomes unstable, in the sense any tiny perturbation drives the system irreversibly towards a new destination (remember stable and unstable equilibria in Chapter 3). Bénard cells are precisely this new destination: by proceeding through the thermal highways described previously, the molecules manage to transfer way more heat across the system than individual molecular collisions could ever possibly do, thus meeting the thermal constraints. This is the new regime which remained silent and hidden under milder constraints: under an increasing thermal constraint the system finds an 'innovative solution'.

How about the temperature profile?

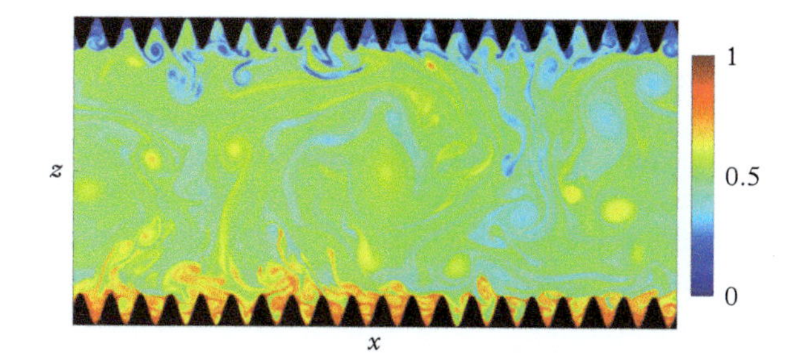

Figure 10.7 *Typical mushroom-like structures resulting from Rayleigh–Bénard turbulence with corrugated walls at a Rayleigh number of about two billions. From S. Toppaladoddi, S. Succi, J. Wettlaufer, (2017). Roughness as a route to the ultimate regime of thermal convection, Physical Review Letters, 118(7): 074503.*

Source: reprinted with permission of aps.org.

In the linear situation, each layer of fluid contributes equally to the heat-flux, the amount of heat crossing a liquid layer per unit area and time. It can be shown that this is proportional the slope of the temperature profile itself, which for a linear profile, is simply constant across the liquid. In the new regime, the linear profile is replaced by a steep drop of temperature, from T_H to T_U, over a thin layer of width h, just above the bottom wall and much shorter than the height H of the overall liquid region. Then, the temperature remains basically constant across the fluid, at a uniform value T_U, until a second sudden drop, from T_U to T_C, occurs again across another thin layer, also of width h, just below the top wall.

Such nonlinear profile is much more effective in transferring heat from the bottom to the top wall, because the entire temperature drop from T_H to T_C is now segregated around a much smaller region, the two layers of width h each, both much thinner than the cell height H. Hence, in the nonlinear configuration, it is just the two thin layers that are basically in charge of the entire heat flux, while the bulk region contributes virtually nothing! Under thermal stress, the system develops a typical 'antifragile' response, i.e. it works out 'innovative solutions' which are almost entirely in charge of an extreme minority of molecules, the ones in the thin layers, the rest just serving the purpose of connecting the two (see Fig. 10.7).

10.8 Where is nonlinearity?

Although discreetly behind the curtain, nonlinearity is again the art director in this plot, as it is always the case far from equilibrium. A detailed mathematical description of the so called Rayleigh–Bénard convection, the general framework which Bénard cells subscribe to, is out of scope here. Suffice to say that the equations which govern the motion of the fluid and its temperature exhibit (quadratic) non-linearities, which are fully responsible for the dynamic instability leading to the convective regime.

The Bénard cells are the outcome of the *nonlinear competition* between various mechanisms: thermal and viscous dissipation, thermal dilation coupled with gravity. A crucial role is played by *buoyancy*, namely the pressure lift experienced by a parcel of fluid, due to the fact that the liquid below is less dense than the previous one, on account the thermal expansion. If you think of Archimedes' Eureka, you're not quite on target, but you're in the right ballpark. Beyond a given thermal delta, buoyancy lifts the parcel of fluid and sets it in convective motion according to an organised pattern: the Bénard cell.

The physical reason for the instability is elementary: if the fluid is heated from above, like the sea receiving light from the Sun, the upper layers are hotter, hence less dense than the lower ones. Light on top of heavy is a mechanically stable situation, and all is fine. In the case of Bénard cells, the fluid is heated from below, hence the upper layers are denser than the lower ones, and we are faced with heavy on top of light, which is a mechanically unstable situation. If the thermal delta is small, the instability remains silent, but above criticality, the genie is released from the bottle. It should not go unnoticed that convection is a *collective* form of ordered motion, in which zillions of molecules behave like one: it this precisely this *Order* that gives groups of molecules the power to transfer an amount of heat which would be totally unsustainable by single molecules in the conduction regime. A bit like Roman phalanxes, at the molecular level .... In other words, the mechanisms which act on (large) *groups* of molecules, versus the ones which act on individual ones.

The aforementioned competition between buoyancy and dissipation is measured by a single parameter, known as the Rayleigh number, from the British John William Strutt, Baron Rayleigh (1848–1919).

The mathematical expression is as follows:

$$Ra = \alpha \Delta T \frac{gH^3}{\nu \chi} \tag{10.1}$$

where α is the thermal dilatation coefficient of the fluid, g the gravitational acceleration, H the height of the box which contains the fluid, ν the kinematic viscosity of the fluid and χ its thermal conductivity. More precisely, the Rayleigh number measures the competition between buoyancy, proportional to $\alpha g \Delta T$ and dissipation, proportional to $\nu\chi/H^3$ (see Appendix 10.1). If buoyancy is small, in a sense to be quantified below, the upward motion of the light fluid and the downward motion of the heavy one, can be arranged at the level of individual molecules, with no need of any mutual coordination. This is the linear regime of thermal conduction.

As the thermal delta, hence buoyancy, get stronger, individual molecules are no longer in a position to sustain the required flow of mass and heat, and a new operational regime must set in. This is the birth of Bénard cells. This happens above a critical Rayleigh number $Ra_c = 1707$, which means that buoyancy is almost

2,000 times stronger than dissipation. Please note that, in terms of mechanical stability, buoyancy would still be in a position to sustain the light fluid below the heavy one, but the point is that, above Ra_c, even the slightest perturbation would compromise this unstable equilibrium.

Since the Rayleigh number scales with the cubic size of the system, no surprise that the critical threshold can be passed under very ordinary conditions, a ten centimetre thick layer of water with just a few degrees of temperature delta across, already features Rayleigh numbers around a million! No wonder that geophysical phenomena easily reach up to the trillions and more .... The study of thermal convection makes up a very active and important front of modern science, with countless applications to energy and the environment, a subject that we leave to the specialized reader.

10.9 Take-home's

For all their Simplicity, Bénard cells unravel a rich sequence of far-reaching concepts. First, the system is driven by external constraints, the thermal delta. Left alone, without any thermal drive, the system is doomed to dull uniformity. Second, weak (subcritical) constraints leave the system close to equilibrium, where the linear regime (conduction) is the only response the system can provide. Third, strong (supercritical) constraints drive the system far from equilibrium steady state characterized by the nonlinear balance between the various mechanisms in action. The resulting nonlinear response materializes via the onset of collective motion (convection) in which zillions of molecules behave like one: the emergence of Order out of Disorder. Fourth, the emergence of Order out of Disorder, in apparent breach of the second principle, is a paradigm of structure formation via nonlinear dynamic instability.

This is commonly denoted as 'self-organization', a term which, much against the prevailing wind, I am not particularly fond of, since this is nothing but the behaviour contained in the physics of the system, which emerges once the latter is subject to a strong-enough constraint. The constraint is still imposed by the exterior, which is why I feel like 'emergent' is more appropriate than 'self-organized'. But that's matter of semantics.

10.10 Tomaso Albinoni

I distinctly remember the sense of enchantment I experienced when I first learned about Bénard cells, and particularly the question that immediately sprang to my mind upon learning how molecules can organize into such kind of concerted motion. How do zillions of molecules 'know' that they have to move together in order to perform the task of transferring the required amount of heat across the system? Who tells them? At the risk of raising your emotional expectations only to dash them, the answer is dry and plain: molecules do not 'know' anything, all they do is to abide by the rules of the game. And the rules of the game are the laws which govern the motion of matter and energy in this specific set-up, which can be given a very precise mathematical formulation. In scientific language, this is called thermo-hydrodynamics, the chicken-egg loop between fluid motion and temperature.

Is this the death of enchantment? No, it isn't. Quite the opposite, that's where the real magic of emergent Complexity is! If you look at the equations of motion, even with the eye of the specialist, I promise that there is no way you could or would infer the *emergence* of such a beautifully organized behaviour. *This* is the deep and intense beauty of emergent phenomena: you cannot predict them by staring at the equations which rule their behaviour, you have to set the show in motion, meaning that you have to *solve* the equations, watch their solution and see the marvellous beauty unfolding before your eyes! To my real regret, I am not a musician, but I have a deep appreciation for music, and sense a strong analogy here: on 3 Jan, 2020, I had the privilege of listening to a New Year's concert in Budapest, and my daughter taught me how to record the music on my cell phone. Thus, whilst in the midst of the deep emotions triggered by Albinoni's Adagio (see Fig. 10.8), I was struck at the dizzying contrast with the (apparently) dull

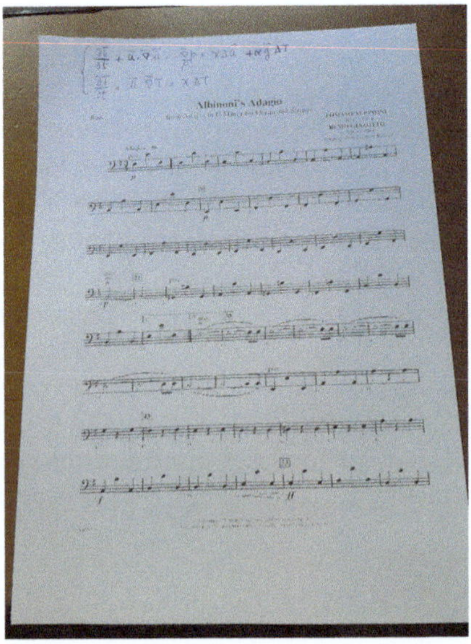

Figure 10.8 *Albinoni's adagio for the professional (not me!), with the equations of motion which govern the physics of Bénard cells scribbled on the top (now it's me and don't worry if you can't read, you can find them in fully display in the Appendix 10.1). If you can foretell the musical paradise in the former, let me know, for I most certainly can't, maybe because I am not a professional musician. But I have some knowledge of the equations scribbled on the top; and yet, I still cannot anticipate the beauty of the physics hidden within, including the Bénard cells.*

signal which was unfolding before my eyes on the screen. In other words, my eyes would give me no clue to the paradise that was being disclosed to me through my ears. Or, even more to the point, there is no way you can tell the immense pleasure of listening to the Adagio by looking at its sheet music, you have to hear the piece played! And if you can play it yourself (alas, I can't), lucky you!

Albinoni is not generally credited as a Complexity scientist but in my view he, fits perfectly, and Mozart, Beethoven, or any other great composer of your choice would do just as well, simply because music also is quintessential emergent Complexity.

10.11 Emergent life?

Although the Bénard cells belong to the context of the physics of thermal fluids, the notion of dissipative structure embraces a much broader class of phenomena, including chemical and biological ones. In particular, Prigogine's work focussed on a specific class of autocatalytic reactions, known as 'Brusselator' (Bruxelles+oscillator) which led to the formation of periodic structures in both space and time. At a late stage in his career, Prigogine (see Fig. 10.9) charged dissipative

Figure 10.9 *Iliya Prigogine, a pioneer of non-equilibrium thermodynamics and a premier advocate of the New Alliance between nature and humankind.*
Source: reprinted from en.wikipedia.org.

structures with a lot of philosophical implications, in an effort to elevate them to the status of a general principle for the emergence of life. For all its fascination, most colleagues felt as if this was a big overstretch. In broad strokes, physicists argued that a general theory of dissipative structures still remains to be formulated. Biologists, on the other hand, contend that living creatures display much higher levels of Complexity and organization than Bénard's cells or the Brusselator. According to Anderson's motto of Complexity, 'More is Different' [3], a huge gap remains between dissipative structures and living creatures. The point is most eloquently taken by Anderson and Stein, [4], where we read that the analogy between non-equilibrium dissipative structures and the equilibrium structures which sustain living creatures 'is clearly out of context in relation to the observed chaotic behaviour or real dissipative systems'. And more recently, Kauffman argues that Bénard cells fail to further escalate the Complexity ladder because, unlike living cells, they do not build up their own boundary conditions [53]. In fact, cells spend actual work to build up and modify their own boundary conditions, a process that, according to Kauffman, proves key in exploring new and more complex states of matter and organization.

Personally, I align with all veins of the previous criticism. This said, I also think that Prigogine's work, together with important contributions by much less-celebrated colleagues, remains a major milestone in the development of modern thermodynamics beyond the 'doom' of the second principle. I met him only once, many years ago in the corridors of his Universitè Libre de Brussels, and I felt very honoured at the chance of shaking hands with him, especially considering the great joy I had experienced in reading his work in my undergrad days. A reading which surely affected the choice of my own professional path, a choice I never regretted.

So much for Bios, now to the Cosmos, Chapter 11.

10.12 Summary: Life on borrowed time

The two cosmic prima donnas are close and fairly distant sisters at the same time. One stands for motion, Order, action, and life, while the other apparently for just the opposite. Based on the second principle, the latter always wins in a fairly bleak picture called thermal death. The Second Principle of Thermodynamics, the one developed in the twentieth century and covered in this chapter, presents a richer and less depressing picture: the second principle can be sidestepped, if only temporarily, by dumping entropy to the environment (sic!). The magic keys to achieve this are: openness and non-equilibrium. This dumping strategy lies at the roots of our own very existence: literally, life on borrowed time from the second principle, with thanks to the rest of the Universe for taking the entropic bill.

So much for the two ladies, but there's a third one in view, which we have already savoured a bit in this chapter: gravity. Indeed, thermodynamics becomes even more fascinating as gravity enters the scene, as it definitely does when it comes to be applied to the entire Universe, as we are going to discuss in Chapter 11. Stay tuned, thermodynamics goes cosmic!

10.13 Appendix 10.1: Rayleigh–Bénard equations

The basic physics of Rayleigh–Bénard convection is rather elementary, basically buoyancy against dissipation. But the quantitative analysis of the Complexity which emerges from such simple physics can be excruciating, as is always the case with turbulence in the room.

The Rayleigh-Bénard equations are rather complicated non-linear partial differential equations, whose numerical solution requires extensive resort to computer simulation. Here we list them for the only purpose of illustrating the physical meaning of the various terms in play.

Equation of motion of the fluid:

$$\frac{\partial u}{\partial t} + u\nabla u = -\nabla p/\rho_0 + \nu\nabla^2 u + \alpha g(T - T_H) \tag{10.2}$$

Left hand side, from left to right:

(i) Change in time of the fluid velocity at a given position in space.
(ii) Change of the fluid velocity along the fluid trajectory.

The second term is the nonlinear interaction of the fluid with itself, and it is responsible for the emergence of fluid vortices.

Right-hand side, from left to right: Change in time of the fluid velocity due to:

(i) Pressure (ρ_0 is a reference density)
(ii) Dissipation
(iii) Buoyancy

In the previous calculation, T_H is a reference to temperature, typically the temperature of the cold plate on top. As discussed in the text, the latter term is the driver of the RB instability.

Equation of motion of the temperature:

$$\frac{\partial T}{\partial t} + u\nabla T = \chi\nabla^2 T \tag{10.3}$$

Left hand side, from left to right:

(i) Change in time of the temperature at a given position in space.
(ii) Change of the temperature along the fluid trajectory.

The latter term is the nonlinear effect on the fluid on the temperature, which is carried along by the fluid. Right-hand side, from left to right:

(i) Change in time of the temperature due to thermal conduction.

As discussed in connection with Albinoni's adagio, you can stare at these equations as much as you wish, but I sincerely doubt you could ever guess the Complexity of the patterns shown in Figure 10.6: the only way to appreciate such Complexity is to solve the equations and watch the beauty of their solution unfold before your eyes. A beauty which remains completely hidden below the critical threshold is $Ra_c = 1707$.

11

Cosmological Escapes

It is a pity that nobody has found an exploding black-hole. If they had, I would have won a Nobel Prize.

(Stephen Hawking)

11.1 The cosmic trio

In the previous sections we have discussed a scenario for the emergence of organized structures in our own planet, with many thanks to the generous low-entropy supply from our star, the Sun. But, how about the biggest game of all, the entire Cosmos, our Universe? How did life materialize out of the primaeval Big Bang 'explosion'? Of course the answer to such kind of grand questions is far from being settled, but one sure thing is that, besides energy and entropy, a third cosmic lady, gravity, must enter the scene [84]. Stay tuned, the energy-entropy duo expands to the EEG (Energy-Entropy-Gravity) trio.

11.2 The living Universe

The escape scenario from thermal death discussed in the previous chapter builds on the ability of dissipative structures to keep themeselves away from equilibrium by reducing their entropy content at the expense of the rest of the Universe. Since the Universe is taking the entropic bill on itself, how could it possibly sustain itself against thermal death based on the same mechanism? Certainly a new actor is needed on stage and, as we shall illustrate, this actor is indeed gravity. Let us therefore begin by revisiting the basic aspects of gravity.

11.3 Basic facts about gravity

Gravity is likely the most familiar force of all and, ironically, the least understood one as well. Of course, we know that gravity keeps our feet on the ground, and thanks to Newton's genius, we also know that the gravitational force which draws the apple to the ground, is the same force that keeps the Moon orbitating around our planet. Two centuries later, Einstein took major leap of abstraction

Sailing the Ocean of Complexity. Sauro Succi, Oxford University Press.
 DOI: 10.1093/oso/9780192897893.003.0011

and showed that, unlike any of the three other fundamental forces, gravity stems from the curvature of spacetime resulting from the motion of massive bodies. This is the basic content of Einstein's 1915 theory of general gravitation, to many the most beautiful and elegant theory ever conceived by the human species. Besides being beautiful and elegant (and mathematically very tough), the general theory of gravitation has met with spectacular success, from the very evidence of the bending of light due to the solar mass, which catapulted Einstein's to rockstar status of the day, up to this day, with the spectacular confirmation of gravitational waves, as well as black holes, the marvelous Big Science we discussed in the very first chapter. Yet, this magnificent theory fails to match with the other pillar of modern science, quantum mechanics. This is another many books, so, here we shall content ourselves with mentioning that when gravity is brought down to very small distances, the kingdom of quantum physics, all sort of problems arise which make the marriage very impervious, if possible at all. This is the problem of quantum gravity, one of the most tantalizing issues of modern physics. For all its charm, we shall not pursue this road in any depth here, and turn instead to some very basic facts about classical Newtonian gravity to begin with.

11.4 Newton's theory of universal gravitation

When speaking of gravity, the first point to appreciate is that gravity is a very peculiar force in at least three crucial respects: *first*, it is always attractive, *second*, it is incommensurably weaker that any other known interaction, *third* it acts on long distances. Let us comment each of the three in some more detail.

11.4.1 The fatal grip of gravity

Newtons' law of universal gravitation states that two material bodies attract each other with a force that is proportional to the product of their masses and goes inversely with the square of their distance (see Fig. 11.1).

In equations:

$$F = G\frac{mM}{r^2} \tag{11.1}$$

where m and M are the masses of the two bodies, r the distance between their barycenters and G is the universal gravitation constant. The force is aligned along the line connecting the barycenters.

This is exactly the same law controlling the interaction between electrically charged bodies at rest (electrostatics), with a crucial twist, though. Unlike gravity, electric charges come with both positive and negative sign and oppositely signed charges attract, while equally signed ones repel each other. The gravitational charge (mass), on the other hand, is always positive, which means that the

Newton's Gravitational Force

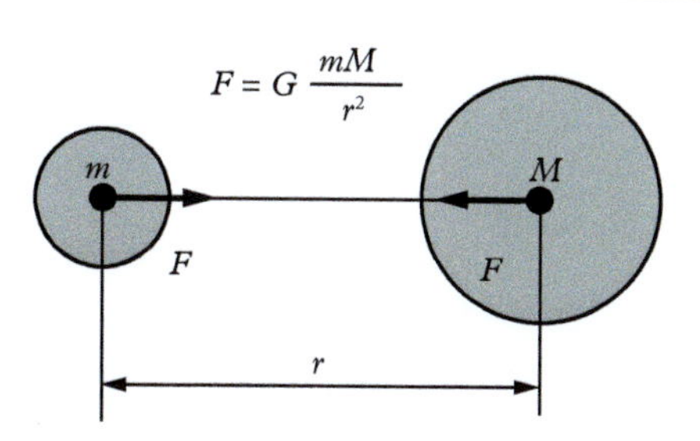

Figure 11.1 *Newtonian gravity. The two bodies attract with a stength proportional to the product of their masses and inversely with the square of their distance. When the distance comes to zero, the attraction is virtually infinite. For two rigid spheres of diameter d and D respectively, the minimum separation is* $r_{min} = (d + D)/2$ *(contact condition), corresponding to a finite value of the gravitational attraction.*

two bodies always attract to each other. This alone reveals the little penchant of gravity to generate equilibrium: since the two bodies attract to each other, they reduce their mutual distance r. But, since the attractive force increases like the square of the inverse distance, this means that upon getting closer the two bodies attract even stronger, in a typical unstable loop, whereby the two bodies collapse on top of each other: the fatal grip of gravity. Not that attraction would come without its benefits: after all, gravity keeps our feet on the ground, literally, preventing us from levitating away out the outer space. But, as we know all too well, too much attraction can be fatal: under gravity alone, material bodies tend to pile up into the same position, a dangerous state of affairs called 'gravitational collapse'. Hardly something that reconciles with life as we know it.

Gravitational collapse promotes Order, and sometimes to the extreme, possibly triggering 'cold' death by untamed self-aggregation, the exact opposite of Clausius's thermal death into featureless uniformity! An interesting flip of roles should be noted here: potential energy, the constructor, now takes the role of the villain, whereas heat, the dreaded carrier of thermal death, now plays the galliant rescuer. In fact, thermal motion is a prime mechanism withstanding gravity's fatal attraction! Again, in a sort of over-indulgence to romanticism, one could say that our Universe takes shape through various forms of resistance to the gravitational grip. For instance, stars are kept together by nuclear forces acting against against the gravitational clamping, and when such forces are no longer strong enough to withstand the gravitational grip, they collapse too. The Indian-American physicist Subrahmanyan Chandrasekhar (1910–1995), Physics Nobel 1983, famously computed that a white dwarf star cannot sustain itself against the grip of gravity above a mass of about 44 per cent more than the mass of the sun. Given that the competition between heat and gravity appears as a cosmic duel of crucial importance for the ultimate fate of the Universe, a few additional words on this subject are worth their while.

11.4.2 Gravity is weak

That gravity is by far the weakest force in the Universe as we know it may come as a surprise at first, because, as we just observed, we experience its very crucial effect on the world around us on a daily basis. How come? The point is that we do not notice gravity feebleness only because we are used to deal with macroscopic bodies, like us, which consist of Avogadro numbers of molecules. Once applied to single molecules, gravity becomes ridicolously feeble. Just to convey the idea, if the electrostatic forces which keep atoms together were replaced by gravity, the atoms would inflate beyond the size of the current Universe! Hence gravity is the master force of the macroscopic world. Being the master force of the macroscopic world, it is clear that gravity has a decisive say on the history and fate of our Universe.

11.4.3 Gravity is long-ranged

As observed previously, the gravitational force decays in space with the square of the distance. This means that upon doubling the distance between the two material bodies, the attractive force becomes *four* times weaker. This may seem a substantial decrease but, compared with the other fundamental forces in nature, it is a much milder one. For instance, the attractive forces between molecules, to be covered in a forthcoming chapter, typically decay with the *seventh* power of the distance, meaning that doubling the separation between the two molecules reduces their attraction by factor $2^7 = 128$! This is what physicists call 'short-range' interactions, to indicate that molecules interact only with comparatively close neighbours, typically less than hundred in an ordinary liquid, because beyond a certain distance, the molecular interactions are small to the point of being totally negligible.

Gravity, on the other hand, is 'long-range', in the sense that its decay in space is too weak to allow the neglect of the interaction even with most distant bodies. Differently restated, every gravitational body interacts with every other one, no matter how distant. This gives rise to global communication network which is drastically different from the local communication patterns resulting from short-range interactions. In particular, it opens up a wider host of collective arrangements, with a profound impact on the dynamics of the system, particularly the relaxation to thermal equilibrium. This global communication is utterly relevant to structure formation in the Universe, particularly to the formation of galaxies out of the collective motion of individual stars (see Fig. 11.2).

Figure 11.2 *NGC 4414, a typical spiral galaxy in the constellation Coma Berenices, is about 55,000 light-years in diameter and approximately 60 million light-years from Earth. A light-year is the distance travelled by light in a year, approximately 9,461 billion kilometers .... For the record, according to 2016 National Aeronautics and Space Administration (NASA) estimates, there are about two billion galaxies in the known Universe totalling more 10^{24}, i.e. one million of billion billions stars, hundred to thousand more than grains of sands on planet Earth. Interestingly, the number of stars is quite comparable to the Avogadro number:* there are about as many molecules in a single glass of water as there are stars in the Universe. *The Universe in a glass of water!*
Source: reprinted from en.wikipedia.org.

11.5 Gravitational thermodynamics

All of the above sets gravity apart from any other known force, which is why it deserves the status of a third primadonna on the cosmic scene. Let us dig a bit deeper into these matters by inspecting what happens when gravity meets thermodynamics.

We are naturally accustomed to the idea that upon absorbing heat, any material body would increase its temperature. This appears natural to the point of verging on truism. Such natural intuition is scientifically encoded in the notion of *heat capacity*, namely the (inverse) ratio of the temperature change resulting from heat absorption, in formulas $C = \frac{\delta Q}{\delta T}$, where δQ is the heat flowing from a hot body at temperature $T + \delta T$ and a cold one at temperature T. A body with large heat capacity can absorb large amounts of heat without any appreciable effect on its temperature (thermal reservoir). Conversely, bodies with small heat capacity experience large temperature variations upon absorbing modest amounts of heat. Either ways, heat capacity is expected to be a positive-definite quantity: upon absorbing heat a material body heats up and upon releasing it, it cools down. As we have discussed before, this is a plain matter of thermal stability. These are the typical rules of the game, but with gravity in town, they can turn upside down, as we are going to discuss next.

11.6 Gravitational balloons

To get a qualitative sense of the story, let us consider a balloon filled with an ideal gas, where ideal means that the molecules do not interact with each other, their energy being entirely kinetic. Next, heat up the balloon: the kinetic energy inside goes up and the molecules hit the surface of the balloon more energetically, thus driving its expansion. The heat absorbed by the gas does not increase its kinetic energy, hence its temperature, because the excess kinetic energy is fully spent to expand the balloon. The bottom line is that the volume increases, pressure decreases accordingly, and temperature is left the same: isothermal expansion, Carnot is back again.

Now repeat the same experiment with a gravitational gas instead, in which the molecules attract each other with a force inversely proportional to the inverse of the square of their mutual distance. Gravitational forces tend to bring the particles together, while kinetic energy favours their expansion. If kinetic energy wins the competition, the system expands, while in the opposite case it contracts, eventually ending into a gravitational collapse. When the two are in exact balance, a thermo-mechanical equilibrium is attained, whereby the system neither expands nor contracts, but just keeps its size constant in time. The textbook case is a rocket of mass m, launched at an altitude h, along a tangential orbit with velocity V. If the kinetic energy of the rocket, $mV^2/2$, is smaller in magnitude than its gravitational energy, $GmM/(R+h)$, R being the radius of the Earth, M its mass, and G the gravitational constant, then gravitational attraction wins and the rocket starts spiralling inwards, until it falls back to the ground, miserably failing the task it was designed for .... In the opposite case, the rocket starts spiralling outwards to outer space and leaves the planet for good; that's precisely what a professional rocket is supposed to do. Finally, if the kinetic energy matches exactly the potential energy, the rocket starts orbiting around the Earth, keeping its altitude h constant in time. That's what satellites do for a job. For the record, this happens at the sizeable speed of about eleven thousand kilometers per hour, some fifteen times faster than standard airliners.

An important point should be raised here regarding the sign of kinetic and potential energy. Kinetic energy goes with the square of the velocity, hence it is always positive or zero if the body is a rest. Gravitational energy, on the other hand, is attractive, hence tends to form bound structures which do not move. Hence, by convention, potential energy takes a negative sign: the more negative it is, the more tightly bound together the two bodies, like in a well. By this convention, at equilibrium, the total energy is zero.

Back to the gravitational balloon. Upon absorbing heat, the balloon still starts to expand. However, at variance with the ideal gas, potential energy is now part of the equation and since the balloon expands, the average molecular separation increases and gravitational energy becomes less negative, i.e. it increases. If the heat absorbed is spent entirely in expansion work, the total energy must be conserved, and since potential energy increases, kinetic energy must perforce decrease, which means a lower temperature!

The math says it best: using the language of Chapter 8, the First Principle of Thermodynamics now reads as follows:

$$dE_k + dE_p = \delta Q - \delta W \tag{11.2}$$

where subscripts k and p stand for kinetic and potential, respectively. As a result, if the heat absorbed is entirely converted into expansion work, the result is $dE_k = -dE_p < 0$, kinetic energy goes down, and so does temperature. This is how, in the very act of absorbing heat from the environment, the gravitational balloon manages to decrease its temperature! In technical terms, one speaks of *negative* heat capacity, as we noted earlier on, a decided hallmark of thermal instability.

11.6.1 Entropy and gravity

So much for gravity versus heat and energy. How about gravity and entropy?

This is a profound theme and one much debated in modern cosmology, especially in connection with the thermodynamics of black holes (BH), to which we shall return in the final part of this chapter. As to our balloon though, nothing has changed: it still absorbs heat at a temperature below that of the environment, hence the total entropy must increase. Fair enough. The main question, however, is whether or not such an increase ever comes to a halt, i.e. as discussed in the very beginning of this chapter, whether the gravitational balloon (soon to become a pretty useful toy model for the entire Universe) ever attains thermal equilibrium. In fact, since the temperature decreases upon absorbing heat, there is no mechanism, in principle, to stop the heat flowing from the hot environment to the increasingly colder balloon! We are clearly facing a thermal runaway scenario. Although somewhat oversimplified, the previous example unveils a very fundamental issue: once gravity takes stage, whether and how the system attains a thermo-mechanical equilibrium, if at all, becomes a fairly non-trivial question. In fact, this is one of the central open questions of modern astrophysical and cosmological thermodynamics, as we proceed to discuss briefly.

11.7 Cosmological thermodynamics

Having pinpointed that gravity is a potential game changer in the energy-entropy relation, the next question is what this means for the second principle in general and of thermal death in particular. Modern cosmology met with paramount progress in the last decades, starting with Edwin Hubble's (1889–1953) fundamental discovery that the Universe is not static, as Einstein's thought in the early days of his theory of general gravitation, but it expands instead. And since the whole field of cosmology was actually sparked by Enstein's 1915 theory of general gravitation, before we plunge into cosmolgical thermodynamics, let us get acquainted with the main ideas behind this theory.

11.8 Einstein's theory of general gravitation

In 1905, his 'Annus mirabilis', Albert Einstein published three epoch-making papers, one on the theory of Brownian motion, one on the photolectric effect (the emission of light from metal surfaces) and finally one in which he formulated the theory of special relativity. The first paper contributed to establish once and for all the atomistic nature of matter, the second made a decisive contribution in establishing the particle-like nature of light as a 'fluid' of photons, and even though the second earned him nothing short of the 1921 Nobel Prize in physics, they are both largely unknown to the general public. Indeed, Einstein's universal fame as the iconic genius in the eyes of the general public rests entirely on his theory of relativity, back then regarded as 'too mathematical' for a Physics Nobel. Indeed, relativity theory shook the foundations not only of physics but on our very view of the world. The main outcome of this theory is that the time interval between events (duration) depends on the state of motion of the observer. More precisely, take Bob standing on the train track and waving to Alice who just caught a train. Alice spots the train officer coming to her and counts say 30 seconds before she's asked to present him her train ticket. If Bob could see the very same scene, nobody doubts that he would count the very same 30 seconds. Actually, what Einstein showed is that this is *not true*! In fact, Bob would measure a sligthly longer duration, where 'sligthly' is a decided understatement, as the difference would count in about one part in hundred thousands billions (10^{-14}). Here, we have assumed that the train moves at 30 metres per second (108 Km/h), but any other reasonable train speed would do, as long as it stays well below the speed light of light, about 300 million metres/second. The reason is that the relative departure between the two durations scales is approximately like the square of the ratio between the train speed and the speed of light. This is what Einstein's relativity predicts, and this simple example says it all on the reasons why relativity remains totally silent for the macroscopic objects we are used to. If Alice was catching an airplane instead of the train, the departure would shift to the 12th digit instead of the 14th, still totally below detectability for our senses. Not so in the microscopic world of elementary particles, which zip around at near-light speed: from our standpoint, they live much longer than they know, a very real effect which is routinely measured in high-energy accelerators.

11.8.1 Entry general relativity

If this sounds totally weird, please, please be prepared for more .... Einstein's 1905 theory of relativity refereed to objects in uniform motion, i.e. moving at a constant speed, a very special situation indeed, whence the namer 'special' theory of relativity. Einstein's was deeply dissatisfied with such stringent limitation, and poised to lift it, i.e. extend it to any type of general motion, not just uniform. He succeeded, and his success led him to the theory of general relativity, possibly the highest triumph of theoretical science ever.

Gorgeous but ..., but what does all this have to do with gravitation? In fact, as we shall see, quite a lot. In formulating his theory of general relativity, Einstein realized that spacetime 'deforms' (again, literally!) under the effect of mass and energy! 'Deforms' means that the measurement of space and time intervals depends on the motion of the observed and in particular on its mass and energy. As a result, a massive body would 'bend' spacetime precisley like a ball on an elastic sheet or blanket. The blanket bends and curves under the weight of the ball, and other balls are attracted as a result of this bending (Fig. 11.3). Energy and matter 'deform' spacetime and the curvature of spacetime directs the motion of material objects: hence, gravity is essentially the curvature of spacetime!

Based on this picture, Einstein predicted that light would be bent by the solar mass, to the point that it should be possible to see stars that, on a straight line, would sit right behind the Sun, hence invisible to us (see Fig. 11.4). In other words, light would literally 'turn around' the Sun! Incredible as this seems, such a prediction was spectacularly confirmed just a few years later (1919) in an experimental

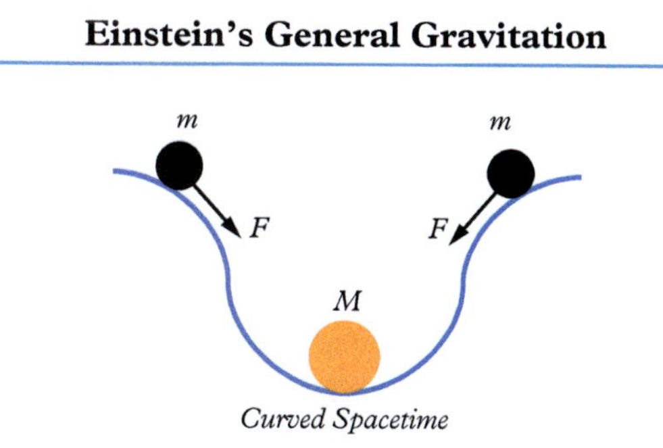

Figure 11.3 *The central ball bends and curves spacetime, and the balls on the sides are consequently attracted to it. This is how gravity works in Einstein's general relativity theory.*

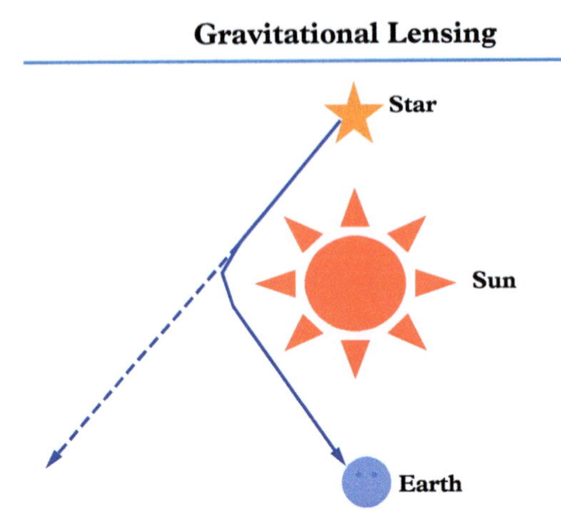

Figure 11.4 *The light emitted by a star 'behind' the Sun is deflected by the spacetime distortion induced by the solar mass and can reach on Earth (gravitational lensing effect). Without gravitational lensing, the light ray would keep going on a rectilinear trajectory (dashed) and the star would not be visible on Earth. The effect is largely exaggerated in the figure, the actual deflection observed in the experiment being of the order of one thousandth of a degree.*

expedition led by the British cosmologist Arthur Eddington, thus conferring Einstein's worldwide rockstar status that would only grow stronger all through his life and beyond.

The general theory gave birth to the science of cosmology from scratch, a wonderful story that is covered in many beautiful books. Cosmology is littered with fundamental breakthroughs, but here we shall focus on the ones which bare most relevance to our target: cosmological thermodynamics, and particularly the entropy-gravity connection.

11.9 Cosmological chronicles

Despite his revolutonary deeds, Einstein acted conservatively in his pre-conceived view of the Universe as a closed and static system. Therefore, he reacted badly a few years later (1922) when an obscure Russian mathematician, Alexander Friedmann (1888–1925) came up with solutions to Einstein's equations predicting a dynamic Universe which would actually expand in time. First, he dismissed Friedmann's solutions, but to no avail, and when he realized that Friedmann was right he changed his equations by adding its by now famous cosmological constant, just for the purpose of making the Universe static again.

But again, to no avail, since just a few years later, he had to surrender to experimental evidence, this time against him, that the Universe *does* expand indeed, as proved beyond question in 1929 by the American astronomer Edwin Hubble (1889–1953). What Hubble did was to measure the recession velocity of distant galaxies, i.e. the velocity of galaxies distance from each other, hence from us as well, and found a strikingly simple law, known as Hubble law and recently justly renamed as Hubble–Lemaître, after the Belgian priest and cosmologist Georges Lemaître (1894–1966), the father of Big Bang theory, soon to come. The Hubble–Lemaître law states that the recession velocity V of a galaxy a distance D from us is imply given by:

$$V = H_0 D \tag{11.3}$$

where H_0 is the so-called Hubble–Lemaître parameter. For the record, H_0 is about 70 km per Megaparsec, where one Megaparsec is about 3.09×10^{19} km. This means that a galaxy a Megaparsec away from us expands at the fairly respectable speed of 70 km/s, about 252,000 km/h. Or, if you prefer, the Hubble–Lemaître law also says that the Universe expands at a rate of about 7 per cent every billion years (en.wikipedia.org/wiki/Hubble). We started from a toy ballon, and now we are placing numbers on the expansion rate of our Universe: this alone gives an idea of the spectacular progress of cosmology in the last century. Moreover, coming back to thermodynamics as the science of slow change, no question that the Universe fits the definition pretty well . . . . This led Einstein to regret his cosmological term as his 'worst blunder', even though, like it is often the case in science, this blunder turned out to be a very productive and useful one, as it gave rise to another family of cosmological models which is still under debate today.

11.9.1 Georges Lemaître

We mentioned Lemaître and here he comes (see Fig. 11.5). He first 'derived' Hubble's law and measured the Hubble constant in 1927 two years ahead of the experiment. He also proposed the Big Bang theory of the origin of the Universe, calling it the 'hypothesis of the primeval atom'. In his own magnificent words (G. Lemaître, The beginning of the world from the point of view of quantum theory, *Nature*, May 9, 1931): we could conceive the beginning of the Universe in the form of a unique atom, the atomic weight of which is the total mass of the Universe. This highly unstable atom would divide in smaller and smaller atoms by a kind of super-radioactive process. . . . . The whole story of the world need not have been written down in the first quantum, like a song of a disc of the phonograph. The whole matter of the world must have been present at the beginning, but the story it has to tell may be written step by step.[36]

It should be noted that the idea of a Universe starting from a spaceless point was already present in Friedmann's solutions, which predicted the radius of the 'balloon Universe' to be exactly zero at the beginning of time. In simple math:

[36] Personal aside: the last two lines of Lemaître's sentence ring like a true gem to me. I am thinking of how many needless and actually detrimental polemics between Science and Religion could be spared if only one would stick to these two lines.

Figure 11.5 *The Abbé Georges Lemaître, the father of Big Bang theory.*
Source: reprinted from en.wikipedia.org.

$$R(t = 0) = 0 \qquad (11.4)$$

But, since such spaceless point should already contain all the mass and energy of the entire Universe, we are clearly faced with an infinite density, what physicist use to call *singularity*. As we discuss in the initial part of this book, singularities are usually interpreted as the signal that the theory has been taken beyond its waters and new ideas are needed. An obvious candidate was, and still is, quantum physics, according to which any material particle is associated with a finite size that cannot be brought to zero without completely loosing information on its velocity, the famous 'Uncertainty Principle' proposed by Werner Heisenberg in 1927 (see Appendix 19.3 on Quantum Mechanics).

However, even before one invokes quantum physics, a simpler possibility is that the singularity is simply due the highly idealized picture of the Universe as a spherical balloon, as assumed in the earliest cosmological models. It took the genius of Roger Penrose to show that singularities are a deep and robust feature of the Einstein equations, that persist even once the simplifying assumptions on the geometry of the Universe are released. This marked the portal towards possibly the most exquisite and popular outcome of Einstein's equations: BHs. Before we venture, if only shortly, into BHs, let's recap the main ideas introduced so far.

11.10 Pause for reflection

First, Einstein's equations show that the Universe is *dynamic*, i.e. it evolves in time. The details of this evolution are crucial to assess its final state: does it ever settle, or does it keep growing, or maybe contracts again, or perhaps oscillate indefinitely?

This eschatological question is still up for grabs.

Second, the Einstein equations predict singularities, i.e. region of spacetime with infinite density of mass and energy, and the initial state of the Universe might well be one such singularity itself (Big Bang). And since the Universe allegedly started from a point-like configuration, it must have had an incredibly low entropy. According to Roger Penrose, the starting entropy of the Universe corresponds to a probability of 1 over $10^{10^{123}}$, i.e. less than one part in a Googolpex (see Chapter 19, Appendix on numbers). Why, still is everybody's guess. Both previous features have profound implications for cosmological thermodynamics and still make the subject of intense research.

11.10.1 Dancing in the dark

The answer to the first question depends critically on the amount of mass and energy contained in the Universe at its inception: it can be shown that in order to recover the conditions of homogeneity observed today, the Universe should have contained much more mass and energy than we can count in the particles and radiation than we are currenty aware of. This is the famous problem of *dark matter* and *dark energy*, a form of matter and energy that we simply cannot detect because it does not respond to anything but gravity. And if you think that this is footnote, you'll be surprised to hear that the count of the missing versus 'ordinary' is 95:5. More precisely, based on the famous Einstein's equation $m = E/c^2$, the numbers for dark matter, dark energy, and ordinary matter are 68:27:5, respectively. This means that about 95 per cent of the energy and equivalent mass in the Universe is still unknown! It is speculated that this unknown form of matter and energy is responsible for the fact that Universe not only expands, but it does at an increasing pace; it accelerates, as clearly demonstrated by Nobel-winning experiments in the early 2000s!

11.10.2 Singularities

Regarding singularities, as observed many times in this book, physicists take them seriously only to the extent to which they signal the need for a better theory. Singularities are already present in Newtonian gravity, but in the Einstenian case they acquire a much deeper meaning, as they involve not only matter but spacetime itself! As we shall see, this has far-reaching consequences, the most spectacular one being the prediction of black holes, as we are going to discuss next.

11.11 Black Hole thermodynamics

We began this chapter by mentioning that, being always attractive and increasingly stronger at short distances, gravity drives matter towards gravitational collapse. Unless, of course, competing effects arise which withstand the gravitational grip, thermal pressure being certainly one of them, the promoting mechanism for the formation of black holes, see Fig. 11.6.

The condition for black hole formation is handily described in terms of the so-called Schwarzschild radius, from the German astronomer Karl Schwarzschild (1873–1916), who provided the first exact solution to the Einstein's equations. The Schwarzschild radius is defined as the distance at which the gravitational energy $E_g = Gm^2/r$ comes to an exact balance with the rest energy of the body, as expressed by the famous Einstein's equation $E = mc^2$ (conventionally divided by two). Equating the two, one obtains:

$$r_S = \frac{2Gm}{c^2} \tag{11.5}$$

A black hole is formed whenever gravity manages to squeeze a material object of mass m within a sphere of radius below r_S.

How likely is it going to be? Since gravity is weak and the light speed is large, even before inspecting number, we get a clue that the Schwarzschild radius must be generally pretty small unless the mass is really large.

A few numbers illustrate the point: the Schwarzschild radius of the Sun ($m \sim 2 \times 10^{30}$ kg) is about 3 km, while for Earth ($m \sim 6 \times 10^{24}$ Kg) it is about 9 mm. For a 70 kg human being it goes down to 10^{-25} metres, a million times shorter than the proton radius (see https://en.wikipedia.org/wiki/Schwarzschild-radius)! The reader may want to notice that in all cases above, the Schwarzschild radius is far below the material size of the object, which is precisely the reason why under ordinary conditions, we do not generally perceive the effect of the curvature of spacetime and live happily with Newton's gravity.

However, extreme states of matter, such as black holes, are totally imbued within Einstein's gravity. Black holes have a number of exotic properties whose description lies far beyond the scope of this book. Here, we wish to content with the most popular one: within a black hole, the gravitational clamp is inexorable: if you fall within it, you will never escape. Not even if you are a massless photon, because, always based on Einstein equations, gravity acts on you through your equivalent mass $m = E/c^2$. This is why light is bent by the Sun in the first place. But with black holes in town, bending turns into an inexorable and unforgiving catch. Black holes are fascinating and somehow threatening entities at a time, the perfect entry for scifi. But they are not sci fi, as ultimately proved in the 2018 experiments in which they were at last imaged and visualized in their full splendor, promptly rewarded with the 2020 Nobel Prize in physics.

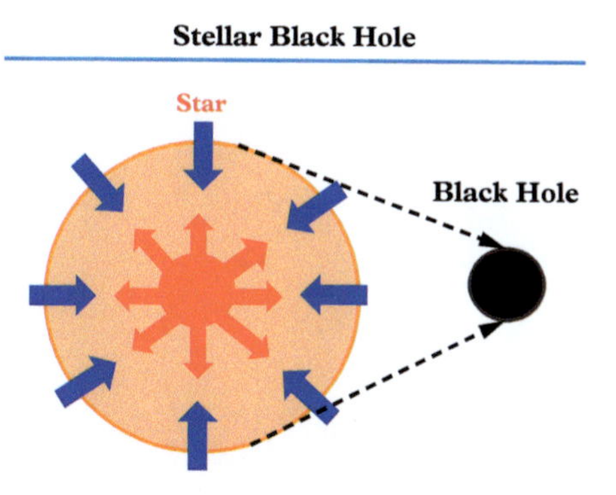

Figure 11.6 *A star is sustained by internal pressure (outward arrows) against the grip of gravity (inward arrows). When the star runs out of fuel, usually nuclear fusion reactions, the internal pressure surrenders to gravity and the star eventually turns into a black hole.*

11.12 Black Holes get grey (not much though . . .)

In this chapter we have written a lot about relativity and said virtually nothing on the other pillar of modern physics: quantum mechanics. To forestall the risk of further digressions, we direct the reader to the short Appendix at the end of this book. Here we just wish to note in quantum mechanics material bodies cannot be assigned a specific position without loosing information about their velocity, a weird property known as Heisenberg's uncertainty principle. Quantum mechanics rules the behaviour of matter at the atomic scale and below, while gravity is the queen of the macrosopic world. However, since, as we have just seen, untamed gravity can squeeze massive objects far below their original dimensions, at some point the two interactions must come in touch. And when they do, a big elephant enters the room, as the two theories just don't get along, spawning one of the most tantalizing challenges of modern physics: quantum gravity (or maybe gravitational quantumness, but it doesn't ring as well). This failed match has far-reaching implications for cosmological thermodynamics as well, especially with concern to the role of black holes on the long-term evolution of the Universe (see Fig. 11.7).

Notwithstanding this missing match, a number of crucial results have been obtained nonetheless. For one, as first shown by Stephen Hawking in 1971 [59], the laws of black-hole physics imply that their area can only grow in time or stay constant. This finding, along with previous work with Penrose on singularity theorems, has been honoured with the Physics Nobel 2020, 2 years too late for Stephen Hawking (1942–2018) who would have most certainly shared it.[37]

[37] I always thought that Hawking's larger-than-life popularity would have made it immune to Nobel longings. Yet, a colleague of mine once told me that this was not the case. He did care, as witnessed in the header statement of this chapter.

Figure 11.7 *A simulated view of a black hole in front of the large Magellanum Cloud.*
Source: reprinted from commons.wikimedia.org.

The area law suggests an immediate and exciting analogy with entropy. At this point, however, a puzzle arises: if black holes are attractive to the point of letting nothing out, including radiation, how can they possibly contribute to the thermodynamics of the Universe, and to its entropy in the first place? This puzzle was solved in 1974 by Stephen Hawking, who showed that, thanks to the law of quantum physics, a black hole can nonetheless exchange radiation with the surrounding Universe.

Standard thermodynamic arguments provide the following expression for the Hawking temperature:

$$T_H = \frac{1}{8\pi} \frac{\hbar}{k_B} \frac{c^3}{mG} \tag{11.6}$$

Interestingly, this formula involves all the three fundamental constants of nature, plus Boltzmann's constant k_B.

This simple expression spawns a number of far-reaching conclusions.

Figure 11.8 *Stephen Hawking and Roger Penrose, the two main pioneers black holes and modern cosmology at large.*
Source: reprinted from pixabay.com and en.wikipedia.org.

The first observation is that black (grey) holes are pretty cold objects: a BH of thirty solar masses features a pretty chilly temperature of two nanokelvins, hundred billion times colder than the standard temperature in our room! Most importantly, this temperature is a billion times colder than the temperature of the surrounding Universe, currently at 2.7 Kelvin. This means that in order to thermalize with the rest of the Universe the BHs must increase its temperature. Based on the expression (11.6), since the Hawking's temperature goes inversely with mass, i.e. energy, in order to raise its temperature the BHs must radiate away its energy, i.e. lose mass, a process known as black hole evaporation. Note that, in this respect, the BH behaves precisely as the gravitational balloon discussed earlier on in this chapter: it raises its temperature by radiating energy and vice versa (see Fig. 11.7).

BH evaporation is however a very slow process: for a BH of solar mass, the evaporation time is 10^{72} seconds, and it grows with the cubic power of the mass. These lifetimes are incommensurably higher than the age of the Universe, about 10^{18} seconds, which speaks clearly for the black hole persistence in time. Such persistence is basically the result of the extreme feebleness of Hawking radiation, billions times smaller than an ordinary houseold bulb! Thus, black holes are not that grey after all ....

11.12.1 Bekenstein–Hawking Black Hole entropy

As mentioned previously, while the merge between relativity with quantum mechanics and gravity are both spectacular success stories, the third side of the triangle, quantum mechanics and gravity still has not come full circle.Even though this is holding back progress in the study of quantum cosmology, fruitful results have been obtained nonetheless. One of the most outstanding one is

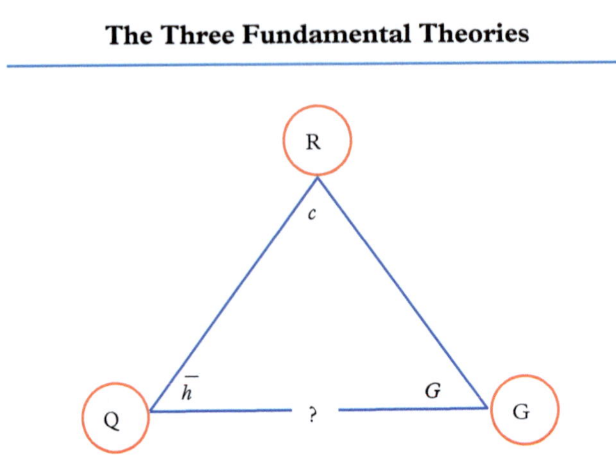

Figure 11.9 *The RQG triangle formed with the three pillars of modern physics: relativity, quantum physics and gravitation, with the respective fundamental constants. Only the RQ and RG sides match so far.*

the calculation of the black-hole entropy performed by the Mexican-born Israeli scientist Jacob Bekenstein (1947–2015), based on the area law discovered by Hawking and Penrose in the mid-60s.

The formula is beautifully simple:

$$S_{BH} = k_B \frac{A}{4L_P^2} \tag{11.7}$$

In the previous formula, A is the area of the black-hole horizon and

$$L_P = \sqrt{G\hbar/c^3} \tag{11.8}$$

is the so called Planck length. The subscript 'BH' might imply 'black hole', but in fact it refers to the originators, Bekenstein–Hawking, a nice coincidence after all (see Fig. 11.9).

Note that the Planck length combines the three major natural constants, the universal gravitation constant G, Planck constant $\hbar$, and the speed of light c. Hence, it brings together three pillars of modern physics, general gravitation, quantum mechanics, and special relativity! Indeed the Planck length is basically the scale at which gravity and quantum physics come with comparable strength, hence gravity can no longer be neglected in spite of its incredible weakness as compared with quantum interactions. The result is an astronomically small number, about 10^{-35} metres: it takes a hundred million of billions of billions of billion Planck lengths to make a single metre! This is no coincidence, but the result of the fact that the gravity is extremely weak at the atomic scale and so is the Planck constant, while the speed of light, which is very large instead, comes at the denominator and with the vengeance of a cubic power!

For the record, these are the plain numbers in MKS (metres-kilograms-seconds) units: $G \sim 6.67 \times 10^{-11}$, $\hbar \sim 1.54 \times 10^{-34}$ and $c \sim 3 \times 10^8$.

The Planck length is the ultimate scale below which spacetime itself loses physical meaning, at least based on what the experts tell us. As such, it offers a natural regulator of the singularities that pervade modern cosmology.

Coming back to the BH formula (11.7), since black holes are definitely macroscopic objects, featuring several solar masses, it is clear that the BH entropy is extremely large. For a black hole with a 10 km radius, it is of the order of 10^{80}. Since there are an estimated 100 billion black holes in the Universe, this makes quite a significant contribution to the overall entropy of the Universe. The odds are open, but it appears likely that in order to form a coherent picture of cosmological thermodynamics, the current tension between gravity and quantum mechanics must be released first. What this will bring to the table of long-term cosmological escape scenarios still is a largely open and fascinating question.

11.13 The thermal history of the Universe

We started with a toy balloon filled with classical gravitational bodies and ended up with a cosmic balloon, filled of ordinary matter (and antimatter), radiation, quite possibly large portions of dark matter, dark energy, and black holes. The thermodynamics of this rich and fascinating 'cosmic balloon' is still an ongoing adventure. Yet, much is known: from the Big Bang on, the thermal history of the Universe is basically an enchanting tale of structure formation at all scales, from elementary particles in the primaeval soup, to the synthesis of nuclei, to heavy elements, stars, planets, galaxies and clusters of galaxies, and the emergence of terrestrial life as we know it . . . .

This thermal history shows that our Universe is still far from equilibrium, hence ageing, although its ultimate fate is still open for grabs.

Although the detailed mechanisms are still begging for a quantitative explanation, it is clear that the ordering power of gravity played an essential role in promoting structure formation in some parts of the Universe at the expense of others. This is after all the 'entropy-dumping' mechanism discussed Chapter 10, and the 'cosmic balloon' discussed in this chapter provides a rich spectrum of possibilities for such a mechanism. For a most enjoyable articulation of these optimistic views, I warmly recommend Freeman Dyson's (1923–2020) delightful book [30].

11.14 Summary

Thermodynamics becomes utterly interesting when gravity enters the scene, as it definitely does when it comes to discussing the evolution of entire Universe. In this case, even the attainment of the most basic cornerstone of thermal death, thermal equilibrium, becomes questionable. This offers a 'cosmological' escape

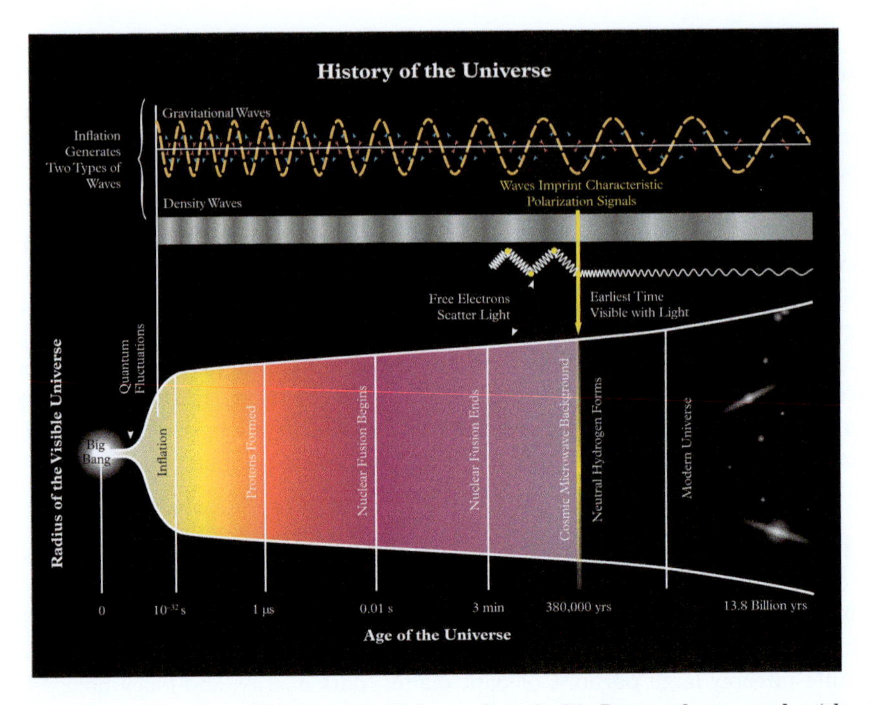

Figure 11.10 *The thermal history of the Universe, from the Big Bang to the present day (about 13.8 billion years). The size of the Universe at the Big Bang is estimated between 17 cm and 170 metres, while it is currently about 90 billion light years, corresponding to about 10^{27} metres (a billion of billion of billions). The temperature went down from $T = 10^{32}$ (hundred and thousands billions of billions of billions) Kelvin to the current 2.57 Kelvin.*
Source: redrawn from https://www.haus-der-astronomie.de/1729890/04cosmo-bbn.pdf.

route to defer the deadly rendez-vous with thermal death. The second principle always downs the last card, but, thanks to gravity, our physical Universe stands new chances to borrow time from the final play before 'the fat lady sings'. And with gravity (and dark matter/energy?) on stage, this time-loan might be pretty substantial on a cosmic scale (see Fig. 11.10).

Whether and how this is going to affect us, the self-appointed *Homo sapiens* species, remains entirely open. But at least we can take some cosmological solace in knowing that our Universe might indeed offer more escape routes from thermal death than we can possibly know.

12

Free Energy

No free energy device will ever be allowed to reach the market.

(Nikola Tesla)

12.1 Energy goes free

In Chapter 11 we have discussed the role of gravity and the potential escape routes from thermal death it may open once thermodynamics is projected to the cosmic scale. In this chapter, we come back on earth and focus our attention on a most important child of energy and entropy prima donnas, a child called *free energy*.

12.2 Nature's spending review

We live in a world in which finance and economy call by far the strongest shots, even above and beyond politics. Although perhaps with less of an obsession, Nature also abides by solid economic principles (I'm not aware of financial ones, though). Now, the expression 'free energy' is vaguely conducive to a sort of magic trick to run your household bill-free. Alas, or perhaps fortunately, free-energy is not free at all in this respect, in fact, quite the opposite, it is precisely the currency in which nature pays the bills that run her show. Formally, free-energy is defined as the difference between energy and the product of temperature and entropy, namely [78, 12]:

$$F = E - TS \tag{12.1}$$

The right-hand side presents three well-known characters at this point, and since we have learned that temperature times entropy is tightly related to heat, we can safely conclude that the free-energy F is what is left of energy once the heat toll is paid off. The attentive reader may observe, along with Atkins [6], that free energy is just an auxiliary quantity, as it derives from the primary characters, energy and entropy. Why do we need another one, then? The reason is that it is not energy and energy separately that 'drive the world', to borrow from Atkins again, but precisely their combination expressed by the simple, yet far-reaching,

Sailing the Ocean of Complexity. Sauro Succi, Oxford University Press.
 DOI: 10.1093/oso/9780192897893.003.0012

relation (12.1). Hence, free energy has plenty of scope as a thermodynamic function on its own.

For the historical record, there are *two* free energies in classical thermodynamics, due to two giants of the nineteenth century, the German Hermann von Helmoltz (1821–1894) and the American Josiah Willard Gibbs (1839–1903). In Helmoltz's case, E is the internal energy of the system, the one that depends on its temperature alone, typically indicated with the symbol U, so that one writes $F = U - TS$. Gibbs's free energy, on the other hand, is defined as $G = H - TS$, where $H = U + PV$ is known as *enthalpy*. The reason for adding PV, the product of pressure times volume, is to exclude energy changes due to system's expansion or compression at constant pressure and temperature, which is particularly useful for chemists, not frequently seen to deal with pistons, cylinders and all that. If, on the other hand, compression/expansion PV work cannot be ignored, as it is the case for engineers, then Helmoltz's free energy is a better choice, which is the one made here.

Whether in Helmoltz or Gibbs vests, the key point remains that free energy is the literal currency of the natural world. Any organized structure delivering a function comes at a cost, and this cost is measured in units of free-energy, there is no other currency in this market. Indeed, like any shrewd investor, nature has been shaped in such a way as to minimize her running costs or, better still, maximize her returns on investment. Here is where the nature of the duet/duel between energy and entropy gets really interesting.

12.2.1 Walls and Doors

Since they come with opposite signs, at first sight it might seem that energy and entropy are poised to fight each other. On a closer look, though, a subtler truth is revealed: they represent two alternative, and not necessarily conflicting, ways of achieving the same goal, namely minimizing the free-energy cost (investment) to achieve a given structure/function (return). However, as the formula (12.1) makes it clear, they contribute in opposite directions, energy has to be at a minimum and entropy at a maximum. What does this mean? It means that nature favours arrangements and structures that cost the least energy and offer the most entropy. Both speak for easiness, but in two different ways: to put it in metaphor, low walls and wide doors, respectively (see Fig. 12.1). The metaphor is far from empty: low-energy is the economy part, as it costs a lot of energy to jump over high bars, think of high-jump sports. High entropy is the wide doors part, places that are easy to reach because they can be accessed through a major doorway or, equivalently, many alternative routes. If you prefer another analogy, energy is the height of the wall; entropy is the width of the door in the wall, or better yet, the number and size of doors in the wall.

But isn't this low-walls-wide-doors approach precisely the route to dreaded thermal death again? Yes, it is! Just like buying at the cheapest prize hardly gets you the best product, likewise, minimizing free energy in an absolute and global

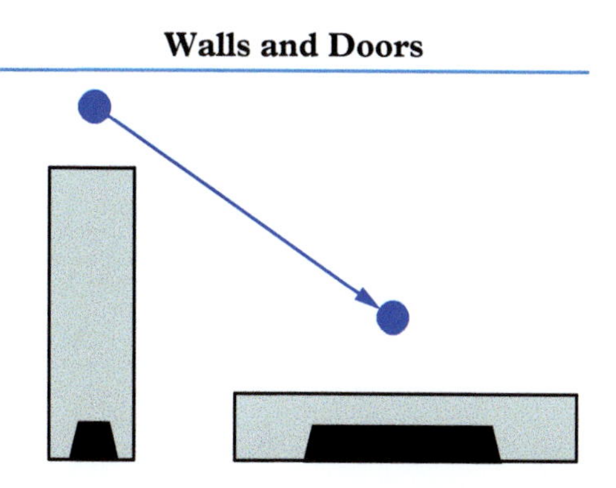

Figure 12.1 *Symbolic wall-door representation of free energy: energy is the height of the wall, entropy is the size of the door. High walls and narrow doors stand for high free energy, places difficult to reach, low walls and wide doors stand for the opposite, low free energy, places that easy to reach. The spontaneous course is to move from the former to the latter, under the effect of entropic forces. Reversing entropic forces, i.e. reaching to difficult places, takes a free-energy investment.*

sense, i.e. the minimum of minima, takes you indeed to thermal death, definitely not the best return on investment!

Energy usually goes with Order, structure and function, while entropy with just the opposite. Hence, at first glance, it looks like we should stick with the former as much as we can. But think a moment: certain peaks might be just too hard to reach for life to prosper there, the energy bar is too high. On the other hand, high entropy means (too) much Disorder, and life cannot prosper in such lowlands either. What life (and any desirable function, for that matter) needs is the right mix of Order and Disorder, and free energy is precisely the gauge of such balance. Granted that the global minimum is not the most desirable destination, the key point is that on the way to thermal death, there are interesting stopovers, oases of life, where free energy attains intermediate *local minima* and the descent to the fatal global minimum is suspended. By local minima, we mean places such that any short-range move is bound to increase the free energy (see Fig. 12.2). If the descent happens to be suspended for long enough to allow the system to do useful things in between, such as reproducing itself and giving birth to offspring, then the wave of life wins, if only in a collective rather than individual sense. This idea in itself is not new, we have commented along the same lines in Chapter 10 devoted to dissipative structures. What we have done here, though, is to place the idea in a more precise context, through the identification of free energy as the most suitable quantity to express the competition/cooperation between the cosmic ladies. In a language closer to the theory of complex systems, one could say that free energy provides a suitable *fitness function* for natural systems. Next, we take the idea one step further, by introducing the key notion of *free-energy landscape* [118].

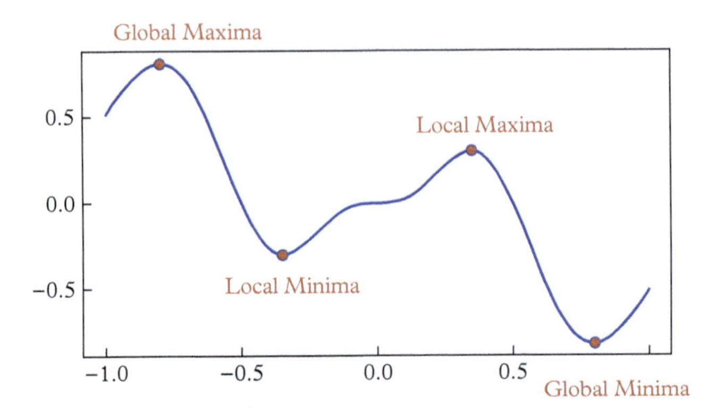

Figure 12.2 *A free-energy profile with two minima and two maxima, one local and global each. The global minimum is the cheapest of all.*
Source: reprinted from https://www.i2tutorials.com. i2tutorials.com/deep-learning-interview-questions-and-answers/what-are-local-minima-and-global-minima-in-gradient-descent/

12.3 Beautiful mountains

Consider the most iconic mountain of all, at least for Europeans, and certainly for this author: Mount Cervino, Matterhorn for the Swiss, scoring 4,478 metres in height (on both the Italian and the Swiss sides (see Fig.12.3).

There are two major routes to the top, the Italian one, Cresta del Leone, and the Swiss one, Cresta Hornli, starting from Cervinia (altitude 2009 m) and Zermatt (1608 m), respectively. Both Cervinia and Zermatt can be reached in many ways through ordinary automotive routes, say from Milan and Geneva respectively, both basically at sea level. But the tip of Cervino/Matterhorn can only be reached in two ways (ok, there are variants for the specialists, but we can hopefully skip this level of mountaineering detail). Thus, the tip of Cervino (Matterhorn) has more potential energy and far less entropy than Cervinia or Zermatt, hence far more free energy than both. And indeed, guess what, it is way less populated than either. In other words, it takes far more free energy to reach to the top of Cervino/Matterhorn than to sit in a comfortable hotel in Cervinia or Zermatt.

The story goes on. One can reach Cervinia from Milan, but you have to drive nonetheless through the mountains, and, for sure, there are far more routes to Milan than Cervinia, including airlines. Same goes for Zermatt versus Geneva. And again, sure enough, we find more people in Milan than in Cervinia, and more in Geneva than Zermatt. By now the reader sees where the analogy is heading; altitude and accessibility are the (concrete) metaphors for energy and entropy. And the number of people goes in inverse proportion to the cost of reaching the place. Thermodynamics works the same way: the probability to attain a given state goes inversely with the free-energy cost of reaching it, more often than not *exponentially* inverse, as we shall detail shortly. The most likely place, the global

Figure 12.3 *Left: Mount Cervino, from the Italian side Plateau Rosa. Right: the same mountain from the Swiss side, with a different appearance and a different name, Matterhorn. Whichever side, beauty can hardly be escaped. Reprinted from it.wikipedia.org and en.wikipedia.org. Below, an early-teen version of the author at the feet of the Cervino.*
Source: reprinted with my own permission!

minimum, is thermal equilibrium, where nothing changes anymore: energy is at its nadir, entropy at its zenith, there is no cheaper place to go. This is thermal death, and the mountaineer analogy only means that thermal death is cheaper than life, no surprise here.

For that matter, aging sings a very similar song: it takes more free energy to keep your hair in its original colour than let them go grey and white, as I learned early in life myself (they are white, no hue) . . .. The analogy works in many natural systems. The tip is the hardest structure to reach energy-wise, and nature won't be found there, unless powerful anti-entropic forces push it over. The natural course is to sit in a nice and comfortable valley, which costs less energy and can hopefully be reached via many routes from all over the country. Since life at the very top is hard because it comes with the highest bill and death at the very bottom is the cheapest location, the only option is to count on intermediate valleys, where various forms of life may be expected to flourish before they run down to the global minimum. So much for the metaphors, but how does the story go in actual practice? What are the actual free-energy landscapes which control the natural world in general and specifically the biochemical ones?

12.4 Chemical mountains

To address this question, let us remind that many basic phenomena in the biological world are controlled by so-called *activated processes*. This means that in order

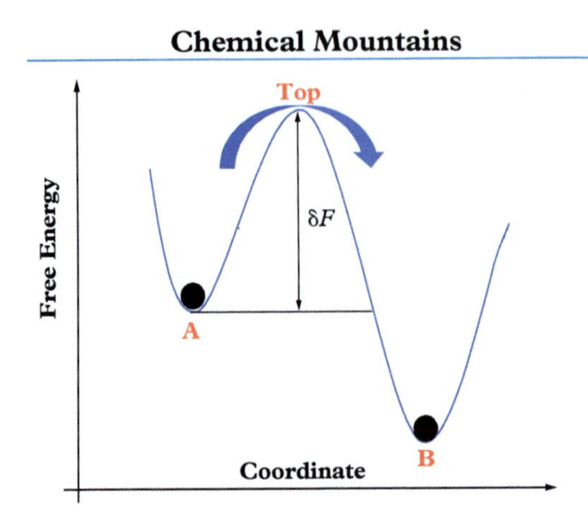

Figure 12.4 *Crossing the chemical mountains from A to B through the Top. The inverse process, B to A, is exponentially suppressed as it faces a larger free-energy barrier.*

to move from state A to state B, the system must typically cross a free-energy barrier, (see Fig. 12.4). Based on the mountaineering analogy discussed before, this is like going over the top of a 'chemical mountain', sitting between the two. The probability of crossing barriers is governed by a most important formula, due to Swedish chemist Swante Arrhenius (1859–1927), chemistry Nobel 1903.

The Arrhenius formula states that the probability of crossing a free-energy barrier decreases exponentially with the height of the barrier, *divided by the temperature.* In mathematical terms, the number of jumps per unit time between A and B of going from A to B, crossing the top of the free-energy barrier between them, is given by

$$P(A \to B) = Ke^{-\frac{F_{Top}-F_A}{k_B T}} \tag{12.2}$$

where T is the temperature, K is a frequency factor and k_B Boltzmann's constant.

Note that since, by definition F_{Top} is larger than F_A, the argument of the exponential is negative, hence the probability is less than 1, as it should. Of course, the larger the difference between the free energies of the top and of the valley, the smaller the probability of crossing. In fact, *exponentially* smaller.

Also to be noted that the probability of going from A to B is generally not the same as the reciprocal one, B to A, unless A and B have the same free energy. To keep going with the mountaineering analogy, take A for Cervinia, B for Zermatt and the top for Cervino/Matterhorn, and identify free energy with the altitude. We then have $F_{Top} - F_A = 4478 - 2009 = 2469$ and $F_{Top} - F_B = 4478 - 1608 = 2870$. The probability to go from Cervinia to Zermatt via Cervino by random jumps of 100 metres on average is $e^{-2469/100} \sim 10^{-11}$, one-hundredth of a billion. The same probability from Zermatt to Cervinia via Matterhorn is $e^{-2870/100} \sim 10^{-13}$, nearly hundred times smaller. That means that, before succeeding, the 'alpinist flea' from Cervinia should try a hundred billion times and her Swiss colleague in Zermatt even a hundred times more! Even if they could jump at a rate of one per second, it would take hundreds of thousand years from Cervinia, and hundred times more from Zermatt .... Now make our alpinist fleas ten times more jumpy, i.e. 1 km on average instead of 100 metres. The daunting numbers previously given turn into way more decent 0.084 and 0.056, respectively, which means about 8–9 and 5–6 successful attempts in hundred. At the same rate of one jump per second, a few tens of seconds would do for both .... And if alpinist flea is a no-show in our world, it certainly isn't in the molecular one, and this dramatic change of numbers is exactly what temperature does for real in the molecular world.

Bottomline: *the temperature is key*: it means that the height of the free energy barrier must be measured in thermal units: if temperature is high, the effective height is small and vice versa. The reason is no mystery, if one looks at the atomistic world, as discussed in Chapter 11. Alpinist fleas do not exist, but molecular ones do! Molecules are indeed like tiny little jumpers, trying all the time to go over the top under the effect of thermal fluctuations. At low temperature they have little kinetic energy, hence they are like dog fleas trying to mount a horse from the ground floor, and most of the time they fall miserably short! As temperature goes up, though, molecules acquire more and more energy, until eventually some of them manage to go over the top, and accomplish the sough transition from A to B. This is literally what happens in the atomistic world, and this is why the mountain picture is way more than just a useful metaphor. In fact, I think it is no exaggeration to state that the landscape is one of the most useful notions in the Science of Complexity, (see Fig. 12.5).

The main message here is the key role played by temperature, the macroscopic face of molecular chaos: due to the exponential nature of Arrhenius law, changes of just a few tens degrees in temperature can turn the timescale of chemical reactions from seconds to years and vice versa (see Appendix 12.1)! This is why life is hosted in a very narrow band of temperatures, in fact less than a hundred degree excursion (250 to 350 Kelvin is a generous bound) over a full range going from a billionth of Kelvin up to nearly a billion billion in particle accelerators!

12.5 Chemical time

This shows the tremendous power of temperature on chemical time, with paramount consequences for biological systems, hence us, in a most real sense, given that aging proceeds mostly through bio-chemistry. Just think of the dreaded 'free radicals', the aggressive molecules which accelerate the aging of our body.[38] The typical chemical clock at the atomistic level is the femtosecond, the time

[38] Oxygen in the body splits into single atoms, with unpaired (free) electrons. Electrons like to be in pairs, so these atoms, called free-radicals, scavenge the body, seeking out electrons to form a pair, a process call oxidation, which causes damage to cells, proteins, and DNA. That is, aging.

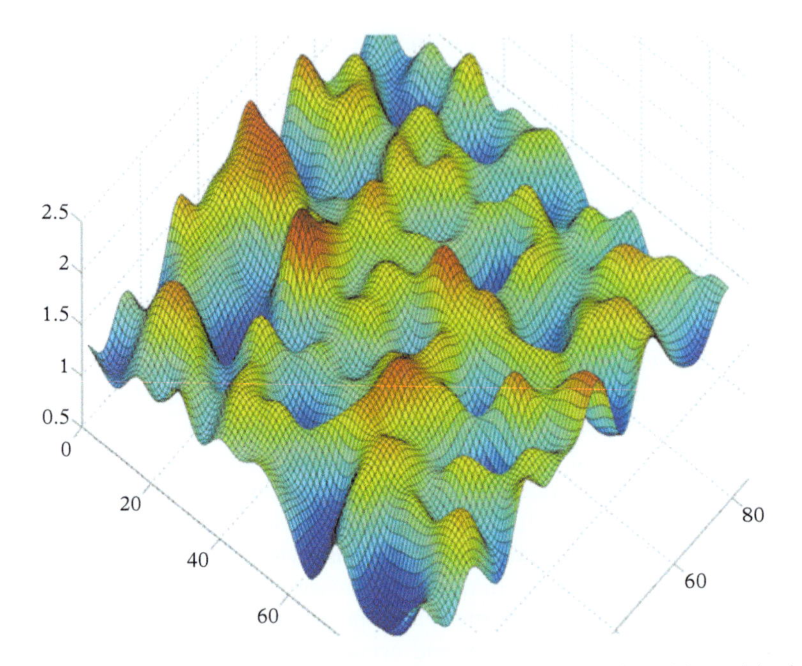

Figure 12.5 *Schematic visualization of a three-dimensional (3D) section of the multi-minima energy landscape of chemical matter.*
Source: courtesy of Christoph Dellago, University of Vienna.

it takes to an electron to jump from one atom to another, leading to a chemical reaction. It takes one million billion femtoseconds to make just one second, which means that if femtoseconds were turned into seconds, we would live hundred millions of billions of years, and cause the Methuselah to be jealous in the process . . . .

'How long is forever?' Asks Alice to the white rabbit to be returned a baffling: sometimes, just one second. The white rabbit must have been a chemist, for it is indeed true that while for us a second is a fleeting beat of the eye, for some high-temperature reaction, a second is a literal close proxy to forever. One wonders whether Lewis Carroll (1832–1898) knew about Arrhenius (1859–1927)!

In summary, at high temperatures, molecules stand exponentially increasing chances to overcome the chemical barriers, eventually it will be too easy, at which time the system can visit all the mountains at a low cost but can't come to a halt on the way. At high temperatures, even the highest mountains become flat hills, no constraints, no life either. Conversely, as the temperature goes down, even small hills turn into unsurmountable mountains: the system is frozen in its initial state, unable to visit the landscape: cold death. I cannot refrain from drawing a psycho-parallel: when we feel tired and blue, even the smallest obstacles seem insurmountable, while when we are in a rosy mood, we can swiftly jump across problems that have obsessed us all through a sleepless night. Thus, the art of travelling across the free-energy landscape at the right pace is the key to success and temperature has a game-changing role on this story.

12.6 Stopovers in the landscape

By now, we have learned that barriers can be crossed in two fairly different ways, either by jumping above the top of the mountain (the energy way), or by sneaking via alternative paths, which steer clear of the top of mountain (the entropy way). Interestingly, life flourishes where energy and entropy come more or less on a par, or, better said, when changes in one are exactly compensated by corresponding changes of the other: see the duet now?

This is a no brainer: peaks and valleys are local maxima and minima (extrema) of the landscape, respectively. By definition, this means that small displacements away from the extrema do not (appreciably) change the value of the free energy.

Some little math says it best. Consider a thermodynamic transformation involving a change of energy δE and a corresponding change of entropy δS. Let us further assume that the transformation occurs at constant temperature (isothermal), so that, by the very definition (12.1), the associated change in free energy is given by:

$$\delta F = \delta E - T\delta S \tag{12.3}$$

As discussed previously, local extrema are peculiar locations such that a small displacement from these locations leaves the free energy unchanged. In math terms,

$$\delta F = 0 \tag{12.4}$$

This condition is met at local minima (valleys) and local maxima (peaks) alike, the crucial difference being that if you move further away from a local minimum, free energy increases, while a corresponding substantial displacement away from a local maximum meets with a free energy decrease. In other words, *local maxima are unstable and local minima are stable.* This condition can be obeyed in many different 'locations' of the landscape and since the changes are small by construction, it says nothing about the absolute value of F itself, which is why they are called *local* extrema. The lowest one is the *global* minimum, and we have given abundant reasons why we'd rather stay away from it.

The expression (12.4) sends another important message: in a local extremum, the changes of energy and entropy are tied up together. The relation is readily derived by combining (12.3) and (12.4), yielding: $\delta E = T\delta S$, or, in equivalent terms:

$$\delta S = \frac{\delta E}{T} \tag{12.5}$$

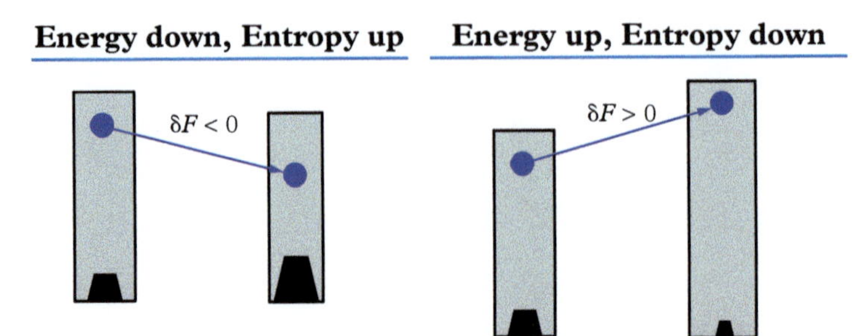

Figure 12.6 *Cooperative energy-entropy scenarios: Down-Up (left) and Up-Down (right). In the DU scenario, free energy (the elevation of the ball) is bound to decrease, indicating that the transformation can occur spontaneously. In the UD scenario the opposite is true: the free energy is bound to increase, and such transformation cannot occur spontaneously but must be driven by external means.*

Let's focus on this latter. What this expression says is that the entropy change required to keep the system at a local extremum upon an energy change δE is precisely $\delta E/T$. This shows again the crucial role of temperature: at high temperature, the compensating change in entropy is comparatively small because T is a large number. At low temperatures, though, such change becomes increasingly larger and in the limit of zero temperature it becomes virtually infinite. This connects to the third principle of thermodynamics: zero Kelvin is a chimaera because even the tiniest change in energy would command an infinite amount of compensating entropy. This also shows the smiling face of entropy. While it is true that, in the end, high entropy leads to the ultimate minimum where no Order can survive, it is also true that entropy gives access to free-energy niches that would otherwise remain unexplored on a purely energetic basis. This is the cooperative part of the duet: if one gets too demanding, the other can compensate and vice versa, (see Fig. 12.6). Away from the extrema, the energy and entropy changes no longer balance each other and free energy either increases or decreases but cannot stay the same.

12.7 The duel and the duet

At this stage, it proves expedient to distinguish between *cooperative* scenarios and *competitive* ones. Cooperative scenarios occur whenever the changes of are opposite in sign, namely energy decreases and entropy increases or vice versa. This may sound weird at first, because cooperation would be expected when two quantities change in the same direction. However, since entropy *subtracts* from energy, in our case cooperation occurs when they change in opposite directions. Let us begin with the two cooperative scenarios, namely, (see Figs. 12.6 and 12.7):

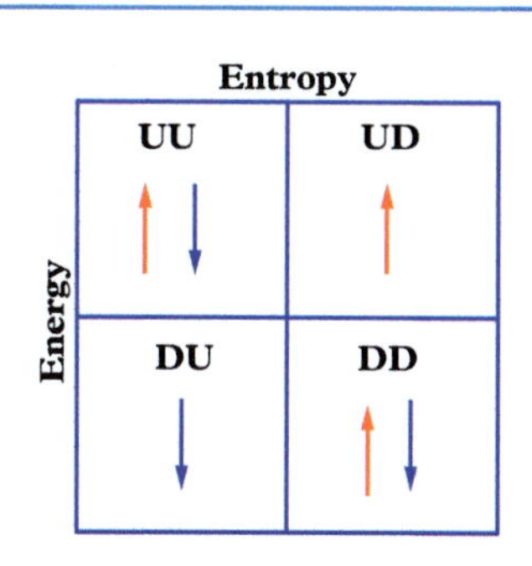

Figure 12.7 *Schematic summary of the energy-entropy competition/cooperation scenarios. Arrows up (down) indicate free-energy increase (decrease). In the DU cooperative scenario free-energy always decreases, hence the corresponding transformation occurs spontaneously. In the UD cooperative scenario, the opposite is true and such transformations cannot occur without external supplies. In the competitive scenarios UU and DD both outcomes are possible, depending on the absolute value of the energy and entropy changes (see text).*

DU Down-Up: $\delta E < 0$, $\delta S > 0$; $\delta F < 0$

UD Up-Down: $\delta E > 0$, $\delta S < 0$; $\delta F > 0$

It is worth reminding that both energy and entropy changes refer to the system under consideration, not the entire Universe, which is why entropy is eventually allowed to decrease, as discussed in Chapter 8. In the DU case, both terms on the right-hand side of the expression (12.3) are negative, hence the free energy is bound to decrease. This means that DU transformations can occur spontaneously. Reciprocally, in the UD case, both terms are positive, and the free energy is bound to increase instead. Hence UD transformations cannot occur spontaneously and must be driven by external sources or constraints.

Next, let us inspect the competitive scenarios (see Figs. 12.8 and 12.7), i.e.

DD Down-Down: $\delta E < 0$, $\delta S < 0$; δF?

UU Up-Up: $\delta E > 0$, $\delta S > 0$; $\delta F > 0$?

Here, the situation is trickier precisely because the two ladies compete instead of cooperating, so that the final outcome depends on which one of the two prevails. For the DD case, the driver to lower free energy is energy, hence a lowering of free energy occurs if the negative change if energy exceeds, in absolute value, the negative change in entropy times temperature. In math terms: $|\delta E| > T|\delta S|$.[39]

[39] We remind that the absolute value of a given number x, denoted by $|x|$ is the value of x itself is x is positive, and $-x$ if x is negative, so that the absolute value is always positive.

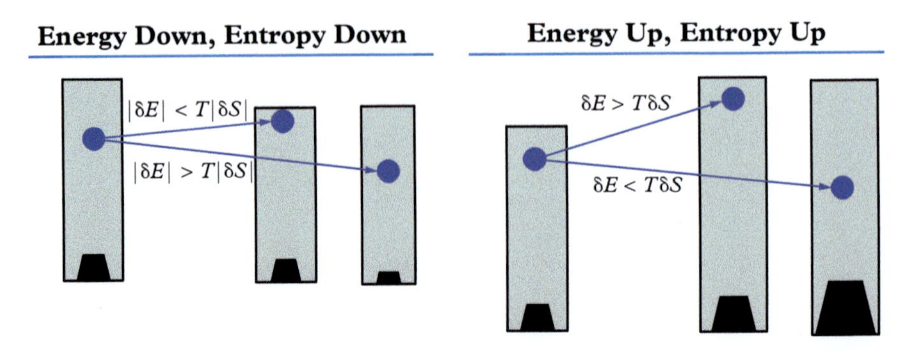

Figure 12.8 *Competitive energy-entropy scenarios: Down-Down (left) and Up-Up (right). In the DD scenario free energy decreases under the condition $|\delta E| > T|\delta S|$, namely the energetic decrease must exceed the entropic one in absolute value. In the UU scenario, the opposite is true.*

In the UU scenario, the driver to lower free energy is entropy, hence the condition for lowering free energy is just the opposite $\delta E < T\delta S$. The positive change of energy must be smaller than the positive increase of entropy times temperature. To be noted is that whenever the two opposing contributions match exactly each other, we obtain precisely the condition for local extrema. This shows that in order to keep the system in such local extrema the two ladies must necessarily compete, see Fig. 12.6!

Finally, we mention two special instances which bear major thermodynamic relevance, the so-called *iso-energetic* and *iso-entropic* transformations, characterized by the condition $\delta E = 0$ and $\delta S = 0$, respectively. Either way, the duel/duet fades away since only one of the two is on stage at a time. For isoenergetic transformations, $\delta F = -T\delta S$, hence they occur spontaneously if the system's entropy increases and must be driven in the opposite case. In the iso-entropic case $\delta F = \delta E$, and free energy is just the same as energy, hence they occur spontaneously if energy decreases and must be driven in the opposite case.

These are the general conditions which govern what we metaphorically called 'duet&duel' of the two cosmic prima donnas. For the sake of concreteness, let us specialize the discussion to a case we been covered at length in Chapter 8, namely the Carnot cycle. Let us consider the hot isothermal expansion, in which the gas expands upon absorbing heat at constant temperature. As shown in Chapter 11, in an ideal gas of non-interacting molecules, the energy is a function of temperature alone. This means that the isothermal expansion is an iso-energetic transformation with $\delta E = 0$. Entropy, on the other hand, grows because the gas occupies more volume, thereby increasing the number of microscopic configurations. Hence, the isothermal expansion features $\delta E = 0$ and $\delta S > 0$, namely $\delta F = -T\delta S < 0$. By absorbing heat, the gas spontaneously expands and frees up an amount $\delta F = -T\delta S < 0$ of free energy, which is exactly the heat absorbed and converted into work done by the gas on the piston! This shows how heat is transformed into useful mechanical work by means of purely entropic forces, i.e. the forces exerted by the gas molecules impinging on the piston head! Next consider the adiabatic expansion, in which no heat is exchanged, so that entropy stays the same, thus qualifying the adiabatic expansion as an iso-tropic transformation. In this case $\delta F = \delta E < 0$ is precisely the work done by the piston as the gas expands. With this, we hope we have spelled out in general but nonetheless concrete terms what the duel/duet of the cosmic ladies actually means.

12.8 In the biological world

Having clarified the terms of the energy-entropy duel/duet, the next question is: how big are the energy and entropy changes in the biological world? As it is often quoted, the biological world rests on '$k_B T$' physics, the so-called soft matter domain, where energy and entropy never win over each other hands down. The reason is again Arrhenius's law: barriers much higher than $k_B T$ are exponentially inaccessible, and because the exponential function decreases very steeply, states 'protected' by barriers much above $k_B T$, remain exponentially unvisited. Just to give a sense of it, the probability of going over a barrier of 10 $k_B T$ is about 40 in a million, and at 20 $k_B T$, it falls down to two in a billion . . . .

So far, we have told a long story about the crucial role of free-energy min-max 'locations', valleys, and peaks, but apart from mentioning cities like Milan, Geneva, Cervinia, and Zermatt, we have remained completely silent as to what such locations really are, out of the geographical metaphor. In other words, what are the concrete analogues of the geographical coordinates in the free-energy landscape of biological systems? This question takes us for a tour to the magic country of molecular hyperland, the subject of Chapter 13.

12.9 Summary

Summarizing, free energy, namely the energy left to produce useful work once the entropic tax is paid-off, is the actual currency of the natural world. This means that Nature tends to select structures featuring the least free-energy cost, while still delivering useful functions. Both energy and entropy usually change during thermodynamic transformations, and whenever such changes come to an exact balance a local extremum of the free energy is attained, in which the system can take a pause, sometimes a blink of an eye, sometimes an entire lifetime or even astronomically more, depending on the conformation of the free-energy landscape and on the actual value of temperature. At low temperatures, changes are exponentially suppressed, so that the system remains frozen: cold death. At high temperatures, changes are exponentially facilitated, so that the system never finds the peace to stop anywhere and get anything useful done: hot death. For biological systems, the borderline between low and high temperature is the standard 300 Kelvin ambient temperature (27 Celsius).

Formally, free energy depends on all the coordinates and velocities of the atoms composing the system in point, hence it lives in a configuration space with Avogadro numbers of dimensions! The natural question then arises: how does the temperature hyper-sensitive duel/duet between energy and entropy succeed in shaping free-energy landscapes with life-friendly free-energy minima? How come, given the monster-dimensional hyperspace they live in? This sounds like a mission impossible, yet it is not. Stay with us, this is the question we are going to explore in Chapter 13.

12.10 Appendix 12.1: Arrhenius law

The Arrhenius law states that the number of successful jumps per unit time, $\mathcal{J}_s$, across a free-energy barrier ΔF at temperature T is given by:

$$\mathcal{J}_s = \mathcal{J}_r e^{-\Delta F/k_B T}$$

where $\mathcal{J}_r$ is a frequency factor, the jump rate, measuring the number of molecular attempts per unit time to cross the barrier and the exponential gives the probability to succeed.

There are typically of the order 10^{12} thousand billion, such attempts per second in the atomistic world, so $\mathcal{J}_r \sim 10^{12}$. At room temperature, the thermal energy is about $k_B T = 1/40$ electronvolts and ΔF can be taken off the order of the electronvolt, hence the ratio $f \equiv \Delta F/k_B T$ (the effective height of the chemical mountain), is $e^{-40} \sim 4 \times 10^{-18}$. Hence, there are about $4 \times 10^{12} \times 10^{-18} = 4 \times 10^{-6}$ successful jumps = reactions, per second. Better restated, the reaction (in fact -one-reaction) occurs once every 250,000 seconds, i.e. once every 3 days.

Now heat the room up to 127 Celsius, namely 400 Kelvin degrees, and the Arrhenius exponential becomes $e^{-30} \sim 10^{-13}$, twenty thousand times faster, one reaction every 10 seconds. Heat it up further to 600 Kelvin and the exponential is $e^{-20} \sim 2 \times 10^{-9}$, ten thousand times faster, a thousand reactions per second. Chemistry is on! Although purely orientative, these numbers speak clearly for the power of temperature over (chemical) time. The dramatic writer would speak of the power of 'chaos' on time, and although it is 'only' a matter of probabilities, this power is very real.

Part III

The Physics-Biology Interface

13

Survival in Molecular Hyperland, the Ozland Valleys

You people with hearts, have something to guide you, and need never do wrong; but I have no heart, and so I must be very careful.

(Tin Man, The Wizard of Oz)

13.1 A tour to molecular hyperland

In Chapter 12 we have discussed the subtle duel/duet between energy and entropy and shown that this is most conveniently described in terms of free energy landscapes in analogy with natural masterpieces like Mount Cervino/Matterhorn. We have also pointed out that the local extrema of such landscape are the places where the system can pause and eventually deliver useful functions. The question is how? In this chapter we shall dig deeper into this question with the aid of two fundamental processes from the biological world: deoxyribonucleic acid (DNA) translocation and protein folding.

13.2 Topography

With reference again to Mount Cervino, we note that its topography, like that of any other mountain on planet Earth, can be characterized by its altitude at each given point on the Earth's sea-level surface, say latitude and longitude. In other words, three numbers, latitude, longitude and altitude provide the position of any given point on the rocky surface of Mount Cervino. Beware: this is just because we happen to live in a three-dimensional world. Now, please bear with me, for we need to make a huge conceptual leap of imagination. Consider a system of ten particles in a box, see Fig. 13.1.

Even leaving aside the particle velocities, an exhaustive knowledge of the system of ten particles requires the specification of 20 independent numbers. These numbers are the coordinates $x_1, y_1; x_2, y_2; \ldots x_{10}, y_{10}$ of the particles in the two-dimensional box. This is the microscopic representation of the system 'ten particles in a two-dimensional box of side L'. As a result, the microscopic description of any thermodynamic function of this systems depends on twenty

Sailing the Ocean of Complexity. Sauro Succi, Oxford University Press.
 DOI: 10.1093/oso/9780192897893.003.0013

Particles in the Box

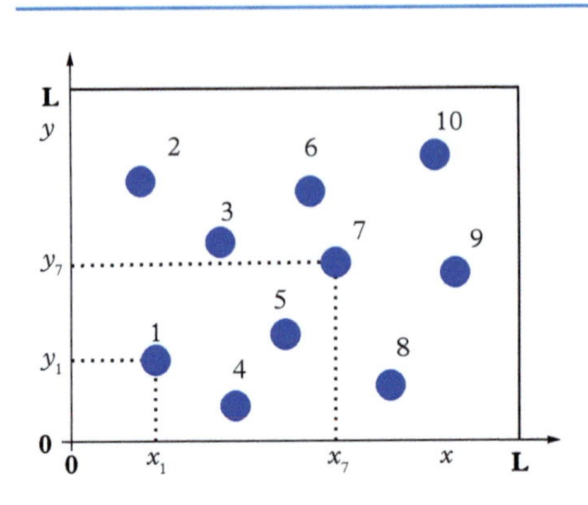

Figure 13.1 *Ten particles in a square box of side L. The position of each particle is characterized by its coordinates along the horizontal axis x and the vertical one, y. For instance, particle* 1 *has coordinates* x_1, y_1, *particle* 2, x_2, y_2, *and so on up to particle* 10. *The dynamic state of the system of 10 particles depends on* 20 *coordinates* $x_1, y_1; x_2, y_2; \ldots x_{10}, y_{10}$, *which means that the free energy is a function of 20 coordinates, i.e. it lives in a 20-dimensional space!*

coordinates, meaning by this that the system lives in a 20-dimensional space! The free-energy landscape of this system is therefore the analogue of Mount Cervino, but with its altitude described by 20 coordinates instead of just two! It is often said that a little drop of water contains a universe inside. This is actually true, as every little piece of matter comes with its own hyperland, namely a free-energy landscape, except that instead of just three, this landscape comes with *zillions of extra dimensions* [35]! Take a cubic centimetre of water: strictly speaking, the energy and entropy of this cubic centimetre depend on the position and velocities of all the water molecules inside, a number around 10^{22}, let's say Avogadro for simplicity. So, if we can still draw a parallel between free-energy and altitude, in the world of molecules there are Avogadro's numbers of latitudes and longitudes: *welcome to molecular hyperland*!

13.2.1 Surviving in hyperland

A question immediately arises: how can nature, i.e. molecules, possibly manage to navigate the monster-dimensional landscape of molecular hyperland and find the right 'oases' where 'things work'? The answer is far from straight, but a few powerful clues are available. First, the $k_B T$ argument discussed before. Since the probability of reaching the top of a chemical mountain decreases exponentially with the ratio of its height over temperature, mountains much higher than thermal ones, height $k_B T$, are likely to be near-empty, hence there is no urge to inspect these dead ends of molecular hyperland. Second, and more subtly, the overwhelming majority of these 'extra-dimensions' are 'smooth and flat', meaning by this the free energy does not change much upon moving for even long stretches along these dimensions. The 'smooth and flat' dimensions, by construction, do not host many local extrema, and when they do, these extrema are not expected to be deep, hence they are easily escaped. As a result, the local extrema of smooth and flat dimensions do not need to be inspected very closely either. They are, to use technical jargon, 'irrelevant'.

The relevant dimensions, on the other hand, are those along which free-energy shows substantial and repeated variations, so that upon walking along them, one meets many high peaks and deep valleys. In other words, along the irrelevant directions the landscape is flat, while along the 'relevant' ones it is rugged and *corrugated.* This resonates with the idea that roughness carries more Complexity than smoothness, as discussed in Chapter 3. With a major smile of Lady Luck, rugged dimensions are vastly outnumbered by the flat ones, which dramatically reduces the dimensionality of the search space. With these encouraging preparations, let's dig a bit deeper into these matters.

13.2.2 Weeding out wasteland

The previous two points alone weed out an immense amount of Complexity from the hyperland landscape, allowing the system to confine its exploration to a comparatively few dimensions, aptly known as *Order parameters* (OPs) in the scientific literature [100]. In fact, not only is the number of OPs astronomically smaller than the Avogadro number, but often even well below its logarithm, 23! Believe it or not, there is much we can learn about nature by exploring just a fistful of OPs, in the most extreme cases even just *one* may suffice to deliver meaningful information![40] In the reader's boots, my reaction at this point would be that this must be a joke, simply 'too good to be true'. First, we made a big deal about molecular hyperland and at the end of the day, we land on a plain one-dimensional landscape, *literally, from hyperland to lineland*! This may indeed look like a joke, but it is not. In fact, it is a wondrous gift we owe a lot of our existence to, so please, wait a minute before you give up. *The point is that nature provides a lot of redundancy, information which is not essential to answer some of the questions which we depend upon.* As a result, redundant/irrelevant features can be 'washed out' without throwing away the baby (us) with the bathwater .... Scientists call this procedure, *coarse-graining*, a less than charming name for the magnificent art of gleaning knowledge and understanding by *removing* information, instead of amassing it! This stands in stark contrast with the Big Data hype of these days, with billionaire gurus informing us that, with

[40] As a wildly unproven observation, the number of OPs is often slightly above the double- logarithm of the number of dimensions. For the case of Avogadro, $loglog10^{23} = log23 \sim 1.36$, i.e. one or two OPs.

more data and (their) algorithms, all the problems of this suffering world will melt down like lemon drops .... Big Data is very likely to help, but coarse-graining is guaranteed to stay.

So, let's take a closer look at the art of surviving in hyperland.

13.3 The art of coarse graining

Let us begin with a very simple instance of coarse-graining: mass lumping. Consider the familiar case of gravity (again!): the force which keeps us down on Earth, is the product of the mass of our body m, say 70 kilograms, times the gravitational acceleration $g = 9.81$ metres per squared second. In math terms:

$$F = mg \tag{13.1}$$

This force, the weight of a material body of mass m, is the result of the interaction between *all* the molecules in our body with every single molecule of the entire planet Earth! Computing the interactions pair by pair is obviously a mission impossible, since it involves a number of multiplications equal to the *product* of the number of molecules in our body, times the number of molecules in the entire planet! For the record, we are talking a number like 1, followed by 80 zeros ....

It turns out that, to an excellent degree of approximation, our weight can be computed in an astronomically simpler way. Here comes the idea. First, lump all the molecules in our body into a single (umbilical) point, containing the entire mass of our body, say m. Second, do the same with all the molecules of planet Earth, located at her umbilical point, totalling a mass M. Third, multiply the two masses, divide by the square of the distance between the two barycenters, namely Earth's radius R plus about half our height, and finally multiply by the gravitational constant G. The result is the Newton's formula for gravitational interactions:

$$F = G\frac{mM}{R^2}$$

that we have discussed at length in Chapter 11. Note that in the previous calculation we have neglected our height since it is million times shorter than the radius of Earth. This formula exactly matches the expression (13.1) upon setting $g = GM/R^2$.[41] The number we just obtained via the previous coarse-graining scheme is not *exactly* the right one, but the point is that the approximation error is quite minuscule.

The calculational savings we obtain in return for this clever approximation are just immense. We do not need to know the exact position of each molecule in our

[41] The interested reader can check the actual numbers: $M \sim 6\ 10^{24} kg$, $G \sim 6.7\ 10^{-11} m^3 Kg^{-1} s^{-2}$, and $R \sim 6.4\ 10^6 m$, give the familiar $g \sim 9.81 m^2/s$.

body, nor in the planet: all we need is to compute their total masses and place them in corresponding barycenters. The name of coarse graining in this game is *lumping*: i.e. replace *zillions* of molecules spread around our body with a *single* 'super-molecule' sitting at the umbilical point, with exactly the same mass as the entire body. And repeat the same procedure with our planet. Fair enough, you still have to sum up all molecular masses, but this is proportional to the number of molecules in our body *plus* the number of molecules in planet Earth, not to the *product* of these two gigantic numbers. With small numbers, the gain is irrelevant, but with large ones, as it is certainly the case here, it is astronomical. Of course, it is plain that in this 'don't throw the baby with the bathwater' scenario, the information to be discarded, or averaged out, *must* be irrelevant, on pain of missing the whole purpose. This is what scientists call 'controlled approximations', indicating that you must be aware of the error you incur by taking the coarse-graining shortcut. In the case of gravitational interactions, one can show that the error is indeed negligible, but this must be checked case by case. In other words, nature has been equipped with enough power to 'do calculations' the hard way, a flight we just can't take for want of computational power. But it is a very happy and generous fact of life, that *nature offers plenty of coarse-graining opportunities* and I think it is no exaggeration to say that our very existence, i.e. the capability of navigating Oceans of Complexity depends precisely on our ability to take advantage of such opportunities. In other words, the ability to tell relevant information apart from the irrelevant one, which, as we shall see, is an art as much as it is a science.

13.4 Order parameters

Sometimes this art comes easily, but most often it doesn't. In the case of our weight, the choice of the OP is very natural: it's the barycenter of the material body. In addition, this OP is particularly simple because it is *linear combination* of the particle positions (see Appendix 13.1). In general the OPs depend in a very non-trivial way on the microscopic coordinates and there is no magic formula to devise such a dependence. This leaves plenty of scope for heuristics and physical intuition, which is precisely the artistic component of coarse-graining [20, 117]. On the campus of a prominent European university, I once read the following sentence 'Science is what we can tell our computers to do, the rest is art . . .'. I agree as much as I don't, for, to me, science and art are part of the same imaginative process, but the statement conveys the idea that in the face of Complexity, fantasy, and intuition cannot be entirely replaced by procedures and algorithms. Again, this flies in the face of current trends of the most aggressive Big Data campaigns, pressing hard for the idea that machine and algorithms can sort everything out on their own [29].

Having introduced the idea of coarse-graining as a general strategy to cut down the cost of otherwise unmanageable tasks without compromising the basic goal, let us proceed to illustrate the idea with the aid of two concrete examples from the biological world: deoxyribonucleic acid (DNA) translocation and protein folding. Incidentally, both processes offer example instances of the use of non-equilibrium statistical physics to the study of *kinetic processes* i.e. processes in which time is everything because their function is either delivered on time or to no avail at all. This reconnects to what we had to say in the Preface, namely that modern physics is by no means limited to the study of infinite-size and infinitely slow systems.

13.5 Modelling DNA translocation

The translocation of the DNA molecule through the cellular membrane is a fundamental biological process which proves key to most vital functions, such as the synthesis of proteins within the cell nucleus. The human DNA is a chain of about three-billion molecules, known as *base pairs* (BP), which come in four 'flavours', A,C,G,T for adenine, cytosine, guanine, and, thymine (see Figs. 13.2 and 13.3). These four base pairs are rather complex molecules, each consisting of about ten atoms, but as we shall see in a moment, for the question we are going to ask, these chemical 'details' can be safely ignored. The DNA molecule is famously composed of two intertwined helical chains which are held together via the binding of complementary base-pairs, namely A with T and C with G. Rivers of ink have been spent on this unique molecule, hence here we shall go straight to our target, namely, how to model the translocation process.

More precisely, the goal we set for ourselves is to understand how long it takes to DNA to translocate from the interior of the cell (cytoplasm) to the interior of the cell nuclei. To reconnect with the language of the previous chapter, DNA translocation is most conveniently regarded as kinetic transition from a state U (for untranslocated) in which the DNA molecule is completely outside the nucleus, to a state T (translocated) in which it is completely inside it. This is an inherently kinetic process, since the first question one asks in this context is how long it takes for the transition from U to T to occur. But before we get started with coarse-graining, let's take a look at the molecular hyperland of DNA.

13.5.1 DNA hyperland

Since there are three billion base pairs in human DNA, each coming in four possible letters A,C,G,T, the number of potential DNA molecules (called *Genotype*) is $4^{3\times10^9}$, a monster number sitting between the Googol (10^{100}) and the Googolpex ($10^{10^{100}}$) (see Appendix on Numbers). The overwhelming majority of these combinations is weeded out by biological constraints and what we observe in real life are only the DNA molecules which are actually expressed in living creatures, the so called *Phenotype*. For human beings, about seven billion at the time of this writing hence, biological constraints serve as a formidable sieve, once again something similar to the *logarithmic* compression from the number of microstates associated with a single macrostate and

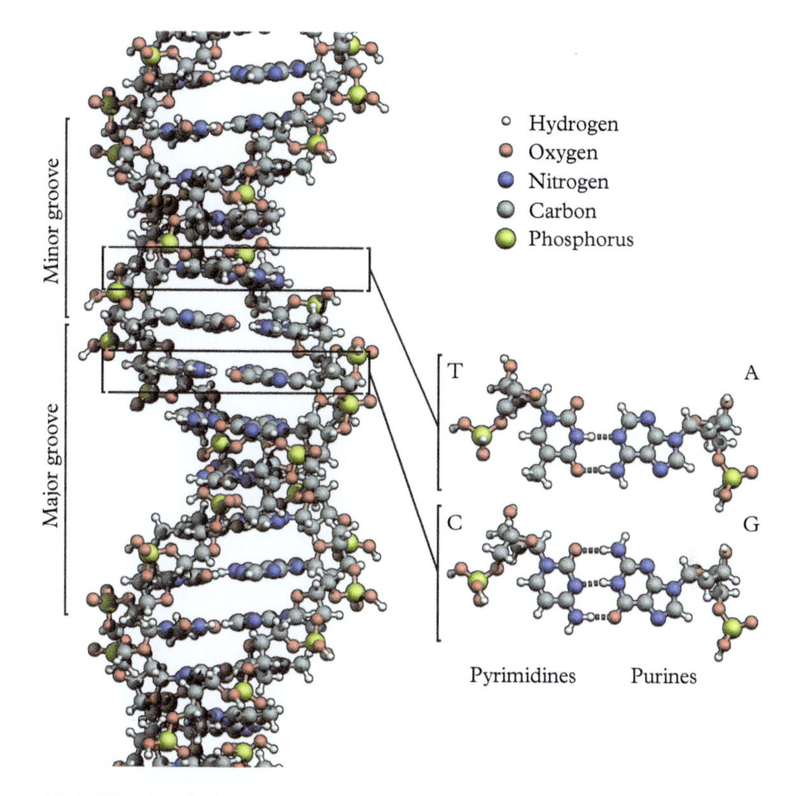

Figure 13.2 *The chemical structure of the DNA molecule with four base pairs thymine pairing with adenine and guanine pairing with citosine to form the famous double helix architecture.*
Source: reprinted from wikipedia.org.

its entropy, according to the Boltzmann's equation $S = k_B logW$, discussed in Chapter 9.

So much for genomic Complexity, how about the conformational one? Even after a single phenotype DNA is selected, which is exactly the case of our exercise, its dynamical state at any given time during the translocation process still requires the specification of three billion positions of each single base pair. And since each base pair consist of about 10 atoms, and each atom features three coordinates in space, this makes a system with 90-billion atomic coordinates! This is the dimensionality of the translocating DNA hyperland! It is worth noting that these estimates do not take into account constraints on the position of the BPs, for instance the fact the bond length, i.e. the mutual distance between two subsequent base pairs, is basically the same for all beads and it does not change in time. Even so, the amount of dynamical information associated with the translocating DNA remains definitely in the class of monster numbers. No way out: to make the problem treatable, coarse-graining is a must.

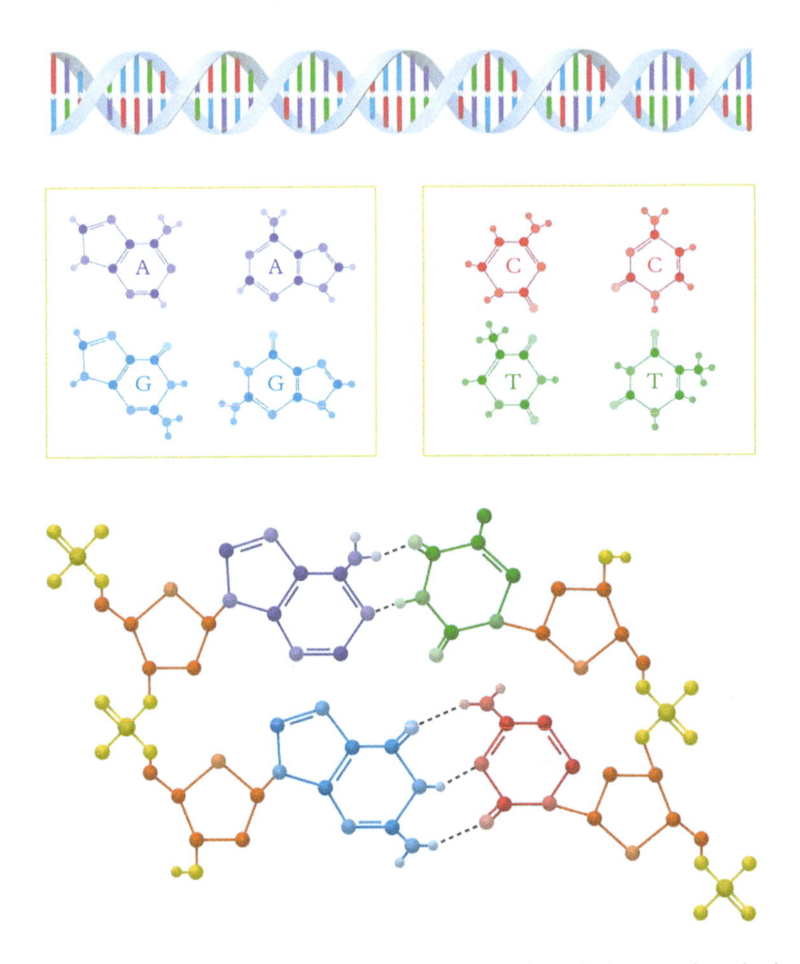

Figure 13.3 *The chemical composition of the four base pairs: adenine, guanine, citosine and thymine.*
Source: reprinted from pixabay.com.

13.6 Coarse-graining DNA

The first coarse-graining move we take is pretty crude: we treat the base pairs as if they were simple spheres of four different colours (see Fig. 13.4). The replacement of BPs with spheres is the stage at which we give up *chemical specificity*: this is already quite a start for coarse-graining, but there will be more coming.

The point is that a number of relevant questions about DNA translocation can be answered to a good degree of accuracy, by replacing molecules with

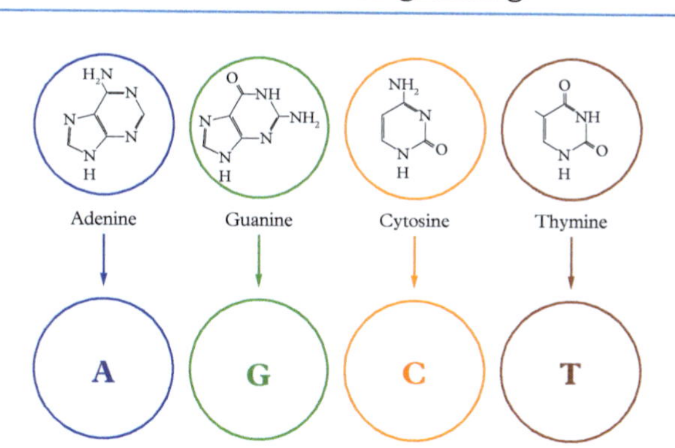

Figure 13.4 *Chemical coarse-graining: the base pairs are replaced by spheres, each with its label (colour), A,G,C,T. With this, chemical specificity is erased.*

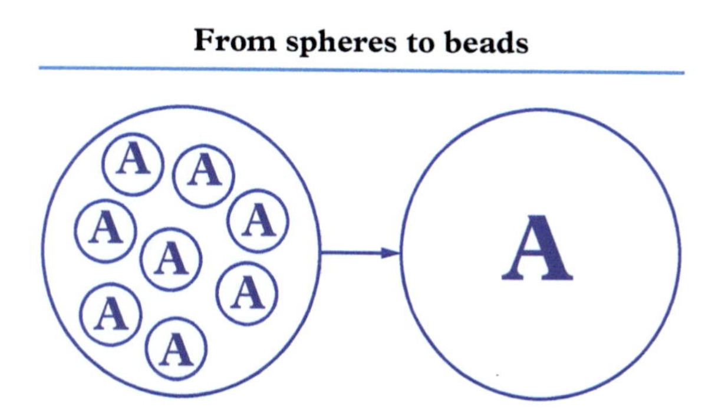

Figure 13.5 *Lumping many spheres of type A into a single bead also of type A. The same procedure applies to the other three types. The figure portrays the case in which only eight spheres are lumped into a single bead, but in actual practice, the number is much higher, between thousands and a million.*

plain spheres and DNA with a sort of coloured necklace![42] Following upon the necklace analogy, we shall call such spheres 'beads', with the important proviso that a single bead represents a large group of spheres, in actual practice easily between thousands and millions of them (see Fig. 13.5). This is our second level of coarse-graining, the lumping procedure discussed earlier on for gravitational interactions.

[42] For the record, in actual DNA the 'spheres' are about 2 nanometres in diameter and they are spaced about 2 Angstroems, i.e. 0.2 nanometres, apart. This means that, unlike an actual necklace, the necklace-DNA spheres exhibit a major overlap with each other.

Having replaced each base pair with a sphere and having lumped thousands of spheres into a single bead, we are left with millions of beads instead of the three billion spheres we started from. This is still quite a lot of information, but something that our best computational science is in a position to handle, the technical name being, coarse-grained molecular dynamics (CGMD) for short, which we now describe in some more detail.

13.7 Coarse-grained molecular dynamics

The task of CGMD is to solve the equations of motion for each and every single bead as it interacts with all the other beads. Even leaving aside chemical specificity, it is clear that having condensed thousands of spheres into a single bead implies that the equation of motion of the bead cannot be the same as the ones that govern the motion of the spheres. This is the key problem of coarse-graining: how to derive the equations of motion of the coarse-grained variables starting from the original ones.

Let me set the record straight from the beginning: this task presents a formidable mathematical problem, one which is generally impossible to solve exactly. Bear with me, dear reader, for this is exactly the point in which coarse-graining becomes an art as much as it is a science. Researchers in the field have indeed worked out suitable coarse-grained forces which serve the purpose of describing the passage of the DNA molecule across the cellular membrane.

Another key point not to be missed is the crucial role of nonlinearity: if the equations of motion of the spheres were linear, coarse-graining would be an empty exercise, in the sense that the beads would obey *exactly* the same equations as the spheres! So, once again, nonlinearity is the premier engine of Complexity!

Notwithstanding the major sources of uncertainty inevitably associated with coarse-graining, CGMD provides indeed a wealth of useful information. In particular, if the question is how does the translocation time depend on the length of the DNA? CGMD provides a very sensible answer in close agreement with the experiments. And the answer is that the translocation time obeys the following law $t_{translo} \sim L^{\alpha}$, where L is the length of the DNA molecule and α is the so called *scaling exponent*, a single number.

The Complexity of the problem has been reduced to a single number, the scaling exponent, and the actual calculations show that $\alpha \sim 1.27$. A brief digression on the physical meaning of this number is in order.

13.7.1 Molecular trains and their delays

If all the beads were moving in sync at a given speed, say V, like a molecular train, the translocation time would be just the DNA length L, divided by the translocation speed V, namely $t_{tra} = L/V$ (we neglect the channel length, on the assumption that it is much shorter than the DNA length.) Note that this is a linear function of the DNA length, i.e. the scaling exponent would be $\alpha = 1$

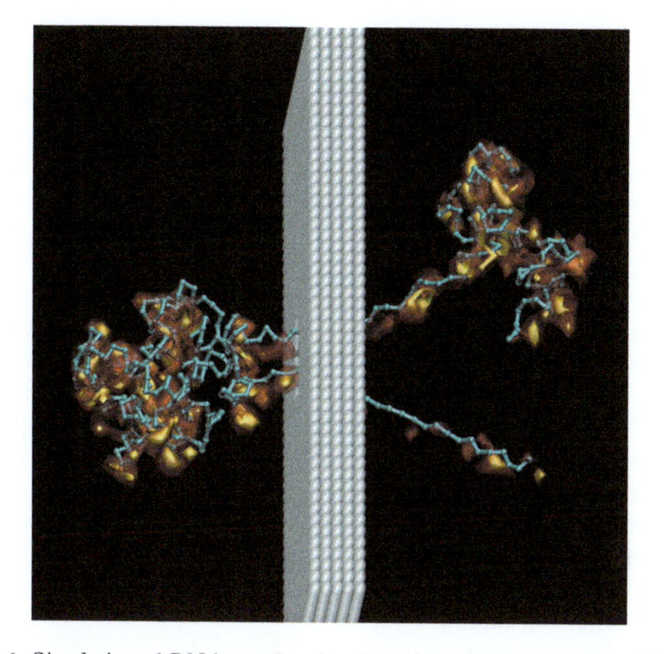

Figure 13.6 *Simulation of DNA translocation through a planar membrane. Note that the 'DNA' consists of two interlaced strands, conferring much more rigidity to the molecule. The simulations consisted of the order of ten thousands beads, i.e. five orders of magnitude below the actual number of base pairs. This means that each bead represents not just a single base pair but hundreds of thousands of them! From M. Bernaschi, M. Fyta, S. Melchionna, S. Succi, E. Kaxiras (2008) Quantized current blockade and hydrodynamic correlations in biopolymer translocation through nanopores: Evidence from multiscale simulations. Nanoletters, 8(4):1115.*
Source: reprinted with permission of acs.org.

But molecular trains don't move straight, as they are subject to a whole series of buffets and shakings due to collisions with surrounding water molecules (thermal fluctuations) and the molecular walls of the channel (see Fig. 13.6). As a result, the process takes longer and the translocation time of the DNA scales *super-linearly* with its length. One might argue that 1.27 versus 1 is no big deal but wait a minute and take a look at the actual numbers: with $L = 1000$, this means an increase of the translocation time by a factor 6.4, which in the context of a finite-time kinetic process is by no means a negligible delay. Just imagine your own reaction at leaving at 9 am and reaching your destination at 3.30 pm instead of 10 am....

With $L = 10^6$ (one million) the delay becomes a factor 41, almost 2 days for a trip which should have lasted just 1 hour! And finally, with $L = 10^9$ (one billion), the delay factor is 269, about eleven days! These numbers speak clearly for the effect of the scaling exponent on the translocation time!

We have emphasized that kinetic processes must occur in due time, on pain of being pointless or even worse, i.e. compromise survival. Most often than not the kinetic constraints can only be met with the help of some form of external assistance.

If we would wait for the DNA molecule to translocate *spontaneously* across the cell membrane, we might just as well be waiting forever, somehow like waiting for the molecules in this room flowing spontaneously all together into a corner. The point is that in order to translocate across the membrane, the DNA molecule has to enter the nanochannel first, align and move all along, until each and every single bead makes it to the exit. No way this can happen spontaneously in any reasonable amount of time: the process of translocation is blocked off by a tremendous entropic barrier. Using the language of the previous chapter, the real game spoiler are the narrow doors. The point is that the highly ordered linear-like configuration is much less likely to occur *spontaneously* than the zillions of disordered configurations the DNA can take either outside the cell before translocating. Hence, in order to translocate, the DNA chain, must necessarily find a way to *dramatically* lower its entropy, i.e. sail against entropic forces.

Such a process would take forever in the absence of facilitating agents, typically electrostatic forces pushing the chain across the channel. Thus, electrostatics provides the anti-entropic force which makes translocation possible in a timespan compatible with biological functionality.

13.8 Extreme coarse-graining: Lineland

As we have just seen, despite being based on very crude approximations, CGMD still involves the solution many thousands, if not millions, of nonlinear equations, a task that can only accomplished by means of the most powerful digital computers. Yet, the answer to our question is basically contained in a single number! This alone raises the legitimate suspect that one could proceed to more aggressive forms of coarse graining. Indeed, a minute thought reveals that the CGMD simulation tells us much more than we asked for in the first place. It tells us not only how long it takes to the necklace DNA to translocate, but also at what time *each and every single bead* goes across the channel, not something we have been asking for! Thus, there must be more economic and suitable ways of answering our comparatively modest question.

This is the prototypical case where another form of coarse-graining comes into play, the strategy being to focus the attention on a much lesser number of variables than the (already heavily reduced) set of CGMD beads. For instance, instead of computing the detailed trajectory of each bead of the necklace DNA, we may content ourselves to figure out just how many beads, *on average*, have passed the membrane at any given time. Even better for practical purposes, we may ask what *fraction of beads* have translocated as a function of time, call it $f(t)$. By definition, at the beginning of the translocation $f(t = 0) = 0$, while at the end of it $f(t_{translo}) = 1$, hence, any model providing this fraction as a function of time would deliver the translocation time, precisely as the time at which the fraction hits the value 1.

This *single number* as a function of time is all we need to know in order to answer our question: how long does it take to the necklace molecules to translocate? Remember Chapter 12 in which we talked about going from A to B across the free energy landscape? That's exactly what we are doing here, with $A = U$ being the untranslocated state and $B = T$ being the fully translocated one. And since we use a single-order parameter (OP), the landscape is just one-dimensional: welcome to lineland! From the billions of real base pairs to the millions of beads of CGMD, now to just a single-OP, from hyperland literally to lineland! How could such a daring ride possibly work? Well, the good news is that as long as the translocation question goes, it really does. Let us see why and how.

13.9 The Langevin approach

To begin with, we observe that there is an enormous amount of molecular configurations which map onto the very same number f at any given time t. This is why the latter is aptly called order parameter, as it collects the contributions of zillions of different configurations, all mapping to the same value of f. In this respect, the OP appears for what it really is: a powerful *information-compressor*. But the point is subtler than just information-compression, and the point is that the distilled information must contain the essence of the physics which controls the answer to the question we are asking. This is the very spirit of coarse-graining!

Let us illustrate the idea with a concrete example. Consider a cartoon-DNA of just ten beads numbered 1 (head) to 10 (tail) from right to left, moving rightwards towards the nanotunnel. Suppose at some instant in time, the first three beads, 1, 2 and 3 have gone through the end of the tunnel, the remaining seven still being left behind. In this configuration the fraction of translocated beads is three to ten, that is $f = 3/10 = 0.3$. This is the 'molecular train' configuration, in which beads translocate in linear sequence. But, as we have just mentioned, in the process of translocating, the beads undergo a series of interactions which cause deviations from the linear chain configuration and corresponding delays in translocation time.

Suppose, for instance, that the DNA folds, like the one in Fig. 13.7, so that beads 3, 4, and 5 translocate out of the tunnel, because 1 and 2 have retracted in between. Although very different from the linear chain, this configuration still has the same OP, $f = 3/10$. We can take an even weirder configuration, whereby the DNA turns around before entering the channel, so that the first three beads to translocate are the 'last ones', namely 10, 9, and 8. Still $f = 3/10$, but again a totally different configuration. This shows that there is indeed a multitude of chain configurations, all giving the same $f = 0.3$. Of course, they are *not* equivalent from the free-energy standpoint, the linear chain is the least likely on a purely combinatorial basis, but also the least expensive energetically because each fold costs energy. Hence, all configurations contribute to the *exact* calculation of $f(t)$, while only the

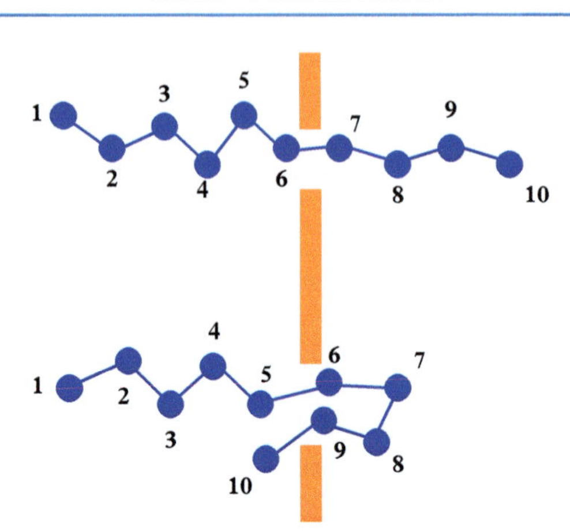

Figure 13.7 *The OP in translocation: the two chain configurations are very different from each other, the one on top goes through the hole in linear sequence (first-in, first-out), while the one on the bottom folds and retract. Yet, they feature the same translocation fraction,* $f = 4/10$.

least expensive from the free-energy viewpoint provide the dominant contribution to the exact solution.

All we need at this point is the mathematical machinery to compute the fraction of molecules that have translocated at given time, instead of tracking the details of all molecular positions (and velocities) in time. As for the case of computing our weight, once the number of beads ranks in the billions, the former task is, in principle, immensely simpler than the latter. We used cartoon-DNA to make the point, but professional models have been devised in the modern literature, in which one formulates *approximate* expressions of the free energy landscape solely in terms of the single-OP f.

With this expression at hand, it is then possible to compute the probability that the molecule translocates within a given time. If only by approximation, the translocation time can be computed by tracking just *one* number in time, the fraction of translocated beads.

The equation for this single variable is simple enough to be put in display:

$$\frac{df}{dt} = F(f) + R \tag{13.2}$$

with the initial condition $f(t = 0) = 0$.

The left hand side is the change per unit time of the translocation fraction. This change is driven by the 'force' term $F(f)$, which depends on f itself at time t. This term is precisely the slope of the free energy landscape, the reason being that, like a ball rolling downhill, the system moves in the direction of decreasing free energy, following the negative slope of the landscape.

The second term at the right hand side is the effect of the random noise resulting from thermal fluctuations. This equation goes under the name of the French physicist Paul Langevin (1872–1946), who first introduced this model equation for the study of liquids. Solving the Langevin equation is a comparatively light task and even if it must be repeated for many different random realizations in order to collect sufficient statistics, it is still far simpler than solving the thousands or millions of CGMD equations. Moreover, it is sometimes even amenable to analytical (paper and pencil) inspection, which is always very insightful. For the sake of completeness, we wish to mention that is also possible to formulate a kinetic equation for the probability $P(f, t)$ that a fraction f of the molecule has translocated at time t. The advantage of this approach is that the kinetic equation, although more demanding than the Langevin equation, contains all the relevant statistical information on the process, hence it has to be solved only once. The reader interested in these professional details is kindly directed to the paper [69].

In the case of DNA biopolymer translocation, the solution of the Langevin equation shows that the key parameter controlling the translocation time is the dimensionless ratio $\phi = FD/k_BT$, where F is the external force driving the process (usually an electric field), D is the diameter of the bead, and k_BT is the thermal energy. Hence, the parameter ϕ measures the work done by the electric field to displace the bead by one diameter (D) in space, versus the thermal energy. When such parameter is smaller than 1, one speaks of 'weak forcing', while the opposite case is denoted as 'strong forcing' (see Fig. 13.8).

It is possible to show that in the strong forcing regime, the translocation time obeys the relation

$$t_{translo} \sim t_0 N^{\alpha} \tag{13.3}$$

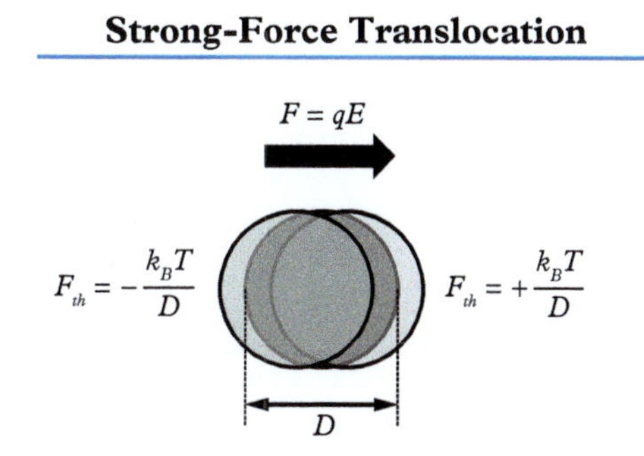

Figure 13.8 *The meaning of the strong-forcing regime: the force pushing the bead across the channel, $F = qE$, is much stronger than the forces due to thermal agitation $F_{th} = k_BT/D$. The translocation time decreases very sensitively at increasing the ratio qED/k_BT between the two. The translocation process is not spontaneous, but a strongly driven one.*

with $\alpha \sim 1.27$. In (13.3), t_0 is the ballistic translocation time and N is the number of beads in the DNA chain. The ballistic time in turn is given by $t_0 = L/V$, where the velocity is $V = qE/m\gamma$, q being the electric charge of the bead (about one tenth of the electron charge), E is the external electric field, m the mass of the bead, and γ the friction coefficient, which has the dimension of an inverse time. Such coefficient represents the dissipative effects due to the interaction of the translocating beads with the surrounding water molecules and with the wall molecules within the nanochannel.

Despite these drastic simplifications, the Langevin equation can still capture the translocation exponent α in fairly good match with the experiment. This is the magic of coarse-graining, the art of exploiting nature's important gift: Redundancy! The strategy is by now clear: the level of coarse-graining can be adapted to the level of detail (Complexity) required by the question you ask! The more you coarse-grain, the less you compute, but the higher the exposure to approximation errors, hence the risk of missing the essence of the phenomenon.

This is the art of coarse-graining, the art of striking the best tradeoff between computational Complexity and physical accuracy (see Fig. 13.9).

We shall return to the general principles of coarse-graining at the end of the chapter, but before we do so, let us turn our attention to a second fundamental

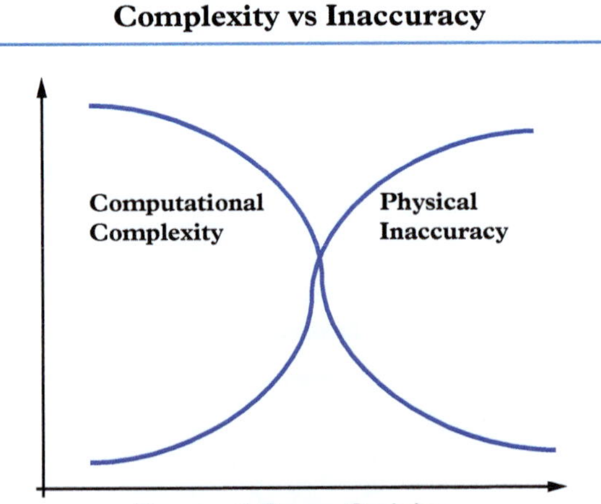

Figure 13.9 *The tradeoff between computational complexity and physical accuracy in the course of coarse-graining. The two opposite trends must be constantly kept under watch and balanced so as to obtain the optimal tradeoff for the problem in point. Such tradeoff is hardly universal and usually changes from one problem to another.*

biological process which benefits immensely from the resort to coarse-graining ideas: protein folding.

13.10 Biological ballerinas: Protein folding

Good players are always lucky

(J. R. Capablanca)

The DNA molecule famously contains the genetic information which is used to code for the actuators of biological life: proteins. Proteins are the life operators, the dedicated workers performing basically all of the vital functions which keep our body going [102]. Originally regarded as 'Cinderellas' at the beck and call of DNA, their paramount role has earned them increasing role and status in modern biology. From a chemical viewpoint, proteins are long chains of molecules, made up by the repetition of a few hundred basic units called *amino acids* (AA), each composed of three base pairs, out of a possible four (see Fig. 13.10). The base pairs are the same as DNA, except for thymine which is replaced by uracil. Hence, symbolically, the amino acids can be regarded as words of three letters out of an alphabet of four. With three letters out of four, one can form $4 \times 4 \times 4 = 64$ three letter words, but it turns out that Nature only employs twenty of them (one less than the Italian alphabet . . .).

Proteins are manufactured in a special piece of supramolecular machinery called Ribosome, from which they exit in the form of a linear chain of AAs, the so-called the *primary sequence*. In such a form, they are gloriously useless. In order to deliver its precious and unique function, the protein must fold into a specific compact shape, called the *native* configuration [102, 63, 42]. The passage from the primary form to the native one proceeds through a series of hierarchical steps (see Fig. 13.10). Without entering the details of this fascinating bio-architectural process, let us just mention that the assembly of the protein goes from the primary structure, basically a backbone of carbon, oxygen, nitrogen, and hydrogen atoms, to form secondary structures, called alfa helices and beta sheets, which further combine into tertiary structures called 'motifs'. This assembly-line entails a highly orchestrated and breath-taking molecular dance, taking from the primary sequence as it comes out of the ribosome, to the native, folded one.

This dance is no joke: failure to fold in the correct configuration may result in major neurodegenerative diseases, such as Alzheimer's. That is why protein folding forms the object of intensive investigation in current biomedical research.

We wish to emphasize that in order to pick up the correct folded configuration, the protein must sail against a monster entropic wind, because there is just one (or very few) native configuration that work, versus an overwhelming multitude of those that don't. This alone tells us that this magical dance must be choreographed by anti-entropic forces docking the protein to the right harbour against the entropic winds. No wonder, water has a crucial role in this plot, as we shall

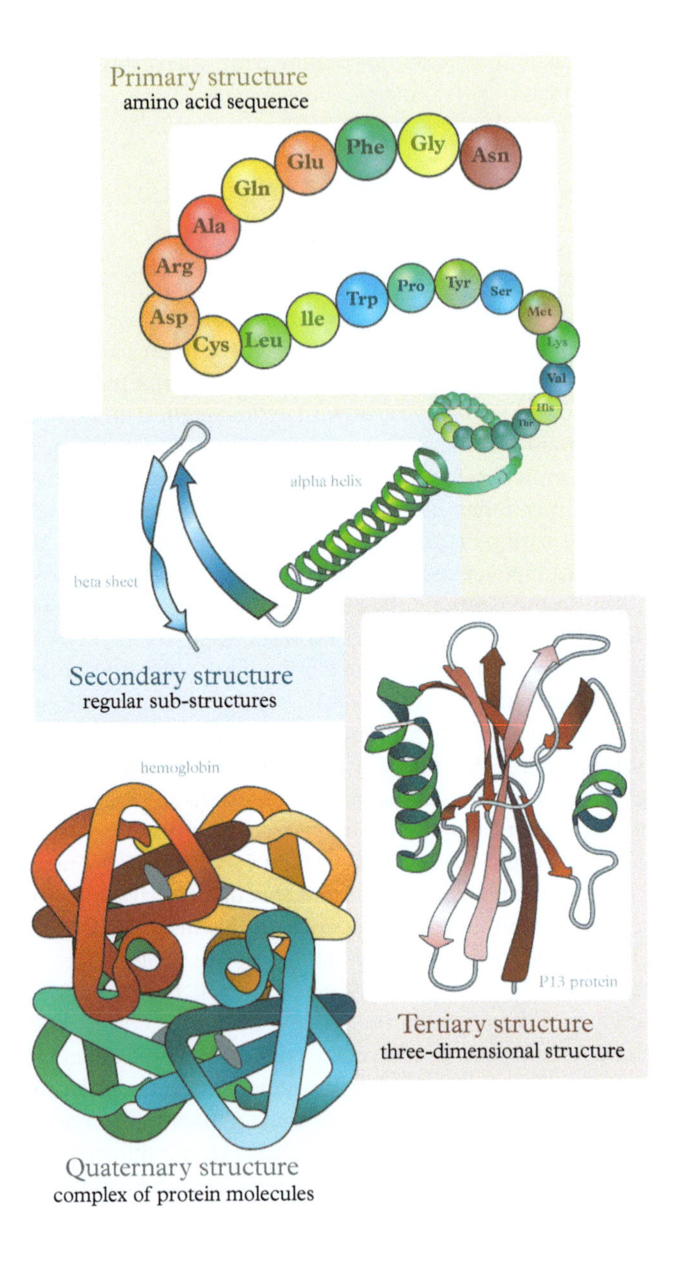

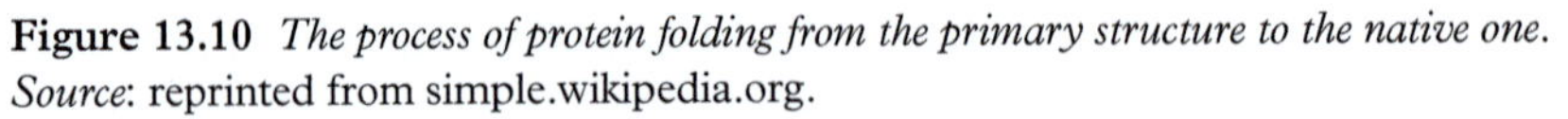

Figure 13.10 *The process of protein folding from the primary structure to the native one.* *Source*: reprinted from simple.wikipedia.org.

discuss in Chapter 16. And when these entropic forces do not succeed, as it is sometimes the case, nature has been provided with extra-help in the form of specialized proteins, aptly named *chaperons*, because they guide proteins along the proper folding pathway, shielding off the hindering effects of other proteins along the way.[43] But let's stick to our mandate and discuss how to coarse-grain the process of protein folding. This is a major topic of modern research; hence we shall only sketch the main ideas.

13.11 Coarse-graining proteins

Like DNA, proteins can be coarse-grained and treated like a necklace of beads, each representing a single amino acid (AA), each coming in 20 possible colours (letters) instead of just four, as it was the case for DNA. The necklace is much shorter than in the DNA case, typically a few hundred instead of three billions, but since each bead now comes in 20 colours, or letters, even a short protein with 100 AAs counts 20^{100} possible configurations, which is basically 1 followed by about 130 zeros, much bigger than the Googol but way smaller than Googolpex. Out of this multitude, however, only about ten thousand pass the test of surviving in our body and doing it good: the logarithmic sieve in action again!

Formally, the coarse-graining of protein folding can be developed similarly to what we did for DNA translocation (see Fig. 13.11). That is, define $f(t)$ as the fraction of beads that have attained their correct position in the native configuration at time t, in a sort of dynamic puzzle. By definition, $f = 0$ in the primary configuration and reaches up to $f = 1$ when folding is complete. Once again, protein folding is a quintessential kinetic process in which the right timing is everything. Based on the previous example, it appears natural to devise approximate free energies for the folding processes, in close analogy with the DNA case.

In practice, f is often taken as the fraction of *native contacts* between any two beads in the actual chain, versus the number of native contacts in the correctly folded one. More precisely, let d_{ij} be the distance between the i-th amino acid and the $j-th$ one. In the folded configuration these distances take a very specific value, call it d_{ij}^{folded}. The order parameter can then be defined as the fraction of distances d_{ij} which match the corresponding folded value (within a given tolerance). This is a popular OP for protein folding, but certainly not the only one. For instance, the diameter of the folding protein also provides useful information on the degree of folding at any given time. We could say that the fraction of native constants is a topological OP, while the diameter is a geometrical-configurational one. Unsurprisingly, the use of multiple OPs enriches the coarse-grained description of the process.

Having emphasized the parallel with DNA translocation, we hasten to add that protein folding is significantly more complicated than DNA translocation. The reason is that proteins are much more flexible than DNA, more active and conformationally richer. For this reason, a single-order parameter is often not sufficient,

[43] The etymology is from the French *'chaperon'*, meaning hood and indicating protection. Later meant as 'protector' especially for the female companion to young women.

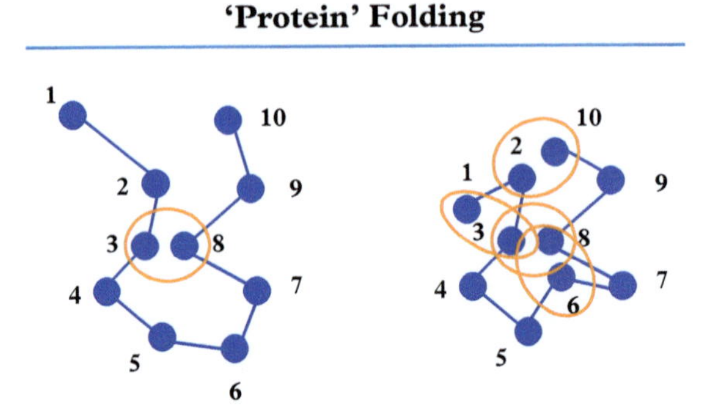

Figure 13.11 *A cartoon of the order parameter in protein folding. The ten-bead biopolymer on its way to the folded state, with one native contact (3–8) out of four native ones, 1–3, 2–10, 3–8, and 8–6. The folding order parameter is $f = 1/4$.*

and more elaborate strategies are needed. For instance, one could partition the protein in a number of sub-domains and define a separate OP for each of them. As mentioned previously, this makes the subject of intense research at the time of this writing and we have to leave it to the specialised literature, focussing instead on a single but crucial question: how does the protein manage to fold within a fraction of a second in the face of the monster entropic barrier which separates the primary sequence from the native configuration?

13.12 Of proteins and monkeys: The Levinthal's paradox

This conundrum, also known as *Levinthal's paradox*, from Massachusetts Institute of Technology (MIT) physicist Cyrus Levinthal (1922–1990), is quintessential Ocean Complexity. And the conundrum is: how does the protein 'know' how to reach its right native configuration? And most importantly, how does it manage do it that fast (fractions of seconds)?

It is readily appreciated that this ordeal takes a breath-takingly clever series of moves, which, if taken at random, would amount to a serious proxy to forever (Alice's question to the white rabbit again). Don't ask me why, but this reminds me one of the Oceans series (no pun, Ocean is the actual name of the main character played by George Clooney!), in which the smart villain (played by Vincent Cassel) performs a series of incredibly clever somersaults to get around a web of rotating lasers, shone across the room to protect the fabulous Fabergé diamond egg. Of course the villain succeeds, and so do proteins. In fact, both have a plan and trained hard around it ....

Figure 13.12 *The monkey on her way to typing Alighieri's Divina Commedia. Dante, left, shows little empathy for his competitor . . .*
Source: reprinted from commons.wikimedia.org and en.wikimedia.org.

Levinthal's question is tightly related to the so-called monkey-on-typewriter paradox, which goes like this. You hire a monkey and let her dance on a typewriter (ok, these days it's rather the keyboard of your desktop) and ask the following question: how long do you need to wait before the monkey types out Dante Alighieri's (1265–1321) *Divina Commedia*, a sequence of slightly over 400,000 characters, (for the record 408,476) (see Fig. 13.12)?[44]

We laugh at such questions and dismiss them as nonsensical, but only until we stop laughing once we realize that if we were willing (should I say able?) to wait long enough, sooner or later, the monkey would actually succeed in reproducing Dante's masterpiece. Strictly speaking, the time it takes for the monkey to type the *Divina Commedia*, is *not* forever. It would be rather later than sooner, though . . . . Just how long this is, can and will be estimated in the next chapter, but for the sake of pathos, let's leave the story here and stay with just plain facts. And the plain facts are that proteins oustmart monkeys by astronomical rates, job done in milliseconds rather than the monkey's eons. How come? How does such magic manage to materialize inside our cells? This is the question we shall address in Chapter 14. But before we do so, a pause of reflection is warranted.

[44] I believe the original version quotes Shakespeare, but let's say I take here the Italian version.

13.13 Why does it work (when it does . . .)?

Coarse-graining is a formidable ally in the enterprise of Sailing the Oceans of Complexity presented by the biological world. And coarse-graining is what most complex systems do all the time on their own to ensure their survival and prosperity in a time-changing environment. In order to do so, they must be equipped with formidable computing skills, and they are indeed. For instance, the cell is a most powerful information processor. But don't be fooled: raw computing power alone won't do. So, the question remains: why does it work?

To attempt an answer, let us go back to DNA translocation: we have described three increasing levels of coarse-graining: from base-pairs to spheres, from spheres to beads and finally from a chain of beads to a single-order parameter, the fraction of translocated beads. We have also commented on the fact that even in the most extreme case of a single-order parameter, we are still in a position of predicting the exponent controlling the translocation time as a function of the chain length. Now, if you think that just a single number cannot get the picture right, you have a solid point indeed.

No matter how smart, a single parameter cannot capture the same information contained in billions of them! The free-energy expression depending on the single parameter f is -not- exact, how could it possibly be? It is just an approximation, but the magic is that oftentimes such an approximation works well enough to provide sensible answers to relevant questions, such as asking how long it takes for DNA to make it through a cellular membrane. The point is that the order parameter is not just any number, it has to be a *smart* one, i.e. one able to capture the essence of the phenomenon in point. Not only powerful information-compressor, but above all a *smart filter*, capable of distilling the relevant information, as we said previously. That's why coarse-graining is an art, as well as a science, for it requires both method and intuition at the same time. One of the two alone would hardly succeed.

13.13.1 The closure problem

Indeed, as detailed in the Appendix 13.1, the task of finding the correct equations for the order parameter is generally one that cannot be solved exactly. No free lunch: the point is that as soon as one endeavours to derive the equation for the order parameter, it becomes immediately apparent that such equation depends on additional parameters not included in the original description. And likewise, as one tries to derive further equations for the additional parameters, new more enter the scene, thus generating a long hierarchy which only ends when the total number of parameters included matches exactly the number of variables one started from! This is a pervasive issue known as the *closure problem*, namely the fact each OP depends on additional ones, along an endless chain that needs to be truncated at some point.

Back to square one? Not really, and here comes another crucial point: *it is not just the number of OPs which matter but their information content* and the way that such information is distributed among them! It so happens that a good OP contains a lot of relevant information, possibly with little left for the additional ones, so that the aforementioned chain can be broken without much damage. In this respect the OPs act literally as the 'information sinews' of hyperland. And missing the right OP may squarely sink the enterprise: if your cat sits on the table and you insist searching for him on the floor, you're doing much worse than the monkey, as you *literally* have to wait forever!

The next crucial point is that the equations which govern the dynamics of the OPs are usually more complicated than the original ones, a natural price to pay for drastically compressing the amount of information. Unsurprisingly, nonlinearity plays a defining role: if the equations were linear, a linear OP would obey the same equations as the microscopic variables. Making the dynamics of the OP as simple as possible, and still realistic, is again part of the artistic component of coarse-graining.

Coming back to the real molecular world, the point is that very often (not always though) a multitude of microscopic details cancel out by the time they cross what we called the 'Boltzmann's bridge', namely as they reach from the molecular to the macroscopic scale, the one (most) relevant to us. If we did depend on the detailed dynamics of the microworld, we would not be here to tell, for we would most certainly drown in Ocean of Complexity. One the reasons we don't is that we are much bigger than the molecules we are made of, hence we don't respond to each of them on an individual basis. In other words, statistics saves us! But there's more to this story, as we shall see in Chapter 14.

13.14 Summary

Biological systems formally inhabit monster-dimensional hyperlands, a dizzying territory which can only be navigated by means of clever information-distilling tools and procedures, which we called coarse-graining. Suitable expressions of free-energy prove key to enable the coarse-graining programme and turn it into a predictive tool to explore the complex frontier between physics and biology. Indeed, finding the free-energy minima in monster-dimensional molecular hyperland sounds like a mission impossible. Yet, it is not, thanks to the fact the macroscopic bodies usually respond to a much smaller number of macrovariables, technically known as OPs, of which we have given a few concrete examples.

This is *Ozland*, the low-dimensional subspace of hyperland described by OPs, where the relevant information is preciously stored and 'dreams come true', like in the magic Land of Oz of Frank Baum's (1856–1919) novel *The Wonderful Wizard of Oz*. Even so, the mystery still lives on, first because Ozland may not exist in the first place, second because even when it exists it may not be easy to find, and third, because even once it is found, navigation across its landscape may still face with a tremendous amount of Complexity, i.e. a very large number of peaks and valleys. Hence, the task of finding the 'Ozland valleys' where things work remains a very challenging one. Yet, Nature has been equipped with 'tools', call them biological GPS, which allow the Ozland valleys not only to be reached, but to be reached on time, often in fractions of seconds, as it is strictly necessary to secure the proper delivery of the biological functions.

This magic GPS is the climax of this book: ready for Chapter 14?

13.15 Appendix 13.1: Order parameter for translocation

Define a score function $H(x)$ like this: if x is negative or zero, the function returns 0: otherwise it returns 1. Note that this is a strongly nonlinear function, a step function jumping from 0 to 1 at $x = 0$. This stands in stark contrast with the barycenter of a material body made of $i = 1, N$ molecules each of mass m_i, which is given by

$$X = \frac{\sum_{i=1}^{N} m_i x_i}{M} \tag{13.4}$$

where $M = \sum_{i=1}^{N} m_i$ is the total mass of the material body. This shows that the barycenter coordinate X is a linear combination of the molecular coordinates x_i, the coefficients being simply the ratio of the molecular mass versus the total mass, m_i/M.

Now consider the position along the x axis of the i-the molecule in the DNA chain at time t, call it $x_i(t)$ and let L be the length of the channel, the mouth being located at $x = 0$ and the exit at $x = L$. By definition, the function $H[x_i(t) - L]$ returns zero if the i-the molecule has exited the channel (translocated) at time t and zero. Next, apply the same test to each molecule in the chain, sum up all the contributions and divide by the number of molecules. In equations:

$$f(t) = \frac{1}{N}\sum_{i=1}^{N} H[x_i(t) - L] \tag{13.5}$$

This is precisely the fraction of translocated molecules at time t, called $f(t)$ in the main text.

The meaning is clear, the compression of information is dramatic, since any coordinates maps to just two values, either zero or one, incidentally a sharply nonlinear operation. Such dramatic compression does not come for free, the price being that while the equations of motion of the particle coordinates are known, the equation for the order parameter is not. Projecting the physical variable by no means implies that its dynamics can be easily projected as well. In other words, the equation which governs the change in time of the order parameter f does not depend on f only, but also on additional variables not immediately apparent from the definition (13.5).

This brings up a very general feature: upon compressing the description to a (much) smaller number of variables, the order parameters, the equations of motion of such variables become *open* and generally more complex. Open means that they depend on additional variables not included in the original compression, thus generating a hierarchical chain of dependencies, in principle as long as the number of the original variables. This looks like a 'Red Queen' stall, a lot of move-around to end up in the same place we started from, but it is not. The point is that truncating the chain opens up a number of imaginative solutions, which is precisely where coarse-graining picks up its 'artistic' component!

14

Free-Energy Funnels

Ignoranti quem portum petat, nullus suus ventus est. (there is no wind for the saylor who doesn't know what haven is heading to.)

Lucio Anneo Seneca (4 BC–65)

14.1 Navigating the free-energy landscape, on time

In Chapter 13 we have illustrated coarse-graining as an effective strategy to navigate across hyperland under the guidance of suitable Order Parameters (OPs). We have also emphasized that, even though smart OPs greatly facilitate the search of the 'Ozland valleys' in the free-energy landscape, this search still is no walk in the park, first because these minima can still be very packed and numerous and also because not only they have to be reached, but they have to be reached *in due time.* In other words, we are dealing with kinetic non-equilibrium processes in which the right timing is essential.

So, time to talk about time!.

14.2 Is time the hero?

Standing on the ledge again. Everybody laughs at dancing monkeys with the typewriter. Not for long though.

(Neil Simon)

Searching for the proverbial needle in the haystack is a notoriously tall order: much tougher, though, is to search for the right *straw* in the haystack! Yet, this is precisely what nature does and in this chapter, we shall discuss how she manages to do so. It is often heard that nature 'works' because it had plenty of time to find optimal solutions through evolution, an idea vividly expressed by Medicine Nobelist George Wald (1906–1997), as he writes:

With enough Time, the Impossible becomes Possible and the Possible becomes Certain.

For all its charm, this sentence is deceptive, as it ignores a big-time elephant in the room, in fact the biggest of all, also known as Mortality .... Let us see why.

Sailing the Ocean of Complexity. Sauro Succi, Oxford University Press.
 DOI: 10.1093/oso/9780192897893.003.0014

14.3 Rare events are rare, not impossible

In the following we shall concern ourselves with rare events which cannot be neglected, because even if they occur very rarely, but when they do, they have a transformative, often devastating, impact. Earthquakes (see Appendix 14.2) are possibly the most immediate examples which springs to our mind, but there are many others, and protein folding is definitely one of them.

So, let us begin from the basics, namely how do we define a rare event? The answer is not unique but let us take a simple route based on the frequency at which it occurs. If a given event occurs *on average* once every t_{ave} time units (seconds, days, years, depending on the phenomenon in question), the number of events recorded within an observational time lapse t_{obs} is, *on average*, simply the ratio

$$N_{obs} = \frac{t_{obs}}{t_{ave}} \tag{14.1}$$

which is strongly reminiscent of the Deborah number encountered before in this book.

Loosely speaking, we can classify events as frequent or rare depending on whether N_{obs} is a large or small number, respectively. By the same token, an impossible event is defined by the condition $N_{obs} = 0$. Based on the relation (14.1), it is clear that this classification depends strictly on the observational time. If N_{obs} is less than 1, no event is observed within the time lapse t_{obs}, which may lead to the conclusion that the event is an impossible one. Is it really like this? Not quite, all this tells us is that the observation time is just too short to detect it.[45] Suppose that, upon doubling the observation time, one such event does indeed materialize: we would safely conclude that the *probability* of observing the event within a time lapse t_{obs} is $p = 1/2$. And if such single event only materializes upon waiting ten times longer, we can safely conclude that the probability of observing the event within a time lapse t_{obs} is $p = 1/10$.

14.3.1 Long enough

You see where I am heading to: if I keep extending the observation time, I may reasonably expect that, upon waiting 'long enough', one occurrence of the event will always be observed. The longer 'long-enough', the smaller is the probability

[45] This point is a bit subtle and warrants some words of clarification. For simplicity, we have referred to the average occurrence time t_{ave}, but clearly there are fluctuations around such an average value, covering a range from, say, a minimum time t_{min} to a maximum time t_{max}. This means that even if no event is expected on average, we might still observe it, due to a statistical fluctuation with t below t_{obs}. Needless to say, such fluctuations are paramount to the behaviour of most complex systems, in which both t_{min} and t_{max} often show dramatic departures from t_{ave}. This is why, in many complex systems, the statistics of extrema is much more informing than the average or the standard deviation, which lie at the root of Gaussian statistics. In light of the this, the notion of an impossible event is more appropriately formulated in terms of t_{min} being infinity, much more stringent than assuming t_{ave} infinity.

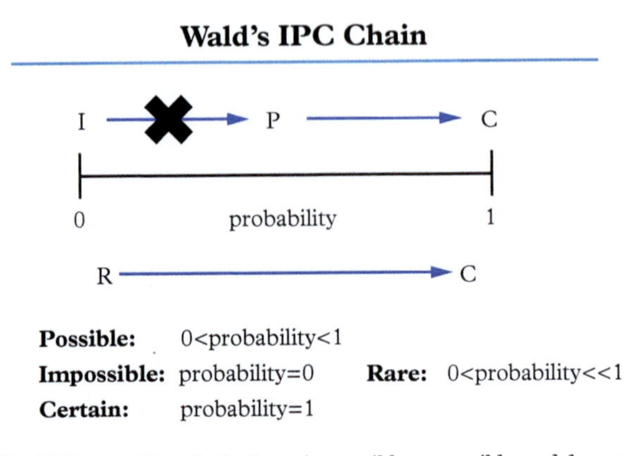

Figure 14.1 *The IPC transition chain from impossible to possible and from possible to certain. How long does it take? Funnels accelerate the PC transition and more precisely they enable the RC transition from rare to certain, but they cannot trigger the IP transition, which remains forbidden in a finite-time world.*

to observe it within the time interval t_{obs}. And when the probability becomes really small, what we have is *rare event*, like winning a lottery once in our lifetime.

But, one story are rare events (very small probability), and a completely different story are impossible ones (strictly *zero probability*). By the previous procedure, an Impossible Event is one that never occurs, no matter how long we are willing (or able) to wait. And, for that matter, we humans, cannot wait forever, remember the Deborah number of Chapter 2? Any observation time shorter than Forever, that is, any *finite* observation time, invariably returns zero.[46] And by zero, we mean a literal zero, not a small number, no matter how small. Bottomline: in our mortal (finite-time) world, rare events eventually do occur, but the impossible remain impossible, period (see Fig. 14.1). For the record, the dire consequences of mistaking rare events for impossible ones are crisply illustrated in Nassim Taleb's books Fooled by randomness and *The Black Swan* [112, 113], with specific reference to Wall Street traders. (See also Appendix 14.2)

As to second part of Wald's sentence, given that things do happen, there are definitely events which turn from possible (probability between zero and one, both ends excluded) to certain, i.e. probability equal one. The point, though, is that they become certain only *ex-post*, i.e. after fact. If we knew for sure that they would happen *before they do*, then they would qualify as certain and not as possible in the first place. What Wald probably meant is the frequently heard 'can-will' statement, namely that anything that *can* happen eventually *will* happen, provided you wait 'long enough'. And, based on the previous discussion, 'long enough' has a very precise meaning, it is just the inverse of the frequency of the event, namely:

[46] I prefer to use forever instead of eternity because eternity is more aptly described as the absence of time than infinite time.

$$t_{long-enough} = \frac{1}{f_{long-enough}} \tag{14.2}$$

If an event occurs 'once in a blue moon', 'long enough' is the blue moon time! As we shall show in a moment, for many crucial biological processes, this 'blue moon time' is a completely impractical wait, the age of the Universe being a fraction of a fraction of an eyeblink in comparison. But let's put numbers where our mouth is, i.e. let us attach a time label to the biological blue moon.

14.3.2 Biological blue moon time

The free-energy landscape of complex systems is fairly non-trivial, multi-dimensional and usually highly corrugated along the relevant dimensions, the OPs discussed in Chapter 13. Had nature been searching for optimal solutions blindly, the entire age of Universe would look like a blink of an eye in the face of the time needed to find the optimal solutions, say the native configuration of a given protein. A back-of-the envelope calculation unrolls the point. Take a modest 150 amino acid chain, each amino acid being endowed with, say, ten conformations, for a total of 10^{150} possible configurations. With a conversion time scale of 10^{-12} seconds between conformations, the time required by an exhaustive search is $10^{150-12} = 10^{138}$ seconds, namely about 10^{130} years. Compare with the 'lowly' 10^{10} years of our Universe, an eyeblink is a close proxy of eternity in comparison .... You now see why, on planet Earth, Wald's 'enough time' is a plain chimaera.

'Time is an ocean, but it ends on the shore', sings Dylan, and that shore comes long ahead of the monkey-typewritten Dante's *Divina Commedia*. Dante Alighieri outsmarted monkeys hands down (it took him twelve years) and proteins far better yet: they accomplish their 'Mission Impossible' in a matter of milliseconds![47] Why? Essentially, because Dante had a plan. At this point, we can deliver the promised calculation for Dante's poem. For the Italian alphabet consisting of 21 letters, one more than the number of amino acids (I discount blanks and other special symbols), the *Divina Commedia* is one out of 21^{408476} possibilities, which we shall call it D, the Dante 'number', for convenience. Like many other monster numbers in this business, D is much larger than a Googol and much smaller than Googolpex. At a rate of one stroke per second, this would exceed the age of the Universe by a factor 1 followed by four hundred thousand zeros ... (the age of the Universe in seconds is approximately 1 followed by less than eighteen zeros ....)

[47] It behoves me to mention modern versions of the monkey metaphor in which the little animal gets a technological upgrade, that is a laptop instead of a typewriter. The key twist is that on a laptop she can do better than typing the full book, i.e. she can write a computer *programme* which would eventually write the *Divina Commedia*. This makes a whole world of a difference, since short programmes can generate awful lots of complex outputs, including perhaps the Divina Commedia. This is at least the thesis put forward by Seth Lloyd in his imaginative and thought-provoking 'Programming the Universe' [67]. I am not aware of any such programme, less so one that would be written by monkeys, no matter how hip their laptop. Yet, the idea is interesting.

As observed previously, Dante did it better, about twelwe years, namely a mere one third of billion (378, 432, 000) seconds, less than nine zeros; and proteins make it in just a handful of milliseconds. Dante and the Oceans villain of Chapter 13, had a plan and trained hard around it: they knew where they were heading to, and they did comply with Seneca's advice. But proteins don't. Unlike Dante and Ocean's villain, proteins do not have a plan, they 'simply' obey the rules of their game, and it is a spectacular smile of Lady Luck that such rules prove capable of guiding their dance through the hyperland landscape by special forces which shape the energy landscape in such a way as to steer them to their target. And pretty fast indeed. In Chapter 15 we shall take a closer look at the physical origins of the anti-entropic forces which make the plan a success against the daunting entropic barriers. For the moment, however, let's indulge a bit longer in a few additional and key features of the landscape.

bear with us.

14.4 Deep corrugations: The funnel

Having glorified the role of OPs as formidable information distillers in Chapter 13, we hasten to observe that, even though OPs might be comparatively few in number, this does not mean that searching for the straw in their haystack is a simple task either. Even in a few dimensions, there could still be very many free-energy minima and finding them, *in due time*, still poses a prohibitive task. In other words, the Fewland landscape might be not smooth but very rugged indeed (see Fig. 14.2). This is a specific instance of the smoothness/roughness duality discussed in Chapter 3. For the sake of concreteness, let us refer again to our folding protein, which must find its way from the unfolded state (U), to the folded one (F), called primary and native structure, respectively. As discussed in the previous chapter, a very natural OP is provided by the fraction f of the folded protein, sometimes also called 'Progress Coordinate' towards the folded state. As discussed in the previous chapter, in the folded state there must be well defined contacts between the amino acids along the native structure of the protein, hence f measures the fraction of such native contacts. Don't miss the key: there are unfolded states galore, while the native structure is basically unique (or a very few), so the sailing from U to F is guaranteed to face an enormous entropic barrier. As discussed regarding Forever, the order is pretty tall: *not only does the protein have to find the right folding path from U to F against the entropic wind, but it also has to move along it pretty quickly*!

In life, as in music, timing is all . . . .

By now, the reader should be inclined to suspect that the reason why nature 'works' is not because it had enough time to search, but because the search is by no means a random one. It's guided, and very cleverly so, by the anti-entropic forces which permit to go around the entropic barriers. But what do such anti-entropic

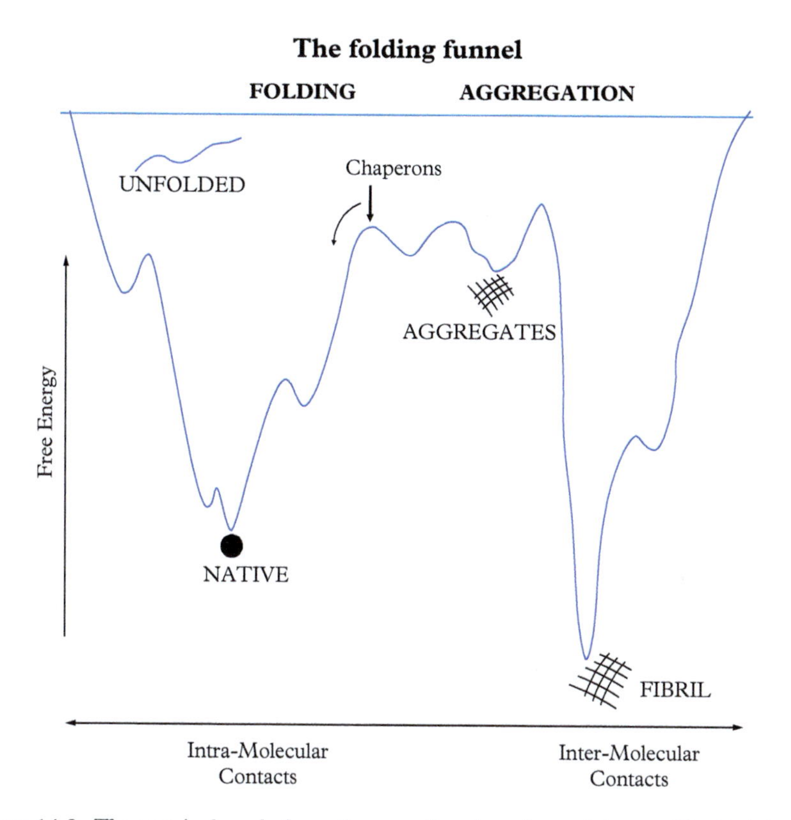

Figure 14.2 *The protein funnel: the native state lives in a deep minimum. The fibrinic state corresponding to mad-cow disease lives in a deeper one. Hence, depth alone does not qualify the 'goodness' of the local minimum. Sometimes folding would not occur spontaneously, in which case special proteins, called 'chaperons' help the protein finding its way to the native state.*

forces look like? Clearly, there must be special features in the free-energy landscape which accelerate the search for 'good' valleys, and do so to an astronomic degree, in order to beat the astronomic slow-down due to the entropic barriers. Simple systems have simple landscapes, with perhaps just a single minimum, like a bowl, and most certainly lack such features.

14.4.1 Fun(nel) time

To convey the idea of the funnel as a portentous accelerator to convergence, let us contrast it with the simple case of a parabolic landscape (see Fig. 14.3) [40]. Imagine convergence as the process of a ball rolling from a given height of the potential down to the minimum, located at zero for convenience. The speed of the ball is proportional to the slope of the landscape; hence convergence is fast far from the target (OP = 0) and gets slower and slower as the target is approached.[48] It is not difficult to show that the time required to reach *exactly* the minimum, $OP = 0$, is infinity. So, perfect convergence to $OP = 0$ on a parabolic landscape takes literally Forever!

Now consider a funnel landscape. Far from the target, the landscape is basically the same, so no difference in the convergence rate either. However, in the vicinity of the minimum, the 'golf-hole' nature of the funnel shows its accelerating power, with a slope which gets increasingly steeper as zero is approached, reverting to zero again only at the very bottom. This means that convergence gets faster as faster as the target is neared and simple math shows (see Appendix 14.1) that by choosing the funnel sufficiently narrow and deep, the ball gets arbitrarily close to $OP = 0$ in a *finite* time: that's how the funnel sidesteps Forever! Despite its patent toy-like nature, this simple example conveys the essence of the idea: *funnels are literal game-changers in the business of accelerating convergence*! Coming back to Wald's I-P-C chain, they provide a tremendous speedup to the P-C transitions, still leaving the I-P one as impossible as ever.

Complex systems typically display very corrugated (rugged) landscapes, with many local minima, each of them potentially hosting a 'good' valley, hence a temporary suspension of the run towards the global equilibrium. The name of the game is quite clear: the ultimate state, the one where free energy attains its global minimum is the state of thermal death. This means that this ultimate state is the one that inevitably wins the race, because nothing can cost less, the ultimate low-cost destination. But in a (deeply) corrugated landscape, the system can lodge on so called metastable states, i.e. local minima, where free energy attains a local balance between energy and entropy, while still preserving enough order/organization to deliver useful functions. So, the art is to navigate through the landscape so as to attain these precious local minima, where Life can flourish long enough to generate further Life. That's how the Life circle goes on in the face of thermal death (see Fig. 14.3).

14.5 Picasso and the proteins

This art takes us to Spain: Madrid, summer 2016. Not that I am not proud of it, but I must admit that I am fond of spending long hours in museums. I am curious but impatient, that's just how I am. My wife and friends in Madrid are much better at that, so I ended up spending almost seven hours in the Prado, not a single minute to be regretted. Among the many masterpieces, my attention was captured by a statement by Pablo Picasso (1881–1973) *Todo lo que se puede imaginar es real*, anything you can imagine is real (see Fig. 14.4). Witnesses, the

[48] This is not exactly what a material ball would do under the effect of gravity, but a good approximation nonetheless if the ball experiences a strong friction proportional to the velocity but opposite in sign.

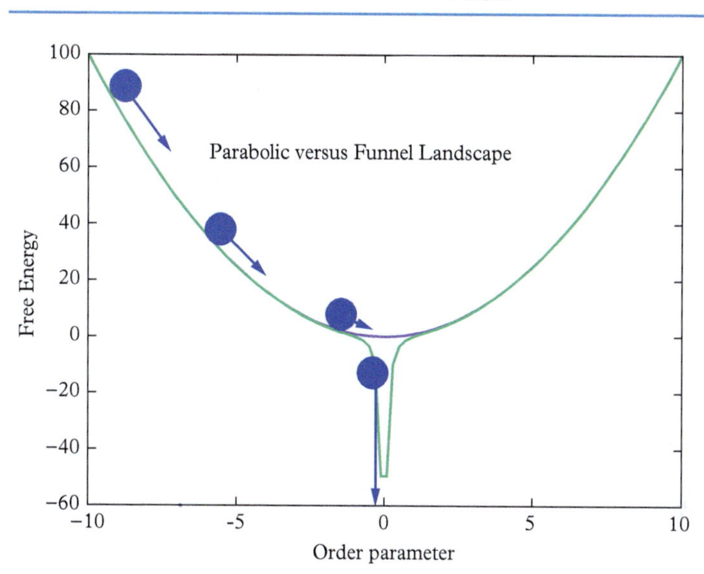

Figure 14.3 *Parabolic versus funnel landscape. An imaginary ball rolls downhill with a velocity proportional to the slope of the landscape. Inside the funnel the ball keeps accelerating until it reaches the bottom, where the slope is zero and the ball stops rolling. In the parabolic landscape the ball takes an infinite time to reach the bottom, while in the funnel landscape this time is finite and actually pretty short, depending on its steepness.*

Figure 14.4 *'Dora Maar au Chat' (left) and Mother and Child (right), by Pablo Picasso. The former is an improbable creature, at least on planet Earth.*
Source: reprinted from en.wikipedia.org and wikiart.org.

beautiful paintings: '*Dora Maar au Chat*' and 'Mother and Child' (see Figs. 14.4). The first is an improbable, albeit somehow fascinating, surreal assembly of shapes, with no parallel in any existing creature on this planet, the latter is a sweet portrait of a mother and her child, as we may imagine many in the real world. *Dora Maar* is the unfolded protein, *Mother and Child* is the folded one. Picasso had many models, but I've my doubts he ever thought of them as proteins!

Picasso probably didn't mean the psycho-virtual reality that drives much of current society, finance in the first place, but I see some truth in his sentence. For one, when we imagine good things happening to us, we improve our feelgood factor, which certainly helps us in many concrete respects. That's very real, at least it works for me, and I think it does for many others. Although I never had the chance to ask him in person, I imagine what Picasso meant was that his mind could fly high, free from physical constraints. Another not quite minor painter seemed to express similar feelings: 'I dream of painting and then I paint what I dream', said once a certain Vincent van Gogh (1853–1890). Dreams and Imagination stand as the epitomes of unconstrained activity, the mind can fly faster than light into new worlds and territories with no counterpart in the physical world. That's how many masterpieces take shape. And if you think that this applies only to genius artists, you may want to listen to Einstein: 'Imagination is more important than knowledge, because knowledge is limited, while Imagination encircles the world'. No limits, the highway to discovery.

But is it really like that? Is unconstrained freedom always the best route to discovery? Maybe, but certainly not in the physical and biological world. Let us spell out the point, by illustrating the constructive role of constraints.

14.6 The principle of minimal frustration

Do not remove a fence until you know why it was put up in the first place.

(GK Chesterton).

'Freedom is just another word for nothing left to lose', used to sing the American song-writer Janis Joplin (1943–1970): if you have nothing to lose, you feel free to go wherever and whenever you want. Cool as this sounds, sooner or later, however, this is going to clash with the very same wish by one of your fellow citizens, for, even if you may have nothing to lose, they eventually might .... This brings up a word we'd rather not like to hear, constraints. Few constraints or many, the fact remains that whenever there are more constraints than available moves, we are frustrated because we cannot take the path we want and must settle for compromises instead. Hence, constraints stand for *frustration*. The same word is used in the scientific context, to express situations where multiple conflicting requirements cannot be reconciled in full, and some must necessarily be surrendered. One line of thought maintains that many biological systems, for instance proteins, obey a so-called *principle of minimal frustration*: given that you cannot match all the constraints you are facing, the best you can afford is to minimize your frustration. It makes perfect sense after all, and it turns out that free energy

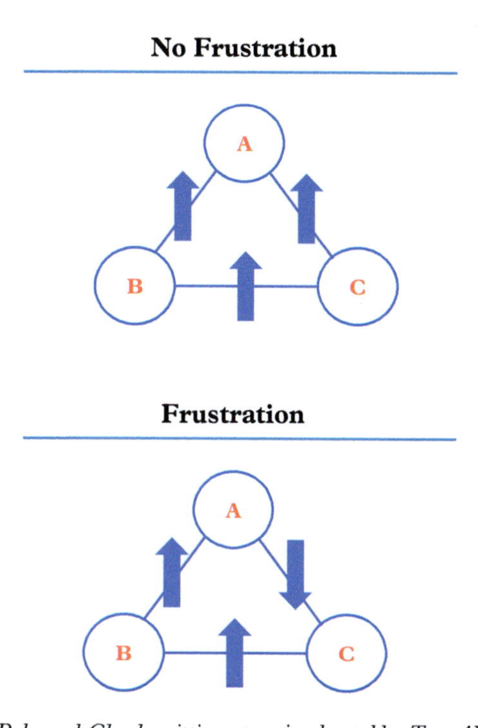

Figure 14.5 *Alice, Bob, and Charles sitting at a circular table. Top: Alice likes to sit next to both Bob (arrow up) and Charles (arrow up) and Bob and Charles like to sit next to each other (arrow up). All constraints AB,AC,BC are met, and everybody is happy. In the following: Alice still likes Bob but not Charles (arrow down), while Bob and Charles still like each other. Bob and Charles are both happy, but Alice is not. There is no way everybody can be happy on this table, frustration in action. Things can get more complicated if we break reciprocity, say that Alice likes Bob, but Bob does not return the feeling. In this case there would be two arrows on the AB link, one up and one down, opening up more patterns of negotiation and frustration.*

provides a very natural means to measure this frustration. Let us clarify with a simple but representative example (see Fig. 14.5).[49] Alice, Bob, and Charles have dinner together and say that Alice likes both Bob and Charles, so she is happy to sit next to both. If Bob and Charles, besides liking Alice, also like each other, they are both as happy as Alice: all is well for everyone, beer is cool, and life is good: zero frustration. But how about Bob and Charles still like Alice as well as each other, but Alice only returns to Bob and not to Charles. It is clear that in this case

[49] This example is often used in a modern condensed matter in relation to spin-glasses, an important class of models of disordered magnetic systems, which has found multiple applications in many different fields, particularly constrained optimization and in the study of neural networks [105]. However, the example was first introduced by the psychologist Fritz Heider under the name of 'Balance Theory' in the context of social psychology (F. Heider, *The Psychology of Interpersonal Relations*, John Wiley and Sons, (1958). I am indebted to Stuart Kauffman for pointing this out to me

there is no way to arrange a dinner for three where everyone is happy. Bob and Charles are both happy, but Alice is not because there is no way she can stay apart from Charles.

14.6.1 Stravinski's constrained freedom

Believe it or not, nature does the same and often to our major benefit. Another remarkable man, Igor Stravinsky (1882–1971), says it best: 'The more constraints one imposes, the more one frees one self. And the arbitrariness of the constraint serves only to obtain precision of execution'. Constraints speaking for constructive freedom, interesting, isn't it? I really like the idea, to the point that, if I could, I would promote it to the status of a principle: the *Stravinski's principle of constrained freedom*. The physical world, as we shall see, fits in well with Stravinsky's principle of constrained freedom, and very happily so. In actual fact, our own existence is an exquisite result of this principle, i.e. the physical constraints which filter *Dora Maar au Chat* away from 'Mother and Child'.

Notwithstanding Stravinski's principle, constraints are not particularly popular, as they are normally perceived as a hindrance to our freedom. Many powerful mechanisms are constantly in action in modern society to remove as many of them as possible, heralding this as a hallmark of progress and liberation. Not so fast, though.

When you drive on a scenic route in Corsica or Monte Carlo and enjoy the breathtaking scenery, you're probably grateful that somebody had the clever idea of laying down a guard-rail along the edge, to prevent your ecstasy from throwing you down the ravine, don't you? This is especially true if you drive, as I did in Summer 2000, counterclockwise, i.e. with the sea on your right hand driving side and the centrifugal force happily pushing you towards the sea, and the ravine .... Well, let's face it, the guard rail *is* a constraint, it limits your freedom, but presumably not in a way you see much point in complaining about (see Fig. 14.6). And the guard-rail analogy sits well with biology: *most diseases are triggered in the wake of a failed constraint.*

For instance, cells are equipped with physico-chemical constraints which inhibit their growth and replication, until a triggering signal is received. A mutation compromising such regulatory constraint may cause the cell to proliferate without receiving any signal, with the dire consequences we can easily imagine. Much of current drug research centres on the search for molecules reinstating failed constraints. So goes the story with constraints; we don't like them, until we realize that we do, because without them, we would run in troubles long before we know, simply because they protect us against our own weaknesses and mistakes (after all, we are all humans)! As mentioned previously, nature and Biology in the first place, are littered with such protection mechanisms, often going by the name of 'feedback loops'.

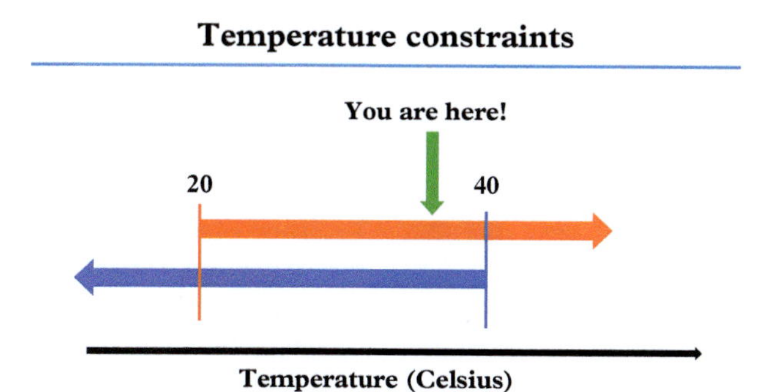

Figure 14.6 *Constraints on the temperature of the human body. Below about 20 degrees Celsius (severe hypothermia) life is threatened by cardiovascular constriction. Above 40 (high fever), the threaten is disruption of the enzymes that control the proper working of metabolic functions. Happily enough, the two constraints match in the window between 20 and 40 degrees Celsius. Frustration of these constraints would close the window and be incompatible with life in flesh and blood as we know it.*

Undertake an action, measure how much off target it went (Error) and correct to undertake a second, hopefully better one. A process called *Learning*, in particular from mistakes, a crucial component of the art of living altogether. For a series of beautiful professional works on protein funnels and the principle of minimal frustration, see the landmark papers [34, 124, 15].

14.7 Broken constraints

It turns out that, as indicated by Stravinski, complex systems improve their performance under the pressure of multiple constraints (sometimes also known as 'anti-fragility').

Simple systems are usually under much less pressure than complex ones and in no need of anti-fragile behaviour. It is well known that when you are under pressure, chances to make mistakes grow pretty fast. Whenever, by intention or not, you break a 'useful' constraint, you make a mistake.

The natural question than is: what is the price of these mistakes? Some mistakes can be deadly, but, happily enough, most are not and can therefore be turned into opportunities to avoid further mistakes for the future. In order to implement such self-improving strategy we need mechanisms to detect errors and correct them. Indeed error detection and correction (EDC) is one the prime strategies by which complex systems manage to adapt to a changing environment, reinforcing themselves in response to new challenges raised by such changes (remember the complex adaptive systems discussed in Part 1).

The perfect world 'no-mistake' condition:

$$Error = Zero \tag{14.3}$$

is the most stringent constraint of all. In social terms we may call it ZT for 'Zero Tolerance' to errors. It is very rare that ZT can be met by any real-life system, if possible at all. The interesting point is that this is often not desirable either, because systems without any tolerance are usually exposed to fragile (no-warning) disruption. Hence, getting *as close as needed* to ZT, is the key to deploying effective survival and anti-fragile strategies. Tolerance is the keyword here.

A major biological case in point is deoxyribonucleic acid (DNA) replication, in which the DNA of the mother cell must be replicated and transferred to the daughter cells. This implies a copy of the entire genome, namely three billions base pairs, each time the cell divides. Clearly, some errors inevitably occur in the process, which can be scored at 1 in 100,000. This may seem pretty close to ZT, but it really isn't, since with three-billions base pairs, it amounts to tens of thousands of faulty insertions in each copying event. Life is way more demanding than that! Fortunately, cells are equipped with sophisticated means of fixing most of these mistakes, some immediately during replication through a process known as *proof-reading* (as we do with our writings, including this book), while some others after replication, in a process called *mismatch repair*. Proofreading can fix over 99 per cent of these errors, which is pretty good, and yet still not good enough.

Without delving further into the details, the point is that the ZT condition is more realistically replaced by a softer one, namely

$$Error\ rate < \ Tolerance \tag{14.4}$$

In the previous calculation, tolerance is the real-life replacement for zero, in fact a number much, much smaller than 1 and yet not strictly zero. Its actual value depends on the specific phenomenon in point, but we are usually talking very small numbers. In other words, errors must be rare events. Coming back to our free-energy landscapes, it should by now be clear why physical constraints take the form of barriers: they reflect the penalty incurred in moving from one place to another in the free-energy landscape. The bottom line is this: since most valleys of the free-energy landscape speak death or doom, *constraints*, rules which negate the access to these regions or make them very hard to reach, are not only welcome, but essential to our survival. Constraints are our guardian angels.

Now we start getting the sense of the story: deep corrugations, typically the outcome of competing molecular interactions, are not only healthy, but essential, in that they build up the fences which prevent the downhill run to death valley! What we need for long-lived stays are deep free-energy pockets where we can prosper, and, as we mentioned previously, it is far from trivial that such deep and *good* pockets should exist at all in the first place! Such deep pockets, both good and bad, are commonly referred to as *funnels*, a self-explanatory name indeed. Only major perturbations can take the 'rolling ball' away from a deep funnel: once the 'ball' enters the funnel, the 'fatal attraction' becomes larger and larger, because the funnel gets steeper and steeper as its bottom is approached. More often than not, the funnel is *hierarchical*: once a proximity of the bottom is attained, the very structure of the funnel is such to progressively facilitate the access to even better positions.

This is what specialists call a *convergent search*, and this is how nature beats combinatorial Complexity and finds not just the needle but the right straw in the haystack!

Coming back to proteins, this means that the native configuration is not reached in a single shot, but through a sequence of structural rearrangements instead. Such rearrangements, also known as 'conformational transitions', mark substantial 'jumps' of the protein configuration towards its native state (see Fig. 14.7).

Convergent search is paramount to biological structures, hence the funnel appears to hold one the most precious keys to the success of biological systems. Thus, in many respects, the *funnel is a made-to-be-found* destination!

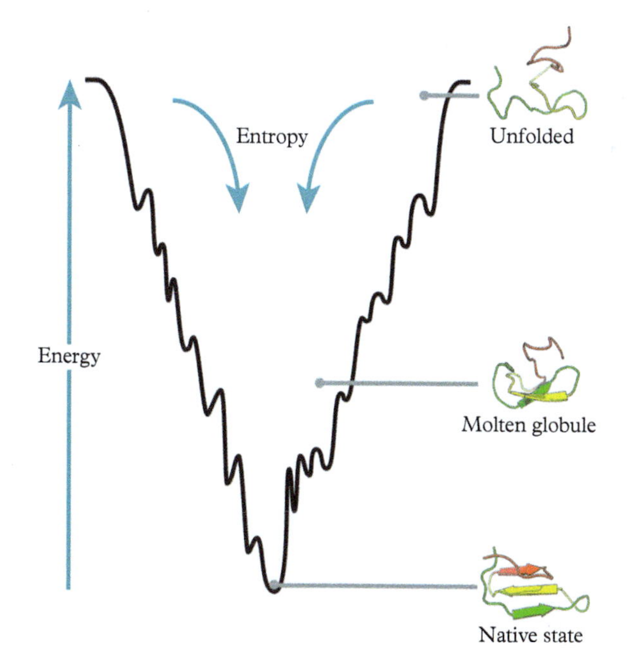

Figure 14.7 *The various stages of the protein folding process.*
Source: reprinted from en.wikipedia.org.

14.8 The good and the bad funnels

Happy end? Not so fast, life is not so easy.

As we mentioned earlier on, not any funnel will do, i.e. not any funnel would take to a comfortable oasis. In fact, most of them don't! Instead, they take to near-death-valleys, mad-cow and other neurological diseases. What distinguishes a good funnel from a bad one? The answer to this question varies from system to system, but in general it has to do with specific values of the Order Parameters. For instance, in the case of Fig. 14.7, a main OP is associated with the ratio between intra and inter-molecular contacts between the molecules which form the protein. Such ratio affects the way proteins interact with each other within the cell. In the native state, each protein preserves its individuality, without forming assemblies with other proteins, and meets the right conditions to deliver its function. This state is described by a 'good' funnel. A particular nasty funnel instead is the one associated with the formation of amyloid fibrils, namely ordered assemblies of proteins which emerge once (to put it simply) the inter-molecular contacts prevail over the intra-molecular ones. Within such ordered aggregates, the single proteins are no longer in a position to deliver their vital function, thus paving the way to serious neurological diseases. As one can see from the Fig. 14.2, the bad funnel is narrower and deeper than the good one. This means that it is harder to reach but also much harder to escape if you have the misfortune of falling within.

14.8.1 Beacons in The Oceans of Complexity

The funnel provides a fairly powerful geometrical picture of what we might call a 'Beacon in The Ocean of Complexity'. Searching in The Ocean of Complexity is not random, it is driven by the energy-entropy duel/duet, materializing in free-energy funnels driving towards the valleys where vital functions can be delivered. The funnel must be 1) *hierarchical*, to be found quickly, 2) *deep*, to offer sufficient stability and 3) *good*, properly located on OP space to deliver useful functions. 'Good' funnels are all but a given, in fact yet another wondrous gift of mother nature, as there is no a priori reason why they should exist in the first place.

Now that we have glorified the (good) funnel, the reader may well wish to make its closer acquaintance, possibly even have a direct look at it. Don't be disappointed, but ... a microscope would not do! We don't see funnels with a microscope, because they do not exist in any material sense, not in the way we use to define things that we can see or touch. The funnel may escape our senses, but not our mind. As mentioned previously, the free-energy landscape is a conceptual construct, a mathematical abstraction which permits us to answer relevant questions without drowning in the ocean of atomistic and molecular Complexity. In Chapter 15, we shall take a closer look at the physico-chemical substrate of the funnel, i.e. the molecular interactions which shape the free-energy landscape in such a way as to support life on borrowed time from the Second Principle of Thermodynamics. Before doing so, a few words on the navigation strategies which allow nature to travel across the landscape and land in the right funnels, are in order.

14.9 Navigation routes in the landscape

Having stated the case for constraints, the next question is: how do biological systems navigate across them? How does the system, our beloved protein, 'know' how to travel towards the hospitable free-energy minima, slalomizing around the crowd of inhospitable and often dangerous ones? Appropriate search strategies must be first learned and then practiced.

14.9.1 Down the steepest slope

A very natural search strategy is the so-called *steepest descent.* Here it goes: you are up in the mountains, in the darkness, and you have to make it home, back to the valley, possibly with the inviting perspective of a warm fire and inviting glass of vin bruleé both waiting for you. It's dark, but much to your relief, you have a torch with you, to illuminate the immediate vicinity, and nothing else. The most natural strategy is to take a step in the direction with the steepest downhill slope and if you're lucky and patient, step by step, the steepest descent policy should take you home. But does it really do this for sure? A minute's thought reveals that this simple policy comes with a number of serious limitations, though. First, it assumes that there always is a descending slope, which is true only if the landscape is smooth and concave, like a bowl with no bumps. In a smooth bowl, there always is a descending direction taking to the bottom, and the steepest descent is guaranteed to take you there. But, how about corrugated bowls, with many local minima? It is immediately apparent that if ever you end up in one such local minimum, the short-sighted steepest-descent strategy will never take you out: you are permanently trapped there (see Fig. 14.8).

This may or may not be a good place to be, but most of the time it is not, because 'unhealthy' valleys vastly outnumber the 'healthy' ones: many valleys are just death valleys, or close proxies thereof. In this case, survival commands you to get out quickly, and steepest descent is helpless at that: a new and smarter policy is needed.

14.9.2 With a little help from our molecular chaos friend

One such smarter policy goes as follows. Instead of insisting on searching a descent direction, which may fail to materialize, best is to accept a temporary loss, namely go *up-hill,* because this is the only chance you have to leave the local

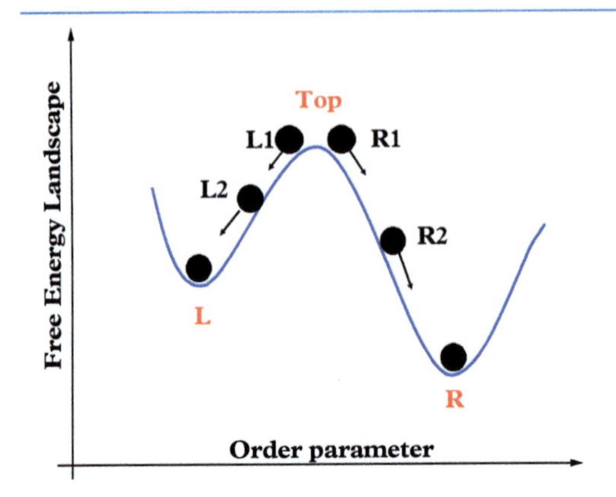

Figure 14.8 *Trapping in a local minimum. A landscape with two local minima, L and R, R being also the global one. Under a steepest descent policy, a ball starting at the top right (R1) ends up in the right local and global minimum R. The steepest-descent policy works, because the path never meets with any corrugation. Under the same policy, a ball starting at the top left (L1) would end up in the local minimum L get stuck there, since any move from L would face an uphill slope, hence would be rejected by the steepest-descent rule. The only way the system can work its way out of the local minimum is to accept uphill moves, bringing it back to the top and hopefully taking the right path therefrom. The downhill move is dictated by the forces acting on the particle, hence going uphill can only be achieved thanks to thermal fluctuations: the constructive face of molecular chaos in full action!*

minimum behind and explore new, hopefully better, regions of the landscape. This strategy is called *ungreedy* because it accepts a temporary loss (the uphill move) in view of a future potential benefit, namely new and hopefully better valleys. Such an ungreedy and far-sighted strategy proves indeed essential to the succesful navigation of rugged landscapes, as opposed to the greedy and short-sighed steepest-descent policy. Fair enough, but who 'tells' the protein to go uphill and actually help it doing so? This is precisely the point where molecular chaos comes to rescue. It is only thanks to thermal agitation, microscopic chaos as we have learned before, that the trapped system stands a chance of leaving local death valleys! If the valley is deep, it will take a 'big kick' to leave it behind, and the system takes a significant time before it can escape and visit some other valley. Conversely, if the valley is shallow, even a small thermal kick will be enough to quit it. So, the temperature must be carefully tuned to serve us well (see Fig. 14.6), and thankfully enough, it really is! And since we have learned before that temperature is the emergent signature of microscopic chaos, the bottom line is that, yes, successful navigation strategies across complex free-energy landscapes, owe a lot to chaos indeed!

This is a concrete example of what the French biologist Jacques Monod (1910–1976), 1965 Medicine Nobel, called *'Le hazard at la necessitè'*, *Chance and*

Necessity [74]. The free-energy landscape is the Necessity and temperature is the macroscopic face of chance. They are both are essential ingredients for biological success.

We have commented at length on the grim face of the deepest valley of all. The fascinating game is to figure out whether there are other sufficiently deep valleys, where we can prosper long enough to get something useful done, like generating offsprings, writing songs or books, playing piano, doing physics, or whatever pleases us best. It is yet another great gift of mother nature that biological landscapes don't look anything like a smooth bowl (see Fig. 14.7). If they did, we would all be stampeding for widest door of all, thermal death. They are complex and corrugated, often with exponentially many local minima, and, most importantly, only a precious few are good enough to host life. That's our chance for life on borrowed time, and, as we just discussed, we owe it to both constraints and chaos alike. You may or may not share the feeling, but I find this moving, for nothing is foregone here: *chaos, constraints, funnels,* they all conspire together to defeat an otherwise unconquerable entropic mountain and accomplish the mission impossible task we call life.

Back to our basics: how does this magic happen in concrete terms? What are the actual forces and mechanisms which carve the funnel out of the entropic mountains? This is the subject Chapter 15. But before we go into this, please, bear with us another little while, since we have another very interesting guest in the room: non-equilibrium.

14.10 Protein folding revisited

As mentioned before, a quantitative understanding of the process of protein folding makes the subject of intense research. Without entering the details, let us portray the main lines.

The first thing to realize is that the conformational changes which take the protein from its primary configuration all the way to its folded state, fully qualify as rare events. To convince ourselves let us just estimate the 'blue moon time' of protein folding. The folding path is ultimately dictated by the interatomic forces between all the atoms in the protein. The typical timescale of these interactions is the femtosecond, namely a millionth of a billionth of a second. The folding time, on the other hand, ranks in the order of milliseconds, which means that the atomic clock must tick about one trillion times before a conformational transition occurs. In other words, the protein spends the overwhelming majority of its time performing little vibrations of no biological significance, only to go through the all-important configurational changes once every trillion moves!

How do we model this?

14.10.1 Multiple-Order parameters

Ideally, one would like to describe this behaviour by means of a single OP, such as the fraction of native contacts, as discussed in Chapter 13. This makes plenty of sense, but with rare events in the room, it often falls short of providing a quantitative description. The reason is that conformational changes imply major structural transitions of the protein configuration, say from a linear chain to a two-dimensional beta sheet. This means that an OP which suits the linear chain may not be equally apt to describe the beta sheet as well. If so, such an OP may fail to detect the transition altogether!

Going back to the cat-on-the-table analogy of Chapter 13: your cat is probably spending most of his time in a few comfortable locations somewhere on the floor of your house. Hence, the coordinates of the floor are good OPs to track him most of the time. Occasionally, however, whether you like it or not (and I doubt you do . . .), he might decide to jump on the table, and the very moment he does, the floor coordinates become gloriously useless. In order to detect the cat's jump, you need an additional OP, the elevation upon the floor. Without such an extra coordinate, your cat is gone, out of your radar! This is why the study of protein folding using multiple OPs is a very active area of protein-folding research.

Another active route is to proceed via coarse-grained molecular dynamics, as we discussed for the case of DNA translocation. This is a pretty intensive task, especially on account of the very large number of timesteps involved, but the last few decades have witnessed enormous computational progress along this direction. Just to give an idea, ANTON, the fastest present supercomputer, built on purpose by Wall Street billionaire David Shaw for protein simulations, computes about 20 microseconds of a protein-water system of about 25,000 atoms, in a day [2]. This means 50 days for a millisecond protein lifetime. Big time, but doable, if you have access to ANTON . . ..

Problem solved? Not quite, and for many reasons.

First, these simulations do not contain the same number of atoms of the real system, in fact just about one tenth. Hence, they must still rely on coarse-grained approximations of the interatomic forces, potentially a significant source of inaccuracy. Second, one single trajectory is hardly enough to collect sufficient statistics, hence the calculations must be repeated for several trajectories, in principle hundreds of them, which takes us from months to several years, even using ANTON. The current tradeoff is to perform a few such long-time simulations and use them to gain knowledge of the energy landscape, a procedure known as 'sampling' in the scientific circle. In other words, the idea is to use the simulation to explore the energy landscape, the way the 'ball' in Fig. 14.8 does, except that the 'ball' explores a much higher dimensional one. And while you explore it, you keep track of its topography, which is what sampling is made for. The sampling of protein free-energy landscapes represents a magnificent example of synergy between computer simulation, applied math and statistical physics: the use of the most powerful computers to track dynamic trajectories and convert this dynamic information into probabilistic prediction of the conformational changes. A very elegant, sophisticated and powerful tool to investigate the Complexity not only of the protein folding process but of many other problems with rugged landscapes.

No relativity, no quantum gravity, we are plain into the Newtonian-looking 'Good-Old-Physics' scenario discussed in the Preface. Yet, this is frontier science no less: simulations, deep statistical physics and sophisticated math, are all in action here! Proteins don't 'think', they don't take any decision but simply abide by the rules given to them, hence they don't qualify as complex adaptive systems (CAS). But, I hope that by now my reader sees that the rules are deep and subtle, hence full of fascinating Complexity, nonetheless.

14.11 The gifts of non-equilibrium

This chapter was all about the glory of free energy, its dizzy landscapes and smart strategies for navigating across dangerous peaks and valleys. Dear reader, disappointment time! Strictly speaking, free energy makes sense only for systems in equilibrium, i.e. whenever competing effects are in (statistical) balance! But didn't we surmise that most complex systems live far from equilibrium? So, what's going on here?

The good news is that free energy is a very robust notion and keeps serving us well as a guiding light also far from equilibrium, provided that the handy landscape picture is revisited and properly generalized. The point is that there is more to biological forces than the slopes of landscapes (the technical term is 'gradient'). Out of equilibrium, the mountaineering analogy we so heavily drew upon in this book, no longer holds.

14.11.1 From mountains to quicksand

And the reason why it does not is simply because, unlike mountains, non-equilibrium landscapes are not static but 'plastic'. They are affected by the motion of the walkers, on a time scale comparable with that of the walker's motion. This is yet another instance of chicken-egg scenario, where cause and effect can no longer be told apart: the walker changes the landscape that guides its motion, which means that cause and effect enter a circular causality loop. Second, and perhaps even more fascinating, the shape of the landscape, hence the ability of walkers to go from place to place, may depend crucially upon the *rate* at which they move, like a sort of quicksand: if you don't move fast enough, you sink! Certain targets can only be reached by walkers who move fast enough: doors that must be passed before they close down forever, trains that cannot be caught twice.

14.11.2 Burnt bridges and cryptic pockets

The upshot is that far from equilibrium, the landscape offers extra features, hence chances, for optimal search strategies. However, these extra-chances are ephemeral, they are available only within a limited time window; like any special offer, they come with a deadline attached! Temporary bridges ('kinetic pathways' is the technical term) which are not available at equilibrium, simply because they stem precisely from the unbalance between the various forces in action.

Hence, once the competing effects come to a balance, these bridges are gone! The language is metaphoric, but reality responds to this call. A concrete example is provided by the so-called *cryptic pockets*, namely dynamic cavities which open up within the protein surface in the course of its 'dance' and close down shortly later (see Fig. 14.9). Energetically, they corresponding to metastable configurations associated with local free-energy minima and their practical interest lies with the possibility of binding them to small molecules (ligands) which may eventually affect the protein function, typically inhibiting undesired (mal)functions, such as binding to an attacking agent (pathogens). This is the way new drugs are discovered and designed.

14.11.3 Trains that don't pass twice

The fragility of non-equilibrium, its ephemeral nature, is clearly a liability but one that also carries a deep charm with it. There is indeed a definite romance to non-equilibrium, a sense of inspired and ephemeral productivity which goes with chances that do not show up twice. As somebody wrote, life is writing on a blackboard without any eraser.

Vasco Rossi, an Italian rock-superstar (the man could collect over two-hundred thousand people in the Modena Park concert, the world record, they tell me, for a one-man show) sings it like this:

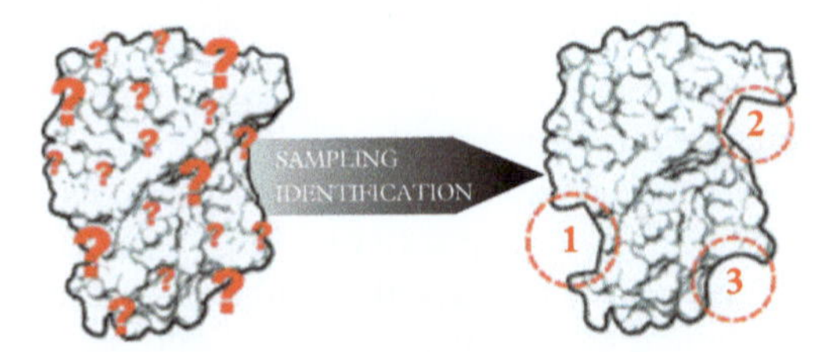

Figure 14.9 *Cryptic pockets on a protein surface (circled regions).*
Source: reprinted from Denis Schmidt, Markus Boehm, Christopher L. McClendon, Rubben Torella, and Holger Gohlke. Cosolvent-enhanced sampling and unbiased identification of cryptic pockets suitable for structure-based drug design. Journal of Chemical Theory and Computation, 2019-04-18, *DOI: 10.1021/acs.jctc.8b01295.*
Source: with permission of acs.org.

> Le nostre canzoni son come i fiori, devi scriverle in fretta, se no spariscono e non ritornano più.

'Our songs are like flowers, you must write them down quickly, otherwise they fade away, never to come back again' .... The same author also sings 'Life is an equilibrium over folly', which goes more or less along the same lines, except that biology possibly works the other way around: life is folly (non-equilibrium) over equilibrium. Equilibrium is about *where* we reach to stay and function, non-equilibrium informs about *how* we get there and actually *how fast*. The former is the key to delivering stable functions, the latter sometimes reveals more of the system's proclivity to imaginative changes: the scout of the group. And, as we have seen in Chapters 12 and 13, biological targets are either reached on time or in vain. As with people, the way they reach to their targets tells a lot about their style and personality, sometimes even more so than the goals they actually achieve. The trip (non-equilibrium) has a deep value in itself, regardless of the destination (equilibrium). It is sometimes heard that the journey matters more than the destination: I am not sure I subscribe in full, but I sense some taste of truth in it. With a little concession to self-indulgement, the way I like to put it down is this: *at equilibrium things work, out of equilibrium they discover new worlds.*

Having, I hope, illustrated concrete examples of the crucial role of free energy in modern research at the cross-border between physics and biology, I would proceed to close this chapter with some general considerations again on the place of free energy in the framework of the Science of Complexity.

14.12 The quark, the jaguar, and the free-energy principle

In the Preface, I commented that the Complexity literature does not pay much homage to the main character of this book, free energy. An adamant example is offered by Murray Gell Mann's deep and highly influential book, *The Quark and the Jaguar* [46]. At page 119, the author comments as follows:

> One of the great challenges of contemporary science is to trace the mix of simplicity and complexity, regularity and randomness, order and disorder up the ladder from elementary particle physics and cosmology to the realm of complex adative systems.

This sentence stands as tall today as it did at the time of its writing, nearly three decades ago. It is therefore intriguing and somewhat puzzling to note that free energy, a most prominent concept in pursuing the ultimate goal envisaged by Gell Mann, does not receive a single quote in his book. The book makes frequent mention of the somewhat related notion of biological fitness, but never free energy, not once. It is perhaps interesting to observe that in just about the same years when the quark and the jaguar was published, a group of inspired researchers, among others Peter Wolynes and Joseé Nelson Onuchic developed a very elegant theory of free-energy landscapes for protein folding and the attendant principle of minimal frustration discussed earlier on in this chapter [34, 124, 15].

And to conclude, free energy has been taken even more radically by the famed British neuroscientist Karl J. Friston, who came to point of elevating it to the status of a *free-energy principle* (FEP) [36, 38]. Friston's FEP formulation is not a walk in park even for the adept, but the idea is fascinating and worth conveying. Using the author's own description [37]:

> We have tried to show that free energy minimization may be an imperative for all self-organizing biological systems and speculate that the attending biological insights may generalize beyond the neurosciences.

These are pretty audacious statements: whether they will lead to lasting advances in the study of the brain and biology in general, it is only for time to tell. Be that is it may, free energy definitely is a milestone concept and tool to advance our knowledge of the kind of Complexity that one meets at the interface between physics and biology.

14.13 Summary

In summary, nature proves capable of bypassing apparently unsurmountable entropic mountains, not because it has been given enough time to do so, but because it has been given a plan.

If you wish, a GPS which guides her through molecular hyperland in ways that time and chance alone would never make it possible. Such a GPS takes the abstract form of funnels in the free-energy landscape, acting as portentous accelerators towards the happy destinations where 'things work'. That's how nature finds her straws within the biological haystack.

I find this beautiful, awe-inspiring and in many respects, even moving. Yet, once emotion and a bit of teleo-philosophy both settle down, the practical question remains: how do such funnels and associated search strategies, *emerge* out the fundamental laws of physics? To borrow from a famous quip by physics Nobelist Isador Rabi (1898–1988) upon hearing of the unexpected discovery of the muon: *who ordered it*? This is the question we shall address in Chapter 15. Bare with us, I hope you'll like it.

14.14 Appendix 14.1 Funnel acceleration

For those willing to follow the math, here is a simple illustration of the way that the funnel turns Forever into a finite-time number. Let x the value of the OP, the equation of motion in the parabolic landscape $P(x) = x^2/2$ is simply:

$$\frac{dx}{dt} = -x$$

The solution is the exponential,

$$x(t) = e^{-t}$$

where we have assumed the starting value $x(t = 0) = 1$ for simplicity. The previous solution informs us that the generic value x is reached at time $t(x) = log(1/x)$, which in the limit x going to zero returns infinity. It takes an infinite amount of time to get exactly at the bottom of the landscape, i.e. $x = 0$.

The same procedure applied to a funnel landscape of the form $F(x) = -1/x$ yields the solution

$$x^2(t) = 1 - 2t$$

This reaches the value x at time $t(x) = (1 - x^2)/2$, which, in the limit of x going to zero, delivers a perfectly finite result, $t(x = 0) = 1/2$. This is how the funnel turns infinity into a finite number. The attentive reader would note that for t larger than $1/2$ the solution loses meaning because the right hand side becomes negative. This artefact due to the fact that $F(x)$ is unbounded at $x = 0$. This can be readily regularized, but these details are of no relevance to the main point of this Appendix.

14.15 Appendix 14.2 Mistaking rare events for impossible ones

As discussed in the main text, a rare event is one which occurs with a very low probability, while an impossible one is an event which occurs, with probability strictly equal to zero. It turns out that, as an effect of our mortality, i.e. the fact that our observation time, even as a species and not just as single individuals, is not infinite, the two are easily mistaken for each other. Unfortunately, in complex systems the consequence of such apparently slight mistake can be truly devastating. In his book *Fooled by randomness* Nassim Taleb kind of mocks this attitude, for instance by blaming people who dismiss things as impossible just because they have never been seen them before. The danger associated with this mistake is simply understood: a single high-intensity rare event is much more devastating than many equivalent low-intensity frequent events. The popular example of earthquakes helps elucidating the point.

It is well known that the frequency of earthquakes is a fast decreasing function of their *magnitudo*, defined as the logarithm in base 10 of the displacement on the seismograph produced by seismic waves. In other words, an earthquake of magnitudo M produces displacements proportional to 10^M.

The relation between the frequency of earthquakes and their magnitudo is known as Gutenberg-Richter (GR) law from the two American scientists who first proposed in 1956. In formulas it reads as follows:

$$f(M) \sim 10^{-bM} \tag{14.5}$$

where $f(M)$ is the frequency of earthquakes of magnitude M or higher and b is a coefficient close to 1. This means that an earthquake with $M = 2$ is ten times less frequent than an earthquake of magnitude $M = 1$, and ten times more frequent than one with $M = 3$ (for the sake of brevity, from now on I shall spare the 'or higher' part). By the same argument, an earthquake with magnitudo $M = 7$ is 10^6 (one million) times less frequent than one with $M = 1$. For the record, earthquakes with magnitudo 8 occur once every 5–10 years, which means that they are rare but definitely not rare enough to be ignored. Based on the GR law, magnitudo 10 or above should occur every 500–1000 years, and, happily enough, we have no record of any so far.

The question is then: is it more damaging to experience 10^M earthquakes of magnitude 1 or just one of magnitude M? For the sake of concreteness, let us take the case $M = 7$: one million earthquakes of magnitudo 1 versus 1 earthquake of magnitudo 7. This question is really a no brainer: everyone would pick the former option in no time. The reason is that an earthquake of magnitude 1 hardly makes any serious damage, hence a sequence of a million such mini-earthquakes distributed in time does not represent any serious threaten and one can live with it, as in fact we do. But a single maxi-earthquake of magnitude 7 sounds like as if the million mini-earthquakes of magnitude 1 would fire in sync all together basically at the same time, which rings definitely way more worrisome even before one does any math. The real point, though, is that doing the math shows that things are much worse than this. The reason is that the energy $E(M)$ released by an earthquake of magnitude M grows exponentially with M, and this exponential growth is stronger than the exponentially decay of the frequency of the GR law. To be sure, the relation is

$$E(M) \sim 10^{aM} \tag{14.6}$$

with $a \sim 3/2$.

The big danger carried by the relation (14.6) is immediately apparent: the coefficient a is greater than b. In other words, the *intensity* of the event (energy released) grows faster with the *amplitude* (magnitudo) than it decreases in frequency. The result is that the *impact* of the event, defined as the product of intensity times frequency increases at increasing magnitude. In formulas

$$I(M) = f(M) \times E(M) \sim M^{a-b} \tag{14.7}$$

Whenever the condition $a > b$ is met, treating rare events like if they were impossible is a literal Trojan horse.

But let us numbers speak their truth. For earthquakes, the coefficient $b = 3/2$ means that upon increasing by one unit in magnitude, the energy released grows by a factor 32. But the frequency decays only by a factor 10, which means that the impact grows by a factor 3.2 each time the magnitude jumps one unit ahead. Or, to make it easier to remember, ten times every jump of two units in magnitude. As a result, the impact of a single earthquake of magnitude 7 is 10^3, a thousand times larger than the impact of an earthquake of magnitude 1! Differently restated, the energy released by a single earthquake of magnitude 7 is 1,000 times the cumulative energy released by a million earthquakes of magnitude 1. And if we scale it up to $M = 9$, this ratio is ten thousand! This means that the energy released by a single earthquake of magnitude 9 is 10,000 times the cumulative energy released by hundred millions earthquakes of magnitude 1! For the record, the largest earthquake ever recorded is the Great Chilean Earthquake, which occurred on May 22, 1960 in Valdivia (Chile) and featured a frightening $M = 9.5$. The energy released in Valdivia amounted to nearly twenty thousand more times the energy released by about thirty-two hundred millions earthquakes of magnitude 1 *all together!* Prior to May 22, 1960, an earthquake of magnitudo 9.5 could have been dismissed as impossible, but on May 22, 1960, in Valdivia the mistaken impossible claimed nearly two thousand casualties. And in more urbanized areas, the impact could have been even more devastating. The most deadly earthquake ever recorded is the one that occurred in on January 23, 1556 in Shaanxi, China, magnitude 8, which claimed over 800,000 casualties ($https://en.wikipedia.org/wiki/1556_{S}haanxi_{e}arthquake$).

These figures speak for themselves on the dangers of mistaking the Improbable for the Impossible.

One last remark to point out how far the statistics of earthquakes is from standard Gaussian statistics. Since the displacement L scales like 10^M, one can rewrite the GR law as follows:

$$f(L) \sim L^{-1} = 1/L \tag{14.8}$$

This expression shows that the frequency of large displacements is astronomically higher than under Gaussian statistics, $f(L) \sim e^{-L^2/2}$. If earthquakes were abiding by Gaussian statistics, events like Valdivia would occur once every hundreds of billions of billion years, hence, they could be safely treated as practically impossible.

That's why for most complex phenomena, reliance upon Gaussian statistics is a close proxy to a sure recipe for major failure.

15

Soft Matter, the Stuff that Dreams are Made of

We are such stuff as dreams are made of.

(W. Shakespeare)

15.1 Back to the ground

In Chapter 14 we have glorified the crucial role of deep corrugations of the free-energy landscape in enabling the convergent search for optimal minima where vital functions can be delivered. We have also emphasized that, for all their elegance and effectiveness, funnels bear no material existence, but reflect instead the highly orchestrated action of the underlying atomic and molecular forces. In this chapter, we take a closer look at such a chain of forces, not only in connection with funnels, but in the broader context of soft matter, namely the state of matter most relevant to the biological world.

15.2 The stuff that dreams are made of

If there is a route to happiness in life, I guess a few would contend that it goes largely across dreamland. The ability of dreaming, i.e. project a pleasant reality out of the actual one, arguably matching our best wishes and aspirations, is a most powerful antidote against the many burdens life presents us with.[50] That is why the ability of abstracting and dreaming is one of the deepest-running threads of human nature. By the very same token, dreams are one of the most sought-after merchandise we can offer to our fellow citizens. The worldwide success and impact of rock stars stands as an impressive witness: with their music and lyrics, they 'can lift you up where we belong', to go with Joe Cocker (1944–2014). A similar story goes for sports, movie superstars, icons of various sorts, including the recent brand known as 'influencers'. And I bet that many would also agree that youth (in the heart, at least) also goes with the ability to dream beyond our bodily limitations. And even though youth by no means equates with happiness,

[50] This statement should be taken with a light spirit, of course. Serious discussions about lightness and happiness pertain to other endeavours than those discussed in this book.

Sailing the Ocean of Complexity. Sauro Succi, Oxford University Press.
 DOI: 10.1093/oso/9780192897893.003.0015

it does not hurt either . . .. So, what is the stuff dreams are made of and, to go with Shakespeare, what is the stuff *we* are made of?

Many readers might sense a regrettable loss of poetic altitude in asking such a 'practical/technical' question, but I hope that by the end of this chapter, the same readers will agree that the opposite is true. Looking for an answer to this question reveals beauty and wonder, hence nothing short of poetry, albeit a very concrete one, the poetry of things that did happen, against the odds. At least this is the way this author feels about it.

And here we go.

In the previous chapter, we discussed the way that free energy guides biological systems across Oceans of Complexity, towards safe harbours, where they meet the right conditions to deliver useful functions. Protein folding was chosen as a concrete and very relevant example in point. We also pointed out that free energy is real but not material, nothing we can touch or see in a microscope. It is a bit like software versus hardware, the latter being the basic constituents of matter, atoms, and molecules, and the former being the rules of the game. Thus, in the end, free energy *emerges* from the basic interactions between these microscopic constituents. The specific mechanism by which this crucial epiphany takes place is by no means a given, nor is it completely understood by modern science, as we speak. This is the poetry we were referring to earlier on, the poetry of things which exist even though, by any reasonable odds, they actually shouldn't. However, certain general properties of microscopic interactions leading to convergent evolution in free-energy landscapes can be identified, and this is precisely the leitmotif of this chapter. This is a comparatively engaging chapter, but we shall make sure that the general concepts are conveyed without any major technical burden.

15.3 States of matter

The first attempt to express a rationale for the structure and functioning of the world around us is usually credited to the Greek philosopher Empedokles of Akragas (a Greek colony in Sicily back then) (494–434 BC), who established the first cosmogony based on four ultimate elements: Water, Air, Earth, and Fire, (see Fig. 15.1). To emphasize their fundamental role, he called them 'roots', which he proceeded to identify with four corresponding mythical names: Zeus, Hera, Nestis, and Aidoneus (best known as Hades in English). The idea goes as follows: the magic four are indestructible and unchangeable and the structures that we observe around us are just different combinations thereof. Empedokles imagines life as the result of the competition of love (*Philotes*) and strife (*Neikos*). Citing Wikipedia: Love and Strife are attractive and repulsive forces, respectively, which are readily seen in human behavior, but also pervade the entire Universe. The two forces wax and wane in their dominance, but neither force ever wholly escapes the imposition of the other. The picture brought up by Empedokles is a striking anticipation of several key ideas discussed in this book. It evokes

Figure 15.1 *Empedokles in the act of 'showing' the four elements of his Cosmogony.*
Source: reprinted from en.wikipedia.org and from www.linnaeus.uu.se, with the permission of Uppsala University.

the energy-entropy duel/duet, as well Boltzmann 'evershifting battle' between nonequilibrium (streaming) and equilibrium (collisions). And even more to the point, it anticipates a most distinctive property of molecular interactions, namely the competition between attractive and repulsive forces, as we are going to discuss in detail shortly. Not bad for his times! Hence, no surprise that this view managed to set the gold standards for the next 2,000 years ....

15.3.1 The magnificent three: gas, liquid, and solid

Modern science has amended this view, but not that drastically after all. Instead of four elements, we get three fundamental states: *gas, liquid, and solid,* (see Fig. 15.2). As you can see, Empedokles didn't miss much, air is still the prototypical gas, water is the prototypical liquid, earth is a good match for solids, and so is fire for heat, as we discussed abundantly in earlier chapters. Of course, by now we know more, thanks, among others, to the kinetic theory of gases developed by Ludwig Boltzmann, also covered in some detail earlier in this book. Kinetic theory informs us that gases are the triumph of kinetic energy, they expand and fill up their containers, no matter how large. Solids, on the other hand, sit just at the opposite end, potential energy dominates and blocks off the atoms at fixed positions in space, around which they can only perform small amplitude oscillations under the effect of temperature. As a result, solids stick to their own shape regardless of the container. Finally, liquids are a kind of intermediate, where kinetic and potential energy come basically on a par, making of liquids the most flexible, hence most difficult, state to handle of the three. Liquids expand little, but easily change shape, hence they adapt to their container but don't fill it up, until their volume exceeds that of the container itself and they spill over. Intuition does well here, just think of air, water, and rock, and you have the essence of it.

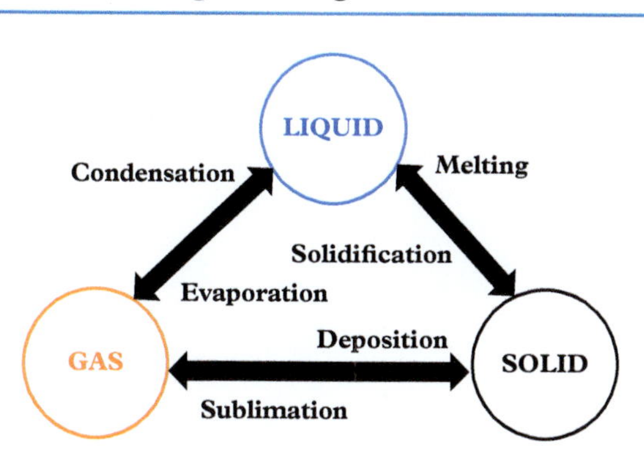

Figure 15.2 *Phase changes between the three fundamental states of matter. Liquid to gas (evaporation) and gas to liquid (condensation); liquid to solid (solidification) and solid to liquid (melting); gas to solid (deposition) and solid to gas (sublimation). These phase transition make the world go around.*

The three fundamental states turn constantly into each other under the effect of heat and work. Upon cooling below a critical temperature, a gas would condense into a liquid, and conversely, above the critical temperature liquids evaporate into gas. Below another critical temperature, liquids freeze into solids and above it, solids melt into liquids. Solids may even turn directly into gases, a process known as sublimation, (see Fig. 15.2). These *phase-transitions*, as they are known in scientific language, make the world go around, in a literal sense.

15.3.2 Hybrid states of matter

Fast forward to the mid-1980s, approximately the time soft matter was born as an independent discipline, the term having been coined by Pierre-Gilles de Gennes (1932–2007), Physics Nobel 1991, unquestionably catchier than its forerunner 'complex fluids'. Interestingly, most materials 'dreams are made of' are neither gases, nor liquids, or solids, but rather a blend between the two or even three of them [90]. Foams are a mixture of liquid (water) and gas (air), emulsions (like mayonnaise) are a delicate blend of liquid (oil) in another liquid (water), gels are made of small solid particles dissolved in water, and aerosols are droplets or small solid particles floating in air. The amazing point is that these 'hybrid' states of matter behave very differently from the three basic states they are built upon and show new properties nowhere to be found in any of the three basic states, (see Fig. 15.3). For instance, both air and water under standard conditions behave like 'Newtonian fluids', i.e. they respond linearly to the applied loads, in different proportion, but still linearly. Yet, despite being made of air and water, foams exhibit a very nonlinear response to stress. Hence, the proportion by which the three basic

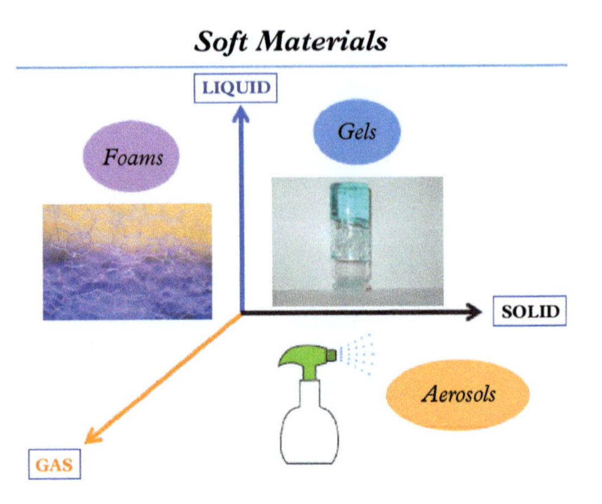

Figure 15.3 *Soft materials result from assorted combinations of three basic states of matter: liquid, gas and solids. For instance, foams are combinations of liquid and gas, gels typically made of solid particles dispersed in liquids, and aerosols are again solid particles dispersed in gases, typically air. Other important soft materials are emulsions, namely liquids dispersed in other liquids, typically oil in water, such as mayonnaise, assorted creams, and many others. The importance of these materials for daily life hardly needs any comment, but they are also vital to the correct build-up and functioning of our own body.*
Source: the insets are reprinted from wikimedia.commons.org.

states are mixed with each other, as well as the temperature at which this mixing takes place, makes a world of a difference as to their spatial arrangement, which in turn dictates their rheology, i.e. how they deform and eventually flow in response to an applied force.

Quantity is quality, another fundamental signature of Complexity, as highlighted in the epoch-making paper 'More is Different' by P. W. Anderson [3]. All of the previous ideas justifiy the birth of soft matter as a new and self-standing branch of condensed matter, itself a modern outgrowth of solid-state physics.

Soft matter is crucial for a large number of industrial applications, food processing, pharmaceutical, and cosmetics to name but a few, and particularly for biological systems. It is fair to say that *soft matter is the physics of the intermediate scale between us and the atoms we are made of*, hence a natural crossroad between physics, chemistry, and biology, physics-chemistry-biology interface [87]. Indeed, it is hard to imagine a living creature based exclusively on just one of the three fundamental states! True, liquids (water) take the lion's share in biology, but water alone would not do either, a good pinch of solid is definitely required to keep our body together. Likewise, breathing totally depends on the availability of air (some cyanobacteria can do away with this, but this is another story). It is therefore a great gift that the three fundamental states of matter can combine in such a way as to sustain living organisms. Unsurprisingly, nonlinearity plays once again a fundamental part in this plot, as these hybrid materials do not respond in proportion to the degree of mixing between the three fundamental states they are made of.

Soft matter is gifted with a lot of practical and intellectual fascination alike, but it also faces some communication difficulties, due to the very different attitudes and goals of the three fields it is based upon. Physics has the cult of universality and makes a point of pride of eliminating all possible details that 'don't matter'. Coarse-graining, as we discussed in the previous chapters, builds heavily upon universality: upon grouping microstates based solely on a given macroscopic property, say energy, we are implicitly stating that other properties of these microstates do not matter, as far as macroscale behaviour is concerned. That is a strong call to universality indeed!

Chemistry, on the other hand, is nearly the opposite: the love for details matters and don't stoop to coarse-graining, on pain of missing essential features of the system. Chemistry is highly specific: proteins and ligands are key and lock, they match only once; they show sufficient chemical affinity, failing which, they simply ignore each other. This so-called 'key-lock' kind of interaction lies at the opposite end of universality; it is the triumph of individuality and specificity instead. Soft matter is the area of modern science where universality and specificity shake hands, in a most profound and fertile way, and it is fair to say that despite the inherent tension between the two, their coexistence is a major linchpin of biological Complexity.

15.4 The physics-chemistry-biology interface

If physics and chemistry show nearly opposite penchants towards universality versus specificity, this is nothing compared with the case of physics versus biology. Historically, physics and biology couldn't be more different. Physics thrives on simplification and models, aims at extracting simple and universal laws from the apparent mess of the physical world and it is strongly reliant on mathematization. Biology treasures the mess, complexity, and diversity of living organisms, details are beloved, and its mainstream still makes little use of mathematics. Francis Crick (1916–2004), the famous co-discoverer of deoxyribonucleic acid (DNA) commented his transition from physics to biology "as if one had to be born again" .... However, if only piece by piece, and not without pain, the two disciplines are slowly coming together. But let's go back to our main goal, i.e. to provide a physical substrate to free-energy landscapes, not just for proteins, but for soft matter in general. In the previous chapters, we have learned about the duel/duet between Order and Disorder and its mathematical encapsulation through the definition of free energy. At low temperature, entropy is quenched, and Order prevails, the kingdom of solid-state physics. At high temperatures, the opposite is true, things are volatile, and this is the kingdom of the kinetic theory of gases. Many complex systems, most notably biological ones, present us with a situation in which the two contributions are tightly balanced. Under these intermediate conditions, matter is called *soft*, in that it supports organized and flexible structures which cannot stand energies much in excess of thermal ones, the proverbial $k_B T$ we have met before in this book. Such fragility is the price for the flexibility required to perform life-sustaining functions, thriving at the border between Order and Disorder. This is precisely the state of interest for soft matter research. But let us proceed by discussing the material substrate of this balance, i.e. the actual physical interactions that lie at the roots of systems for which energy and entropy are in tight competition, as cleverly envisaged by Empedokles back then.

15.5 Molecular interactions

Nature presents us with four fundamental types of forces: *strong*, *weak*, *electromagnetic*, and *gravitational* (see Fig. 15.4). To be sure, they are in fact three, since weak and electromagnetic forces are known to be part of the same electro-weak family.

These forces act at very different scales: strong interactions rule the behavior of matter at nuclear and subnuclear scales, namely a million times below nanometers, gravitational interactions, on the contrary, are most relevant at macroscopic scales, all the way up to the entire Cosmos! The most relevant forces for soft matter are the electromagnetic ones, and more precisely, the *electrostatic* ones, which is the name of electromagnetics for bodies which move much slower than light. In practical terms, soft matter is also (apparently) unaware of the other major pillar of modern physics, i.e. quantum mechanics. I write 'apparently' because quantum mechanics plays an essential role on the molecular and supramolecular interactions that rule soft states of matter, for, ultimately, everything traces

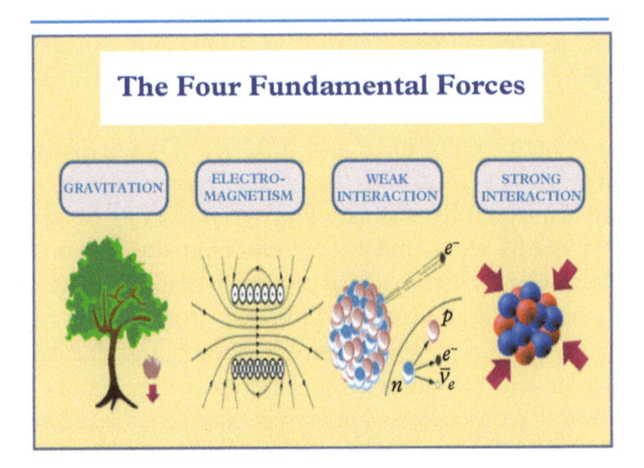

Figure 15.4 *The four fundamental forces in nature.*
Source: reprinted from commons.wikimedia.org.

back to interactions between nuclei and electrons, which are most exquisitely quantum. The point, though, is that by the time these interactions *aggregate* to reach the scale of interest to soft matter, say from fractions of nanometres to centimetres, the quantum details can be comfortably hidden under the form of classical-looking forces and potentials. In a way we could say that soft matter is formally 'de-quantized' , but not quantum free.

15.5.1 What a bore?

No relativity, no quantumness, I can see the little cloud forming over your head, dear reader, with the label 'what a bore' inside .... Not so fast, just wait and see. The reader may indeed wonder what's new here, given that the basic laws of electrostatics were laid down many centuries ago, with the pioneering work of several scientists, most notably the French Charles-Augustine de Coulomb (1736–1806). The point, though, is that in biological systems, and soft matter in general, electrostatic interactions take place in fairly non-trivial geometrical environments, where the distribution of matter and charge gives rise to effective interactions which are very different from electrostatics in idealized geometries, such as point charges, spheres, and similar. No place for spherical cows here ....

To put it distinctly, once light and matter interact in densely packed complex geometric environments, their interaction takes on very different dresses than in empty space. Physicist call these 'dressed' forces *effective*, because they embody the complex interactions with the environment which determine the effective forces experienced by matter. For reasons to be detailed shortly, I shall occasionally call effective interactions *unfundamental*, precisely to emphasize the contrast with the four fundamental ones and to argue that the former are no less important than the latter, as they carry a much more direct bearing on our own existence. So, let's take a look at the basic physical mechanisms which turn fundamental interactions, namely electrostatics, into unfundamental ones, without loosing a bit of their importance. In fact quite the opposite. Before venturing into this task, some simple recap of textbook electrostatics is in order.

15.6 Electrostatic forces

The basic law of electrostatics, named after Augustine Coulomb, states that two point-like charges attract(repel) each other with a force proportional to the product of their charges and inversely proportional to the square of their distance.

That is:

$$F = \epsilon \frac{q_1 q_2}{r^2} \tag{15.1}$$

where q_1 and q_2 are the two charges, r their distance, and ϵ a property of the material in which the charges are embedded. In vacuum we can conventionally

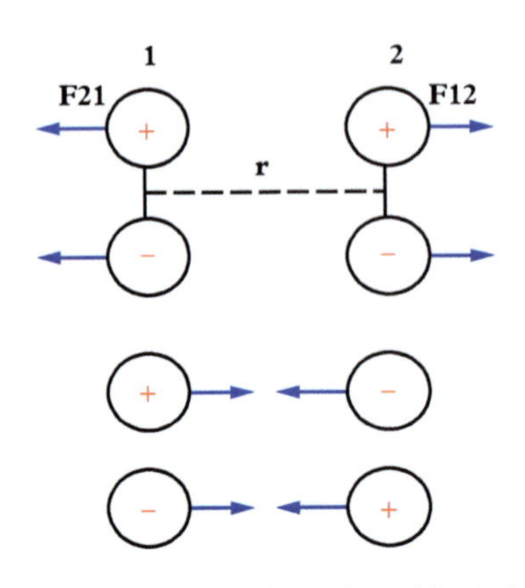

Figure 15.5 *Electrostatic forces between two charged particles 1 (left) and 2 (right). Charges of equal sign (++ and −−) repel each other, while charges of opposite sign (+− and −+) attract. Either way, the force F12 exerted by particle 1 on particle 2 is equal and opposite to the force F21 exerted by particle 2 on particle 1, that is F12 + F21 = 0. Both forces are proportional to the product of the charges and decay with the square of the distance r between them.*

take $\epsilon = 1$. The force is aligned along the segment joining the location of the two charges. If the charges are equally signed, F is positive which means repulsion, and vice versa, oppositely signed charges give negative F, which means attraction. Finally, if F denotes the force exerted by charge 2 on charge 1, the reciprocal force exerted by charge 1 on charge 2 is $-F$, hence the sum of the two is zero. This is sometimes called the principle of action-reaction, (see Fig. 15.5).

The attentive reader might have noticed that this is exactly like gravitation, with a crucial twist though: while gravity is always attractive, electrostatics is attractive for oppositely signed charges and repulsive for equally signed ones. As said by Empedokles, gravity is pure love while electrostatics is both love and strife (see Fig. 15.6). And again as strikingly anticipated by Empedokles, both are needed in a living Universe. Let us see why.

15.6.1 Electrostatic instability

If two bodies attract, they tend to merge into a single one. If, they repel instead, they tend to depart at infinite distance (in free space, where there is unlimited room). Neither scenario is suitable to life as we know it, which tolerates neither unbearable overcrowding nor unsustainable rarefaction. It is a real blessing that charges come in both positive and negative, because if there were just one of the two, we would be utterly unstable, and the tiniest butterfly would blow us away! We can make the argument a little bit more compelling. Indeed, it is easy

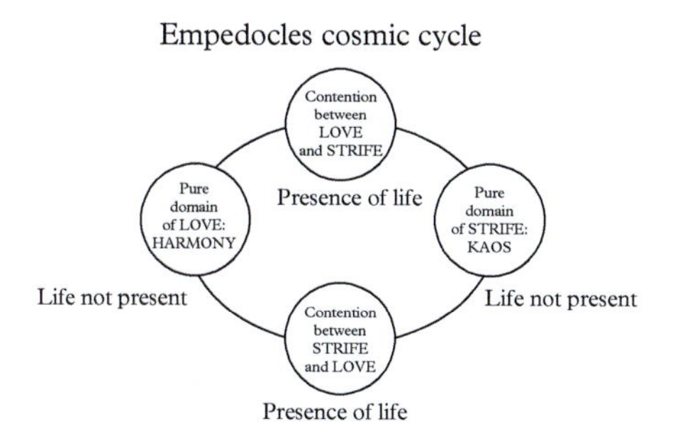

Figure 15.6 *Empedokles connection between love and strife, whose competition promotes the emergence of Life.*
Source: reprinted from en.wikipedia.org.

to see that a material body constituted only by equally signed charges is mechanically unstable; any minor perturbation would quickly dismantle it! Here is the point: take three positive charges aligned and equally spaced from each other (see figure 15.7). The central charge is in equilibrium because, by the principle of action and reaction, it feels the same repulsion from both neighbours along the opposite direction. As a result, it experiences no net force, which is the definition of mechanical equilibrium.

Suppose now that our familiar butterfly (perturbation) pokes the central charge a bit above or below the line. In the new position, the central charge experiences a repulsion from both its right and left neighbours, which add up to push it further up away from the original position. Hence, the central charge will never return to its original position: three equally signed charges on a line form an unstable configuration. Now take again three equispaced charges on a line, but the central positive and two negatives on the sides. The central charge is equally attracted to the left and right neighbours, hence it still experiences zero net force: equilibrium again. The butterfly is still around (by now we have learned that she never quits, haven't we?) and pushes the central positive charge up or down again. At variance with the previous case, both negative charges on the sides pull back the central charge, which regains its initial, pre-butterfly position. The unequally charged linear chain is now stable, (see Fig. 15.7).[51] For all this simplicity, this elementary example illuminates a serious point: the spatial distribution of positive and negative charges, what physicists call *dispersion*, is key to the stability of the structures they form, hence to the very stability of our own world, as we know it. The take-home message is that Coulomb alone in its 'fundamental' form, won't work: more sophistication is needed to sustain functional structures.

Now, please bear with me, as I anticipate the game-changing trick. Suppose that by some magic, two charges repel at short distances but attract at long distances. A minute's thought reveals a new opportunity which does not exist in a

[51] The actual calculation would reveal oscillations around the initial positions, but not run away.

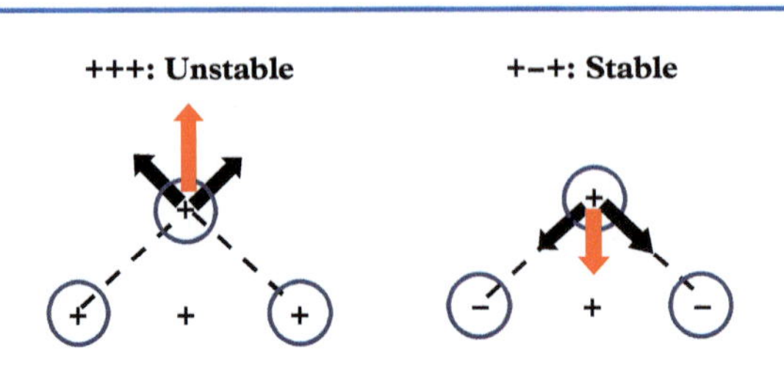

Figure 15.7 *Electrostatic forces on three charges on a line. Left,* $\{+++\}$ *unstable configuration. Right,* $\{-+-\}$ *stable configuration.*

purely fundamental Coulombic world, namely that the two charges find a stable arrangement at some intermediate distance, where the short-range repulsion and the long-range attraction come to an exact balance with each other.

Happily enough, *this is exactly what happens to electrostatics once it is embedded in atomistic matter*! Somebody said that a masterpiece of art should not be looked at too closely, for you would be lost in the details and be at risk of missing the overall picture. If you're far, I'd like you closer, but if you get too close, I'd rather want you further away. No excessive crowding and no excessive rarefaction either. A slightly whimsical attitude which we seem to inherit all the way from our very molecules. Empedokles really got it right!

15.6.2 Long-range attraction and short-range repulsion

The next question is: why do molecules behave like this? The reason is encoded in the very structure of atoms. Repulsion is readily explained: when two atoms get very close, the electrons which surround their nuclei come in close contact, and since electrons are negatively charged, they repel each other, giving rise to short-range repulsion between the molecules. Hence, repulsion is an exquisitely electronic property. This happens at distances comparable with or shorter than the atomic size, typically a fraction of nanometre, the scientific unit being known as Angstroem, from the Swedish physicist Anders Jonas Angstroem (1814–1874), a founder of modern spectroscopy.[52]

Incidentally, this offers an exemplar case of what we previously called 'de-quantization', since the interactions between electrons and nuclei are definitely governed by quantum mechanics. However, as explained earlier on, once the quantum mechanical calculation is completed, their interaction can be expressed in a classical (non-quantum) form, as we are going to detail shortly. So much for

[52] The reader might just as well argue this is another word for a tenth of a nanometre .... But scientists are like this, like to honour their peers who made decisive inroads into new territories of knowledge.

short-range repulsion. The reason for long-range attraction is less straightforward, and it amounts to the way molecules deform in response to charge distribution in space, an effect known as *polarization*, which we proceed to discuss next.

15.7 Polarization

The expression (15.1) applies to charges in vacuum, i.e. no material substance in between and around. This is an ideal condition hardly met in real life, and certainly not in biology, where charged molecules (ions) are typically dispersed among other charges. A positively charged particle placed somewhere in space attracts a cloud of negative charges around it, so that from a distance much larger than the size of the cloud, the effective charge is basically zero, the cloud is *neutral*. However, within the cloud, the charges no longer compensate, and neutrality is broken. This mechanism is called *screening*, since the original charge is literally screened by the neighbouring charges (see Fig. 15.8). As a consequence of screening, the force emanating from the charge acquires a very different space dependence as compared to the plain Coulomb law.

Closely related to screening is the physical effect known as *polarization*, namely the mechanism by which molecules develop a reaction against an external electrostatic field, (see Fig. 15.9), by deforming under its effect in such a way as to generate an antagonist electric field acting in the direction opposite to the external one.

Consider an external electric field pointing West-to-East acting upon a so-called polarizable media. Positive charges within the polarizable molecules are displaced in the direction of the electric field (eastward) and negative ones in the opposite direction (westward). We say that the medium is polarized because

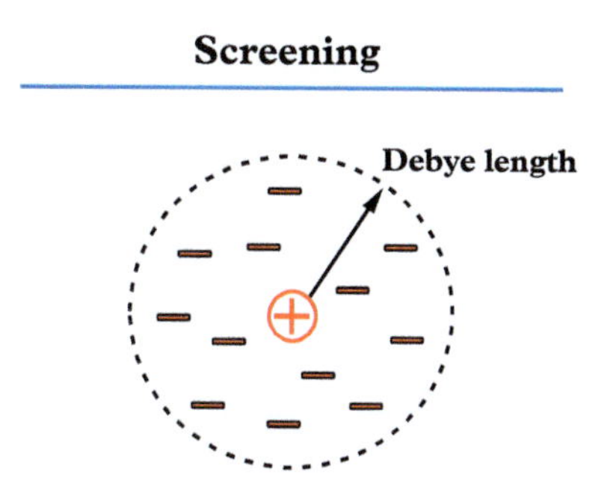

Figure 15.8 *The mechanism of screening. A central positive charge attracts negative charges around. The result is that the effective charge seen by a particle at a distance r from the central charge is reduced as compared to the original charge and beyond a given distance, known as the Debye length, it becomes practically zero.*

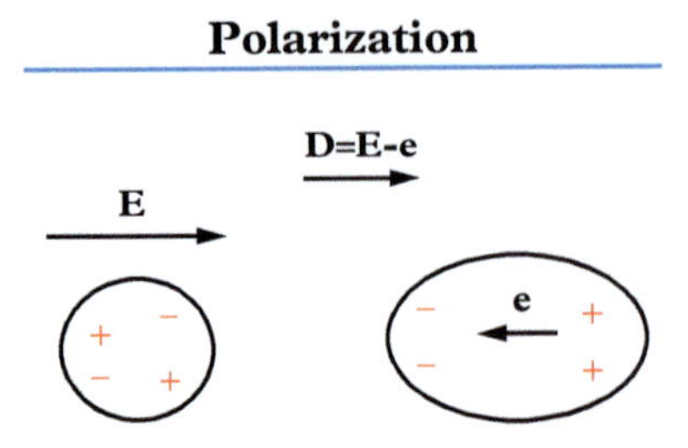

Figure 15.9 *The mechanism of polarization. A neutral molecule containing two positive and two negative charges is subject to an external electric field E. The electric filed pushes the positive charges rightwards and the negative ones leftwards, giving rise to charge separation in space. The positive charges on the right end of the deformed molecule and the negative charges on the left repel each other, giving rise to an internal electric field e in the direction opposite to the external field E. The net electric field experienced by the polarized molecule is D = E − e.*

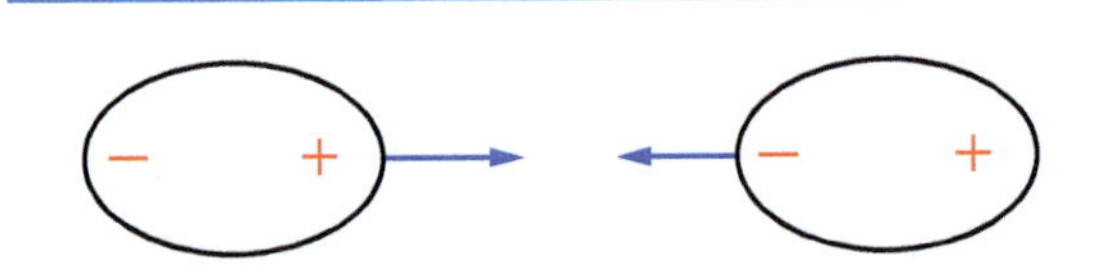

Figure 15.10 *Attractive forces between polar molecules. The positive charge of the left molecule is attracted by the negative charge of the right molecule and vice versa. The result is a net attraction between the two molecules.*

all the molecules align and deform in the direction of the external field (see Fig. 15.10). Being spatially separated, the charges generate an internal field pointing from the positive to the negative charges, hence in the direction opposite to the electric field which caused the displacement in the first place. The total electric field within the material is thus the sum of the external field and the internal one generated inside the molecules in response to it. Being opposed, they sum up to a smaller field than the external one, the ratio of the two defining an important property of the material called dielectric permittivity.

Next consider two polarized molecules sitting next to each other: due to the deformation the positive charge of a given molecule gets closer to the negative charge of its neighbour, with the result that the two molecules attract each other. This attraction lies at the root of the so called dispersion forces, namely screened electrostatic interactions resulting from polarization effects. Since such forces are key to soft matter and biology, it is worth taking a closer look at this crucial phenomenon by discussing polar molecules, namely molecules with a high propensity to polarization.

15.7.1 Polar molecules

In Section 15.7, we discussed the mechanism by which molecules get polarized in the presence of an external field. But how about when there is no such external field? To answer this question, let us recall that atoms are made of a central nucleus, consisting of positively charged protons and uncharged neutrons. The nucleus is surrounded by much lighter (ratio 1:1836) negatively charged electrons, zipping around at near the speed of light, thus forming a sort of negatively charged cloud around the nucleus. Since the electrons exactly match the protons in number, from a distance, all atoms are neutral, because the nuclear and electronic charge cancel each other. A similar argument goes for molecules, which are assemblies of atoms. Hence, from a distance, molecules are indifferent to each other, Coulomb is silent. But inside the molecule it is a different story, the positive and negative charges are not necessarily symmetrically distributed. In this case the molecule is called *polar*, as it displays a permanent mismatch between the barycenter of positive and negative charge. Such mismatch is the source of an internal electrostatic field known as *dipole* in technical terms. Now consider two such polar molecules sitting next to each other; the positive charge of the left molecule sees the negative charge of the right one, whence their attraction. A new interaction is born through the mechanism of polarization. These are called *dispersion forces* because they result from the spatial dispersion of the electric charge. Given their crucial role in building up supra-molecular structures, a little more general discussion on dispersion forces is in order.

15.8 Dispersion forces

The qualitative considerations on the emergence of dispersion forces out of polarization effects were placed on a quantitative basis in the early days of the previous century. The basic picture can be conveyed without any mathematical burden. Let's collect the basic facts: due to the spatial distribution of their charges, the screened electrostatic interaction between molecules feature short-range repulsion and mid-range attraction. Here, short-range means a distance below the size of the molecule, hence a fraction of nanometre and mid-range means larger than the size of the molecule up to a few nanometres. Beyond such distances, which we classify as long range, dispersion forces fade away because of charge neutrality. This is another key difference with gravity. A closer look at these matters takes us first to Holland and then to England.

15.8.1 Van der Waals and Lennard-Jones

As surmized by its very name, soft matter is not supposed to involve high energies, the typical scale being thermal energy $k_B T$ at ambient temperature. Within this framework, forces can be categorized as weak or strong depending on how they compare to the thermal one. Some small numbers won't hurt. The thermal energy $k_B T$ at the standard temperature of 300 degrees Kelvin, corresponds to about $1/40$ electronvolts (eV), the energy acquired by an electron traversing a potential of one volt. Just to get an idea, Italian houses are wired to work at 220 volts, hence each electron contributing to light the bulbs in an Italian house has an energy of 220 eV. For the record, there are about one billion billion of them passing through the wire every second . . . .

Compare with nuclear forces, whose scale can be taken as the rest energy of the proton, according to the celebrated Einstein's equation, $E_p = m_p c^2$. The proton mass m_p is tiny, it takes an Avogadro of them to make just one gram, but this is more than compensated by the speed of light, about three hundred million metres per second, which comes -quadratic- into this stage, to deliver a final 938 million eV, namely forty billion times larger than thermal $k_B T$! And to make it even more humbling, the superstar Higgs boson goes up by another hundred and finally escalates to the largest energies currently achieved at European Council for Nuclear Research (CERN) gives you another extra hundred.

Disappointing?

Not at all, we stated right upfront that Complexity, hence soft matter, does not inhabit the land of extremes. On the contrary, it peaks where competition is fierce because competing mechanisms are close to a tie. Now we can touch it by

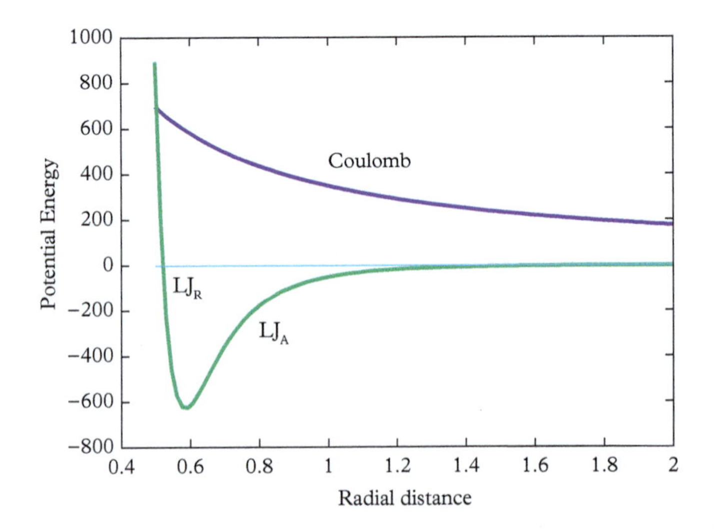

Figure 15.11 *Coulomb and Lennard-Jones potentials, with both attractive (LJA) and repulsive (LJR) branches. The depth of the valley is a typical measure of the energy strength, and the location of the bottom provides the range of the interaction. The former is of the order of the thermal energy $k_B T$ at room temperature, while the latter is typically a fraction of a nanometre.*

numbers, that's all. But don't forget, many small items acting together may make a big one, which is the main point about soft matter . . ..

Back to the task at hand. Forces that feature coexistence of both repulsive and attractive character are named after the Dutch physicist Johannes Diderick Van der Waals (1837–1923), VdW for short, whose name is historically associated with the study of condensation of gases to liquids. The credit for its microscopic underpinning goes rather to the British mathematician John Edward Lennard-Jones (1894–1954), who postulated a specific expression for the potential energy associated with VdW interactions. More precisely, the Lennard-Jones short-range repulsive energy decays like $1/r^{12}$, while the attractive one like $-1/r^6$, respectively. In equations

$$V_{LJ}(r) = \frac{B}{r^{12}} - \frac{A}{r^6} \tag{15.2}$$

where A and B are numerical parameters, (see Fig. 15.11).

By convention, the plus term stands for repulsion and the minus one for attraction. Note that the former is much steeper than the latter: by doubling the distance repulsion goes down by a factor $1/4,096$, while attraction decreases 'only' by a factor $1/64$, i.e. 60 times less. And both are immensely steeper than pure Coulombic $1/r$, which upon doubling the distance goes down in linear proportion, i.e. a factor two, see Fig. 15.12.

As already mentioned, Coulomb interactions are long range because, in vacuo, the charges are unscreened. The coexistence of repulsive and attractive interactions is vital to the stability of matter. Indeed, most molecules settle at the distance where the two effects come into balance, aptly known as equilibrium distance, with

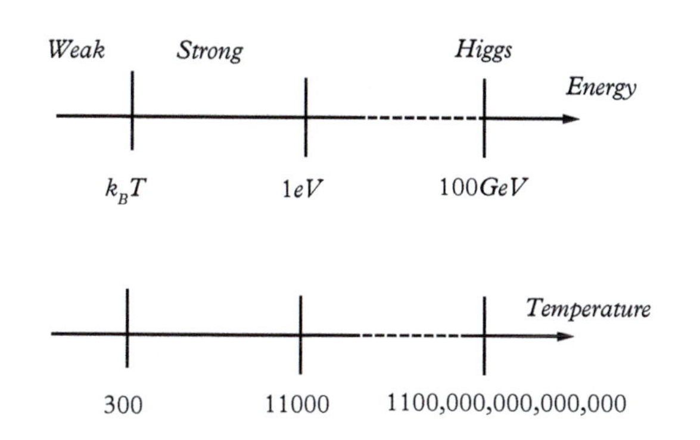

Figure 15.12 *Weak and strong interactions in soft matter. The border is marked by the thermal energy at room temperature, $T = 300$ (Kelvin) degrees, corresponding to about $1/40$ of electronvolts. The tiny electronvolt is a huge energy in soft matter compared with the energy of the Higgs boson, which sits about a hundred billion times higher on the energy scale . . .. (The figure is clearly out of scale.)*

only comparatively small departures due to thermal fluctuations. The equilibrium distance is readily computed as the location at which the Lennard-Jones (LJ) potential attains its minimum. Simple algebra delivers:

$$r_{eq} = (\frac{2B}{A})^{1/6} \tag{15.3}$$

This shows that, as expected, the equilibrium distance depends on the ratio between repulsive and attractive forces. Given the exponent $1/6$, such dependence is pretty weak: to double the equilibrium distance, the ratio B/A must increase by a factor $2^6 = 64$. Without delving into the math details, we simply observe that in the case of pure repulsion, $A = 0$, the equilibrium distance goes to infinity, i.e. total rarefaction. In the opposite case of pure attraction, $B = 0$, the equilibrium distance goes to zero, the overcrowding scenario discussed earlier on. It is only when both A and B are non-zero that the system can find its peace at an intermediate distance, typically a fraction of nanometre. This is an absolutely key aspect of VdW interactions.

The second key aspect, as noted previously, is that they are short-range, namely the equilibrium distance is comparable with the size of the molecules. The far-reaching consequence is that each molecule only interacts significantly with a comparatively small number of molecules around it, typically no more than fifty or so in a 3-dimensional piece of matter, *regardless of its size*. This is what physicists call *locality*: the microscopic interaction is unaware of the global size of its environment. This is an enormous simplification as compared to the coulombic case, in which, due to the much slower spatial decay, $1/r$, basically all molecules are in simultaneous interaction with each other.

But let's dig a bit deeper into the LJ potential. The typical force associated with LJ interactions is $F_{LJ} \sim k_B T / r_{eq}$ and scores in the order of pico newtons, namely thousand of times smaller than a billionth of a newton. This is the typical unit for macroscopic forces (I remind the reader that the force acting upon 1 kilogram of our body due to gravity is about 10 newtons, to be precise, 9.81). This says it clearly on how weak VdW forces are on a macroscopic scale.

Although *weak and short-range*, these interactions are nonetheless vital to the equilibrium and stability of most soft matter systems. Moreover, despite their weakness, they are responsible for rather spectacular effects: if you are small enough, they easily outdo gravity, as any gecco walking upside-down on your ceiling would be happy to attest, (see Fig. 15.13)....[53] VdW interactions are responsible for another major family of dispersion forces, known as capillary

[53] I can't help a private communication regarding what I consider as one of my major achievements in life. On a late Sunday evening, my wife and I were on the verge of closing our summerhouse in Tuscany, ready to drive back to Rome. All of the sudden, she spotted a gecco on the ceiling and cried out, claiming that we could not leave him alone in the apartment, for he would surely die by the time we would be back (would he?). Brief, I was mandated to capture the gecco and set him free. Don't ask me how, but I managed. If this happens again, I am sure I would fail a million times in a row . . . .

Figure 15.13 *Gecco on the ceiling: an (upside-down) walking advertisement for the power of VdW interactions.*
Source: reprinted from commons.wikipedia.org.

forces, which we now describe in some detail because they are responsible for important organizational effects such as molecular segregation.

15.8.2 Capillary forces

In ordinary life and society, not to mention politics, walls do not enjoy a good press, as they stand for separation and segregation. However, at the molecular level, walls are not necessarily bad: in fact they are not bad at all, since they are crucial to the proper build-up and functioning of organized structures. We have already touched at this item when discussing the beneficial role of constraints. A cell needs to isolate itself from the environment and have some peace and quiet to perform its vital functions. Yet, just like any other complex system, the cell just can't live in isolation, hence the wall (membrane) must be permeable, in order to trade mass and energy with the external world. Remember Chapter 1 again: complex systems are typically open, and before long, all close systems die out. In the case of the cell, this isolation/communication function happens through a flexible nanometric wall, called cellular membrane. Given that the capability of matter to aggregate in supramolecular structures is vital to biological functions, the natural question is: what are the forces which make organized aggregation possible?

The answer is multifaceted and still in flux, but a few basic actors can be clearly identified. It turns out that nature has been equipped with a specific mechanism to keep groups of molecules together, typically in the form of droplets and also more sophisticated supra-molecular structures. This property bears the rather cryptic name of *surface tension*, but don't be scared, all will become clear soon (at least, I hope so ...). From a macroscopic point of view, surface tension is a measure of the energy required to increase the area of an interface, say an oil droplet in water or a liquid droplet in its vapour, (see Fig. 15.14). Substances with high surface tension offer more resistance to shape deformation. Conversely, low surface tension speaks for easy deformability, hence longer and topologically complex interfaces.

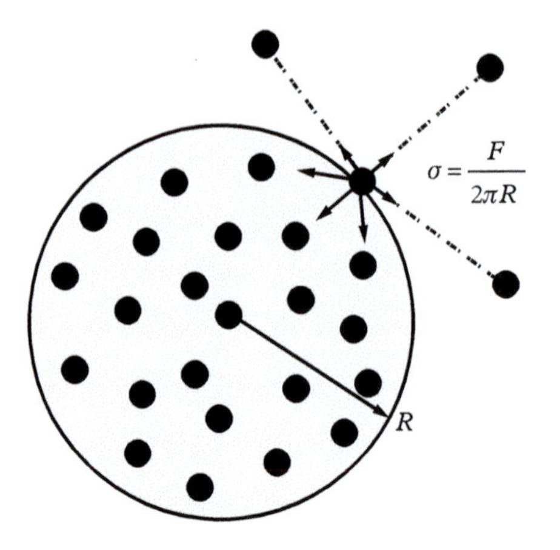

Figure 15.14 *The microscopic origin of surface tension. The liquid droplet inside has higher density that the vapour outside, hence the boundary molecules feel a stronger attraction to the interior than to the exterior, and this difference defines the surface tension. Surface tension is highly sensitive to temperature and vanishes at the critical temperature at which the liquid droplet evaporates.*

15.8.3 The wall builder

Microscopically, surface tension is readily explained by considering 'frontier molecules' sitting at the interface of a liquid droplet, say water surrounded by its vapor. Such frontier molecules experience a net attraction towards the interior of the liquid droplet, simply because due to the lower density of vapor versus the liquid, there are more molecules inside than outside. Remember Lennard-Jones: these forces decrease with distance and since the average distance between two molecules in the liquid is smaller than in the vapour; attraction to the liquid is stronger than to the vapour; the droplet is safe, long live the droplet! This force deficit leads to a corresponding deficit between the pressure inside and outside the droplet.

It is easily checked that surface tension is an energy per unit area, hence it measures the energy it takes to deform and increase the area of the droplet. Low surface tension means easy to deform and vice versa, large surface tension indicates resilience to shape changes.

Water is renonwed for its large value of σ, which is the reason why it can lift high up in thin tubes (capillaries) in the face of gravity. To get a sense of it, in a capillary tube one tenth of millimetre in diameter, the capillary force stemming from surface tension can lift water up to an elevation of 14 centimetres, namely 1,400 times larger than the radius of the tube, (see Fig. 15.15). This is because of the inverse dependence on the radius (see Appendix 15.1): that's how groundwater feeds metres-high trees.

The role of surface tension as a wall builder and, more generally, structure builder in soft matter and biology is paramount. In particular, since surface tension

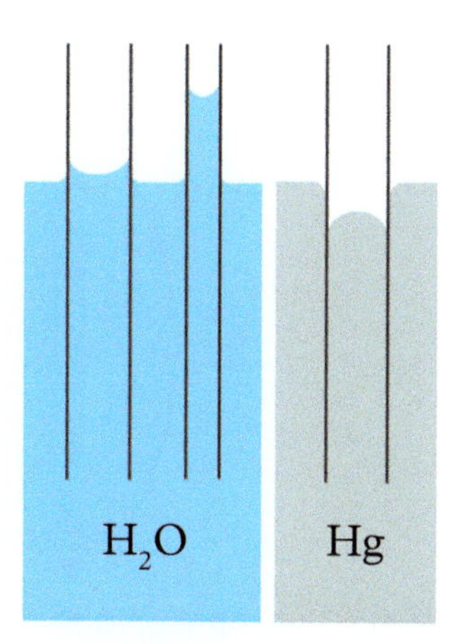

Figure 15.15 *Capillary forces in action. Water does not like to lie on solid walls (hydrophobic glass) and experiences an upward force inversely proportional to the radius of the tube, which is why it lifts higher on the thinner tube on the right. That's how ground water feeds the trees. Mercury, on the other hand, does not like to lie on the walls and consequently it gets pulled back. When the wall temperature is increased, the density of mercury goes down hence to keep the weight the same, the column must increase its height. That's how thermometers work.*
Source: reprinted from en.wikipedia.org.

measures the cost of building interfaces, it provides a natural driver towards configurations which minimize the surface to volume ratio. In other words, surface tension promotes coalescence of small droplets into larger ones, as discussed earlier on in this book. There are indeed several self-structuring molecules in biology but venturing into these matters would be another book. Two important lessons to be retained are that VdW interactions give rise surface tension and surface tension promotes coalescence of small structures into larger ones. The details can be hairy, but the informing principle is crystal clear.

Before leaving this section, it is appropriate to inform the reader that in the presence of multiple species, particularly oil and water, other types of dispersion forces arise which withstand the coalescence effects due to surface tension.

15.9 More dispersion forces

Dispersion forces are not necessarily related to electrostatic interactions, but usually arise whenever macroscopic bodies come in near contact, i.e. at distances within the range of molecular interactions. The forces arising from the coupling between these two disparate scales are largely affected by nanoscale fluctuations and are usually attractive.

A typical example in point is the so-called 'depletion' forces, experienced by colloidal particles immersed in a water solvent. The colloidal particles are usually

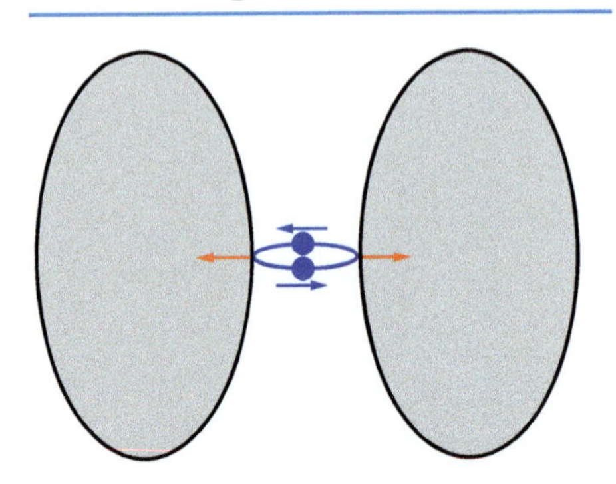

Figure 15.16 *The origin of depletion force. A small molecule bounces back and forth between the surfaces of two colloidal particles. The force transmitted by the molecules to the surfaces of the colloids is equivalent to an effective repulsion between them. Hence, whenever the number of 'trapped' molecules is decreased, the repulsive force is also decreased, which is equivalent to an effective attraction. Less repulsion means attraction!*

spheres of say one micron in radius, hence much larger than the water molecules they are immersed within, whose effective size is about one third of a nanometre. The water molecules constantly hit the surface of the colloidal particles, hence exerting a pressure on both, which keeps them apart. Now suppose that the water density undergoes a negative fluctuation, i.e. a local decrease of the number of water molecule per unit volume (depletion). Such a local decrease leads to less pressure on the surface of the near contact colloids, which amounts to an effective attraction between them, whence the name of depletion forces (see Fig. 15.16).

The pressure associated with such depletion forces is typically of the form

$$p_{dep} = C\frac{k_B T}{d^3}$$

where d is the distance between the closest points on the surfaces of the two colloids and C is a geometry-dependent numerical coefficient. Since $k_B T$ is a small amount of energy, such forces become relevant only when the distance d reaches down to the nanometre scale, i.e. comparable with the range of the attractive branch of the Lennard-Jones potential. As mentioned previously, these forces are typically attractive, but under special conditions, depending on the geometrical details of the surfaces and also their chemical composition, they may occasionally turn repulsive. This has important technological applications, for instance to minimize molecular friction in nanodevices. This is a very exciting frontier of modern nanoscience, a subject that lies beyond the scope of this book.

And the picture can get even richer than that: the interaction between charged colloids can turn Lennard-Jones upside down, i.e. short-range becomes attractive and mid-range repulsive! This means that if two colloids are sufficiently far apart, they repel each other even further apart, but if by some reason they manage to get in close touch, they eventually merge together to form a rich variety of macromolecular structures, including polymers. Obviously, this has major impact on the large-scale aggregation properties of the corresponding soft and biological materials. We could go on with the 'exotic' properties of soft materials, but this is not this book: the reader interested in digging deeper into this fascinating topic is kindly directed to the excellent and highly enjoyable book by Roberto Piazza [90].

15.10 Pause of reflection: The ouroboros

This chapter has straddled across a significant amount of material, much of which can be appreciated in full only by plunging into the details, which is not the purpose here. But aside from details, there is a conceptual underlying thread that should not be missed, namely the *bootstrap* nature of dispersion forces.

By bootstrap, we mean that dispersion forces determine the charge distribution that actually generates them in the first place, whence the metaphor: they lift themselves up by their own bootstraps. In a more erudite version, one may invoke the mythological ouroboros which eats its own tail, often taken as a symbol of self-feeding structures or entities. Coming back to dispersion, charges generate forces, which in turn tell the charges where to go in space. This is the typical chicken-egg loop, in which cause and effect can no longer be told apart. Remember Chapter 1?

For the sake of better clarity, let us denote by $\mathcal{R}$ the set of positions of the various molecules and by $\mathcal{F}$ the set of the forces acting on them. The force-structure relations can be symbolically represented as a sequence in which, given the initial position, one computes the corresponding forces, which in turn determine the new positions, which determine the new forces, and so on down an endless 'Ouroboros chain' :

$$\mathcal{R} \to \mathcal{F} \to \mathcal{R} \to \mathcal{F} \to \mathcal{R} \to \ldots\ldots\ldots? \qquad (15.4)$$

The main question is whether such bootstrap sequence converges to anything sensible, say a supramolecular structure, a folded protein, or perhaps a misfolded one, or maybe it just blows up altogether!

This is basically the same question we discussed in Chapter 14 using the powerful but somehow abstract language of free-energy landscapes. We argued that the funnel drives the is chain towards useful destinations. Here we make the same claim, by speaking however the language of real molecules instead. In principle, 'all one has to do' in order to answer the main question is to solve the equations of motion of every single atom or molecule, subject to classical forces, a task which requires no physics beyond Newton! Compare this to the glamour of string theory or quantum gravity and you may feel like shrinking down to the atomic scale yourself. Not so fast, though . . . .

15.10.1 Die-hard prejudices

There are two deep pitfalls (shall I call them prejudices?) hidden in the innocent-sounding previous statement: the first is 'all one has to do' and the second is 'no physics beyond Newton'. A prominent Italian experimental particle physicist is credited for stating that 'once the equations are known, physics is over'. In my modest view, the opposite is true: once the equations are known, the physics actually *begins*, because the physics is not in the equations, but in what we learn from them, namely their solutions!

Computing the structure of large biomolecules, such as proteins, is a walking advertisement for this point. But there are others, and modern soft matter and biophysics are littered with such instances. The equations are known, but they are just far too many and too hard to solve. So, what we do with them? One way to go is to build ever larger and faster computers, an excellent strategy which helps a lot but cannot solve the problem alone. The other way is to make sensible coarse-grained models. And the third is to combine the two previous examples, which is what current frontier research in the field is actually doing. And it is indeed coarse-graining that best reveals the second prejudice. The point is that by the very act of coarse-graining, genuinely new forces arise, which have no counterpart in the Newtonian world. And since they don't, they have no place in the empireum of fundamental forces. But they matter, a lot, that's exactly what I mean by Unfundamental with capital U. Of course, if we could solve the fundamental equations which govern the motion of atoms, nuclei, and electrons, there would be no need of coarse-graining. But, as we have emphasized several times in this book, besides being unviable this is often not desirable either, for it would provide more information than we need. This said, as computers grow faster and coarse-grained models improve, we can do an increasingly better job at describing the 'ouroboros chain', a fundamental task that has been recognized twice by the Nobel Committee in Chemistry (1998, 2013).

15.11 Beyond dissipative structures?

In Chapter 10 we have discussed dissipative structures as a remarkable example of natural systems capable of displaying organized coherent motion beyond a critical threshold of nonlinearity. We have also observed that in the late stage of his career, their main advocate, Iliya Prigogine, charged them with far-reaching implications for living organisms, and that such a projection did not really meet with much consensus from his peers. In particular, Stuart Kauffman argues that dissipative structures fail to attain the level of Complexity of biological organisms

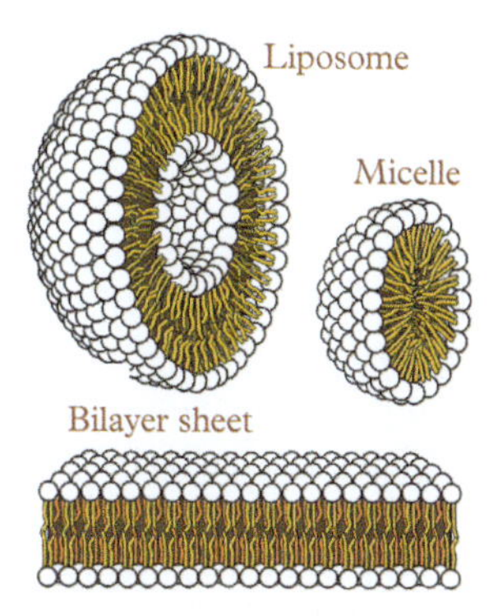

Figure 15.17 *Supramolecular structures: Liposome, micelle, and lipid membranes.*
Source: reprinted from en.wikipedia.org.

because they are subject to fixed boundary conditions. Hence, unlike biological structures, they do not build up their own boundaries in a self-consistent fashion, a property which represents an integral component of *Kauffman's World Beyond Physics* discussed earlier on in this book [53].

Based on what we have learned in this chapter, and most notably the Ouroboros structure of dispersion forces, I think it is fair to say soft matter makes a number of strides in the direction of bridging the gap between physics and biology. Indeed, soft matter often presents situations in which supramolecular structures arise from molecular interactions and set up dynamically adaptive boundary conditions for the very molecules which form them in the first place. This is how a variety of organized macromolecular structures of major biological relevance, such as membranes, vescicles, and micelles, take form out of the underlying molecular physics, (see Fig. 15.17). These macromolecular structures are still far from attaining the functional complexity of the living cell, but they represent nonetheless a major leap towards *Kauffman's World Beyond Physics*.

15.12 Summary

Let us wrap up the broad conceptual ride we took in this chapter. Soft matter is the state of matter most relevant to biological systems. It isn't about gases, liquids, or solid, but rather about a blend of them whose properties cannot inferred from any of the three, because they combine nonlinearly. The main characters are *unfundamental* forces which do not apparently require physics beyond Newton and yet display a great deal of Complexity, due to self-consistent coupling between the geometry that they determine and which they depend upon at the same time. These unfundamental forces, more seriously known as 'dispersion forces', exhibit coexistence of short-range repulsion and mid-range attraction. At distances much longer than the molecular size, they vanish altogether. Even though they are *weak and short-range*, they act cooperatively to the point of promoting the growth of supramolecular structures in the face of the entropic wind.

15.13 Appendix 15.1: Surface tension and Laplace's law

Surface tension gives rise to so-called capillary forces, whose expression is summarized by Laplace's law, expressing the difference in pressure ΔP, between the liquid (inside the droplet) and vapour (outside) as a function of the droplet radius, R:

$$\Delta P = \frac{2\sigma}{R}$$

where σ is the surface tension.

From the previous Laplace relation, it is seen that surface tension is a force per unit length or, equivalently, energy per unit area. Hence, the mechanical work (energy) needed to increase an interface of area by the amount δA is $\delta W = \sigma \delta A$. This is the analogue of the work done on the volume, $\delta W = p\delta V$, as we met it in Chapter 8, devoted to Thermodynamics. The Laplace formula shows that the capillary forces scale inversely with the radius R of the droplet, hence they are particularly relevant to 'small' objects.

Just to give an idea, water features $\sigma \sim 0.07$ newton/metre. Consider now a droplet of water 1 millimetre in radius. The capillary force due to surface tension is

$$F_{cap} = \pi R\sigma \sim 2.5 \times 10^{-4}$$

Newtons, to be compared with the gravitational force

$$F_g = \frac{4}{3}\pi R^3 \rho g \sim 4 \times 10^{-5}$$

Newtons. In other words, for such a small droplet, capillary forces far exceed gravity. This is the mechanism by which water can flow from the ground up to the top of tall trees .... Make the droplet radius 10 cm, and the balance would revert completely in favour of gravity, no way trees could be fed by groundwater with such fat droplets .... Small definitely works better than big in the world of surface tension.

16

Water, the Wonderfluid

If you don't understand water, you don't understand biology.
(Gene Stanley)

16.1 Introduction

In the previous chapters, we have discussed molecular forces and their ability to promote structure formation at the molecular and supra-molecular levels. These forces alone, however, are not sufficient to account for biological processes like protein folding, which are crucially affected by the way the protein interacts with water. For instance, it is known that in-vacuo most proteins simply do not fold at all. In this chapter, we shall convey an idea of the vital role played by water in shaping complex biological processes, including protein folding.

16.2 The magic fluid we call water

Water is often called the fluid of life, and deservedly so.

Fluids are a pervasive presence across virtually all walks of science and life, air and water being arguably the most immediate examples in point. But while air, the champion of ideal gases, is comparatively simple to master, water remains a trove of puzzles [33, 70, 24]. Here goes the story. One oxygen and two hydrogen atoms, the famous H_2O, is the chemical formula of water, as simple as that, (see Fig. 16.1). It is just amazing what kind of marvellous, and world-shaping power emanates from such an innocent-looking chemical formula. Indeed, there's no need of a PhD in chemistry to appreciate how essential this magic substance is for our own survival.

As noted previously, water is often described as the fluid of life, a well-earned definition given its crucial role in most biological phenomena, let alone the fact that it contributes more than 60 per cent of our bodies! We can keep going several days or even weeks without food, but no more than a few days without a water refill. Water is yet another example of magic that the most privileged among us take for granted simply because it is around us all the time. The same thing isn't true in

Sailing the Ocean of Complexity. Sauro Succi, Oxford University Press.
© Sauro Succi (2022). DOI: 10.1093/oso/9780192897893.003.0016

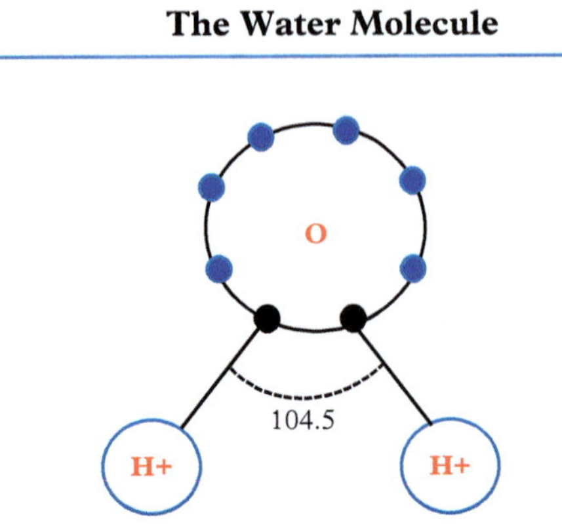

Figure 16.1 *Sketch of a molecule of water. The oxygen atom has six electrons on the outer shall, which means that two are missing to complete the eight required for chemical stability (this comes from quantum mechanics). The two missing ones are kindly provided by the two hydrogens, to form a very stable covalent bond between the three atoms. Covalent means that the two electrons are shared between the oxygen and the two hydrogens and glue them together. The water molecule is born.*

many other parts of the world: billions of people live where rainfall is scarce and a moment's conversation with one of them leaves little doubt as to the immense value of this wonderfluid. The fact is that nothing is trivial about water, it is a most fascinating fluid, still full of mysteries and anomalies, most of them with far-reaching implications for the development of life at all scales, from biomolecular all the way up to geophysical.

16.2.1 Water anomalies

But what are exactly these anomalies? You might be surprised to hear that they number no fewer than eighty, yes eight-zero! But relax, we won't need to go through the full list, we shall just stick with the main ones.

First, water is the only natural substance that exists in all three states, gas, liquid, solid, at temperatures normally found on Earth. This is called triple point, and it is located at about 0.01 degrees Celsius in standard conditions.

Second, water is one of the few fluids which 'loses weight' on becoming solid: that is why ice floats on liquid water.

Third, the density of water peaks at a temperature of 4 degrees Celsius, which means that liquid water can survive below ice in wintertime. Fishes and any other marine creatures are very pleased at this on a freezing winter's day.

Fourth, water is highly polarizable, hence it provides a strong shield to electrostatic interactions, which, as we have seen, is central to the stability of matter.

Fifth, water has high surface tension, which means that it can walk uphill against gravity in small capillaries, a property trees owe their existence to.

The list could continue up to the credited eighty anomalies, but the previous five will suffice us.

16.2.2 Water molecular networks

Next, we turn to nanoarchitectural considerations (for a handy introduction see www://chem.libretexts.org). As mentioned previously, chemically speaking, the water molecule looks pretty innocent: a tiny V-shaped molecule, with the chemical formula H_2O, consisting of two hydrogen atoms attached to a sixteen times heavier oxygen atom. The water molecule is electrically neutral but polar, with the centres of positive and negative charges located about a half Angstroem apart. The hydrogen atom consists a nucleus made of a single positively charged proton, surrounded by a single negatively charged electron. The oxygen atom has a nucleus consisting of eight positively charged protons and eight uncharged neutrons, surrounded by a cloud of eight negatively charged electrons. Based on the laws of quantum mechanics, two electrons 'circle around' in a close orbit around the nucleus and the other six circle around a second orbit at a larger distance from the nucleus. Based on the laws of quantum mechanics, in order for the oxygen atom to achieve maximum stability, another two electrons would be needed in the second orbit.

Hence, the oxygen is said to be 'electronegative', to express, so to say, its 'thirst for electrons'. The oxygen's thirst of atoms is appeased by the two hydrogens, which share their electrons, thus closing the electro-stability count. This is called a O-H *covalent bond* and provides the chemical cement of the newborn water molecule. I speak of cement because in units of chemical energy, that's what the covalent O-H bond really is. The number is 492 kilojoule/mole, which converts to about 5 electronvolts, which in turn corresponds to about 200 thermal $k_B T$ units. Now recall what we wrote about the chemical mountains of hyperland two chapters ago: mountains classify as high or low based on how they compare to $k_B T$. In this respect, the O-H covalend bond is a sort of Mount Everest! I find it really amazing that such a simple toy-like structure holds the key to most of our vital functions! But the magic is not over, bear with us.

16.3 The third thing

Back in the day, the chemist D. H. Lawrence (1885–1930) wrote, 'Water is H2O, hydrogen two parts, oxygen one, but there is also a third thing, that makes it water and nobody knows what it is.' It would be utterly reckless of me to attempt an answer, given that this question still occupies the best chemical minds around the world. However, as far as I can judge, the 'third thing' is not chemistry alone, but it has to do with a distinctive trait of Complexity, namely *connectivity and networking*. Modern research in statistical chemistry has unveiled the richness of the way water molecules manage to team up in space and form complex structures. In a word, water has major networking skills. For instance, it is known that ice comes in a typical tetrahedral configuration, with each oxygen H-bound to another four. This configuration can form large-scale ordered structures which prove lighter than more disordered configurations characterizing liquid water. This is the basic reason why solid water, best known as ice, is less dense than liquid water. So, the richness comes from the chemical structure, sure, as combined though with the networking properties of water molecules, i.e. the way they pack together to form large-scale structures, (see Fig. 16.3). Before we dig a little bit deeper into the chemical basis of water networking abilities, a few general words on the packing business are in order.

16.4 Matters of packing

Ice floats on liquid water, this is possibly the most striking property of water. As noted previously, this has to do with the way that water molecules pack together. To appreciate the point a little bit better, we turn to a very practical daily life problem: packing solid objects within an empty box (an exercise we all did when moving to a new location). We all know that the way how you lay objects down has a decided impact on how many of them you can fit into the box.

For the sake of simplicity, consider a collection of equally sized spheres to be placed in a cubic box and ask the following question: what is the most effective arrangement of the spheres, namely the one ensuring the highest number of spheres fitting within the box? Apparently, the best way is to lay down the spheres along a uniform and ordered array, precisely like a crystal, in which each sphere is inscribed in a cube of side equal to the diameter of the sphere, with six equidistant neighbours, left-right, front-back, up-down, (see Fig. 16.2). Let L the side of the big box, say one metre, and D the diameter of each sphere, say 10 centimetres, we can place $N = L/D = 10$ spheres in a row along each dimension, for a total of $10 \times 10 \times 10 = 1000$ spheres. Physicists measure the efficiency of arrangements of solid objects in terms of the so-called *packing fraction*, in our case the volume occupied by the thousand spheres versus the total volume of the cubic box that contains them.

For the cubic arrangement, the calculation is easy: the volume of a sphere of radius D is $V_{sph} = \frac{4\pi}{3}\frac{D^3}{8}$, and each sphere is contained in a cube of side D, whose volume is $V_{cube} = D^3$. Hence, the volume fraction achieved by this crystal arrangement, known as simple cubic, is $\phi = \frac{4\pi}{3}\frac{1}{8} = \frac{\pi}{6} \sim 0.523$. In other words, the spheres occupy just a little above *half* of the available box space, which sounds a bit disappointing, doesn't it?

The natural question is: is this the maximum packing fraction crystals we can achieve? Detailed studies show that the answer is a definite no! Other crystal

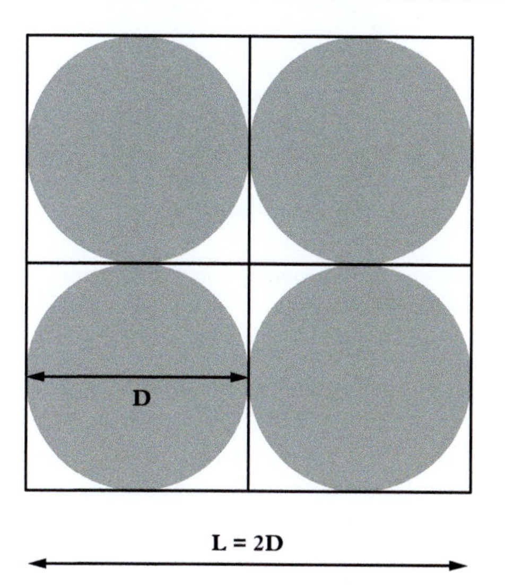

Figure 16.2 *Simple cubic arrangement of four spheres (circles in the plane) in two dimensions. Each of the four circles is contained in a box of side equal to its diameter D, hence the area fraction (packing fraction in two dimensions) is* $\frac{\pi D^2/4}{D^2} = \pi/4$.

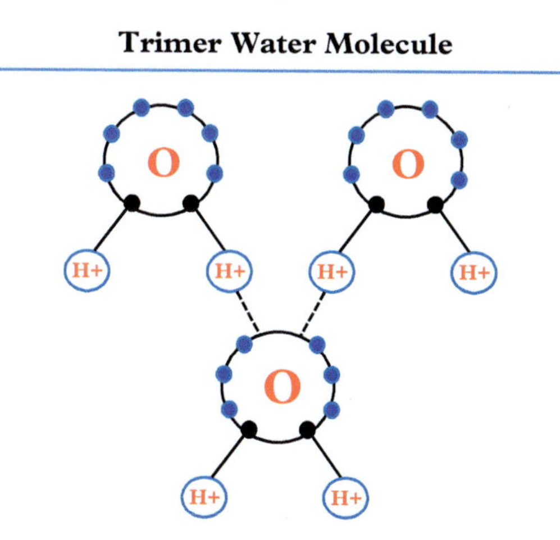

Figure 16.3 *A simple network of three water molecules (water trimer) via HB's (dashed lines), one with the upper-left molecule and one with the upper-right.*

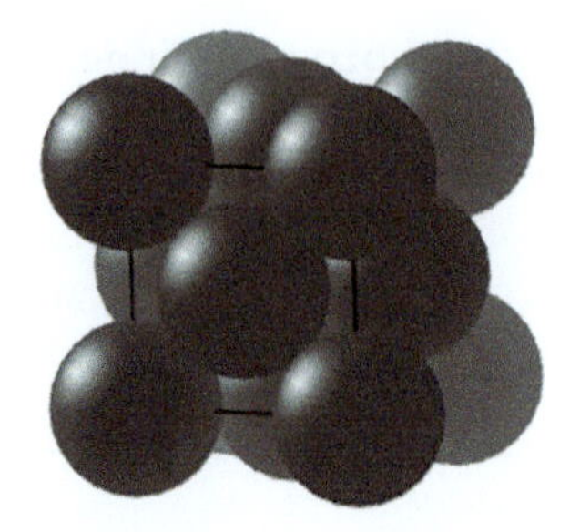

Figure 16.4 *Face-centered-cubic arrangement in three dimensions.*
Source: reprinted from commons.wikimedia.org.

structures exist (face-centered cubic and hexagonal close-packed for the specialist) which manage to reach up to $\phi = 0.74 \sim 3/4$, namely about 75 per cent of the available box volume, respectively. With boxes (cubes) in a box, instead of spheres, we could obviously attain a full 100 per cent, but, whether we like it or not, with spheres in a box, 75 is as good as we can get, (see Fig. 16.4).

Now comes a truly interesting point, and the point is that even disordered configurations ('liquid') can outdo the simple cubic arrangement, with packing fractions up to $0.64 \sim 2/3$. It so happens that liquid water is indeed a disordered configuration featuring $\phi \sim 0.4$, while ice is a tetrahedral network featuring a lesser $\phi \sim 0.34$! This is why the liquid water is denser than ice, letting ice float on liquid water. Secret explained! Even though we have not explained *why* solid water arranges in tetrahedral configurations, I hope my reader sees neatly why the networking properties of water, i.e. the way different water molecules connect to each other, has such a paramount effect on the macroscopic properties of water. The chemical bonds provide precisely the glue for this assembly procedure, while the specific topology depends on the free energy of the overall configuration.

Before we delve into further details, a short comment on the relation entropy-disorder is appropriate.

16.4.1 Entropy is not (just) Disorder

In chapter 8, we briefly mentioned that even though entropy is a measure of microscopic Disorder, we should not hasten to identify the two. Water offers a very clear example of the reason why we shouldn't. We just mentioned that the ordered tetrahedral arrangement features $\phi \sim 0.34$ against the $\phi \sim 0.4$ of disordered liquid water. This means that the tetrahedral arrangement leaves some 6 percent further space to be filled by proper (in fact random) moves of the spheres. If we interpret entropy not merely as a measure of Disorder but more generally as *propensity to change*, [90] it is clear that the entropy of the tetrahedral crystal is *higher* than that of the liquid, because in the tetrahedral state, the water molecules can still be displaced to attain a higher packing fraction. And since tetrahedral states have higher entropy than disordered ones, below the critical temperature of zero degrees Celsius, water spontaneously crystallizes into a lighter tetrahedral solid! This is pretty counterintuitive and indeed, for a good while, it puzzled scientists as well, before they could come up with a rational explanation. Having brought up this very counterintuitive but crucial point, let us proceed to discuss the chemical mechanism which promotes the formation of tetrahedral networks. Believe it or not, if only in humerous modes, this takes us to the most famous secret agent of all: James Bond!

16.5 My name is Bond, H-Bond

My name is Bond, James Bond. Who has never heard the famous catchphrase of the most legendary secret agent in the world? But, given that water was definitely not high in the list of Bond's favoured drinks,[54] what does he have to do with water? Well, the fact is water features another bond besides the covalent one discussed earlier on, named hydrogen bond, (see Fig. 16.5). The hydrogen bond, (HB) for short, is a measure of the chemical affinity of a given molecule to bind to the two hydrogens of the water molecule. This applies to any molecule, including water itself, and even though HB is definitely less known than James Bond, it is hardly less important.

So, how strong is the HB? To answer this question, let us return for a minute on the Van der Waals (VdW) forces discussed in Chapter 15. They are called *non-bonding* because they do not originate from chemical bonds between atoms, but rather from electrostatic forces due to charge dispersion, as we discussed at length in Chapter 15. VdW forces are of the order of $k_B T$ or even less, hence they are 'small hills' in the chemical landscape. Yet, as we have commented before, they

Figure 16.5 *Hydrogen bonds between water molecules give rise to extended complex networks. Icy water is one of them. The most iconic James Bond, Sean Connery (1930–2020), in action. Instead of water, he preferred 'Vesper Martini', for the record, 'Three measures of Gordon's, one of vodka, half a measure of Kina Lillet', as described in Fleming's 1953 book. And of course 'shaken, not stirred'!*
Source: reprinted from en.wikipedia.org and it.wikipedia.org

[54] The iconic Bond's 'Shaken, not stirred' hardly referred to water, but to Martini instead!

do make themselves pretty much felt at the supramolecular scales, because they are pretty good at cooperating. The HB lies somewhere in between VdW and the covalent H-O bond, with energies in the range between a few to about hundred thermal units. Hence, the HB bond is stronger than VdW forces but at least ten times weaker than the covalent H-O bond.

16.6 Hydropathy, the molecular builder

As we said earlier on, the HB largely controls the affinity of any molecule to water: molecules with high affinity tend to link up to water and are called *hydrophilic* while those that don't are called *hydrophobic*. Literally from the Greek, the former are 'water-friends' while the latter abhor it. This attraction(aversion) to water is also called *hydropathy*. There is a third, very important, family of molecules called *amphiphylic*, for they manage to accommodate both sentiments (*amphiphylic* means somehow 'friend of both').

Given that water is the dominant substance in biology, including in our own body (roughly sixty per cent), it is not hard to see why hydropathy controls so much of the Complexity of biological systems. In particular, it is key to the 'dance of proteins' mentioned in Chapter 15, which we shall shortly return to. Water lovers like to mix with it, to the point that when you put them together, they happily mix with water all the way down to the molecular level. Salt and sugar, the two champion opposites in taste profiles, both fall under the water-lover's umbrella. In a more serious language, they are soluble. Among those who don't like water, pride of place is definitely taken by oil, notoriously reluctant to team up with water. When forced together, oil forms droplets, just to make sure it stays as far apart as possible from the surrounding water and minimize its contact with it.

Such contact, however, cannot be avoided because, chemically speaking, oil means hydrogen and carbon, two elements also crucial for life, as they are the basic constituents of lipid membranes. Hence, if not a happy marriage, some form of compromise is necessary, as we shall note shortly.

Finally the amphiphilic, those who love both, although in different regions of their structure, typically head and tail. These molecules, also known as *surfactants*, typically consist of a hydrophilic head and a hydrophobic (lipophilic, for oil-lover) tail. By this very definition, it is quite clear that these molecules are just made for living at the interface between water and oil, head to water, tail to oil. Hence they are ideal 'middlemen' to ease up the 'impossible' but necessary compromise of keeping oil and water together. They are the 'molecular diplomats' in the game.

There is no need of any deep math or chemistry to appreciate the immense building power of the simple rules described previously.

If you put oil in water, the reaction is to segregate and expose the minimum amount of surface, so as to minimize the free-energy cost of building the water/oil interface (remember the role of surface tension as a wall builder discussed in

Chapter 15). This self-segregating attitude lays down the seeds for structure formation, for the case in point, droplets. But it also plays a crucial role in protein folding, as we shall see shortly. The droplet is a simple structure: further Complexity is thrown on the table by introducing surfactants. Indeed, by lowering the cost of building and maintaining interfaces, surfactants allow complex and extended interfaces to survive much longer, thus paving the way to the emergence of significantly more complex long-lived supra-molecular structures, such as membranes, vescicles, and micelles, (see Figs. 16.6 and 16.7). This is how oil, water, and surfactants promote organized growth from the molecular to the supramolecular level, laying down the premises for the emergence of proto-cellular structures.

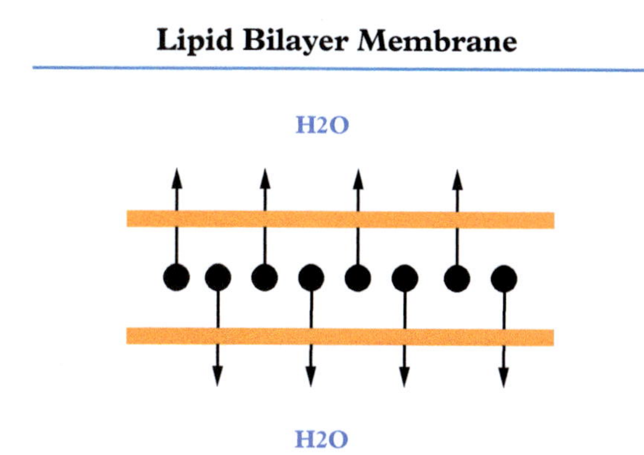

Figure 16.6 *Two layers of amphiphilic molecules (lipids) form a membrane, separating inner water from the outer one. The hydrophilic head of lipid likes water, while the hydrophobic tail stays away from it, so that an alternating sequence of up-down lipids gives rise to a planar membrane separating water on the top from water on the bottom.*

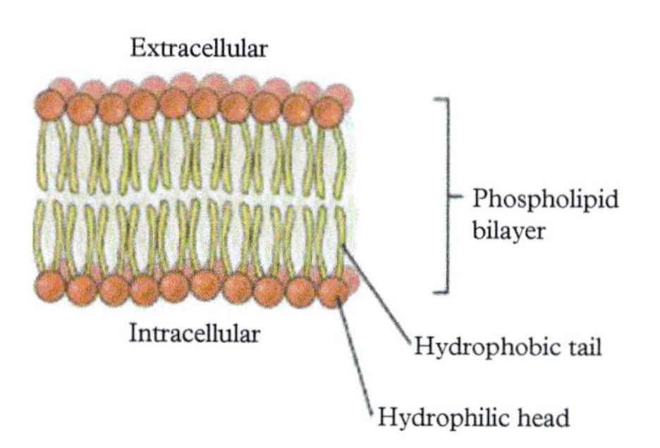

Figure 16.7 *A lipid bilayer, namely a biological membrane consisting of two layers of lipid molecules. Each lipid molecule, or phospholipid, features a hydrophilic head and a hydrophobic tail. Being repelled by water and slightly attracted to each other, the tails match together and form the biological membrane.*
Source: reprinted from commons.wikimedia.org.

This opens up an entire new chapter, in which soft matter gets really close to biology, but again, this is another book [90]. We'll close it here and go back to proteins instead.

16.7 Back to protein folding

As we discussed in Chapter 15, proteins consist of a sequence of amino acids, each formed by a sequence of three basis out of four, adenyne, cytosine, guanine, and uracil. Like any other molecules, amino acids come with their own *hydropathy index* (HI), a positive number for hydrophobic molecules and negative for hydrophilic ones. For instance, Alanine, symbol Ala, features $HI = 1.8$, while Asparagine, symbol Asp, has $HI = -3.5$. A full table of the HI index for the twenty amino acids can be found at page 175 of Piazza's book [90]. It is therefore clear that the dance of the protein in water moves to a very different drum as compared to the case of empty space ('in vacuo' as scientists use to say). In empty space, many proteins would never be able to progress from their primary configuration to the native one: no life in a vacuum! The mechanism is plain: hydrophilic regions of the protein tend to expose to water, while hydrophobic ones tend to hide away from it, (see Fig. 16.8). This provides a powerful driving principle towards the compact globular structures with hydrophilic regions mostly on the surface and the hydrophobic ones hidden in the interior of the native configurations. Hence, this philia/phobia competition/cooperation is a crucial component of the abstract free-energy landscape described in the previous chapters. This is how oil, water and surfactants promote organized growth

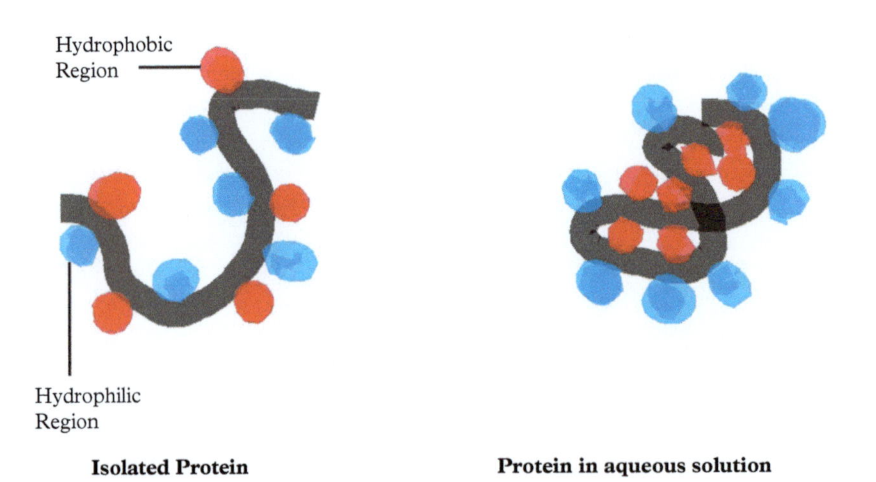

Figure 16.8 *A cartoon of the effect of water on the process of protein folding.*
Source: reprinted from commons.wikimedia.org.

from the molecular to the supramolecular level, laying down the premises for the emergence of proto-cellular structures.

As anticipated when discussing the funnel idea, the folding dance is a pretty touchy one. It is touchy to the point that sometimes the difference between the correct and a faulty fold is in the hands of just a few hydrogen bonds! For instance, it is known that misfolding is often associated with a few hydrophobic residues left on the protein surface. Nice and also a bit scary, isn't it?

In passing, we note that this raises a notable exception to the argument we made before in this book, according to which being made of very many atoms, we are kind of protected against the individual vagaries of the atomistic world. No longer true in this case, where minute errors, such as a misfold, may result in truly dire consequences. This is a typical hallmark of biological complexity: details which make the whole difference.

16.7.1 Molecular storms

We're not done yet with dancing proteins. We should note that, besides experiencing various types of molecular forces, biological bodies are constantly exposed to thermal fluctuations, the vagaries of the environment. To macroscopic creatures, like us, these are peanuts: remember that $k_B T$ is a ridiculously small amount of energy on the human macroscale, as it takes Avogadro's of them to make just one single kilocalorie, of which we need a couple of thousands per day. But proteins are small, and at their scale, $k_B T$ is a lot of energy, nothing short of a molecular thunderstorm! Such thunderstorms come with enough energy to destroy the protein, which they do indeed, thus spoiling their precious dance towards the native structure. Here another wonder takes place: if our car gets seriously damaged, if not written off altogether, in an accident, there is no way to get it back in shape in any reasonable amount of time (let alone money ...). Proteins, on the other hand, do that all the time: they disassemble and reassemble in the blink of an eye! Cars don't self-repair, but proteins do, and pretty fast indeed! This self-healing capability is again an amazing outcome of the various forces acting on the protein, including water. The details of this magic funnel-assisted recovery remain to be fully understood, but there is a general consensus that the hierarchical structure of the funnel is a key ingredient of this prodigious self-recovery ability. Much of current biochemical research is focussed on the derivation of the proper (coarse-grained) force fields which control the intra-molecular interactions within the protein as well as their interaction with water. In the sequel, we provide a few further details on these intra-molecular interactions.

16.8 Intra-molecular interactions

Intramolecular forces are responsible for proteins structure: they shape them and hold them together precisely in the form that is needed to deliver their specific functions. The chief source of structure is the interaction between the amino acid

Intramolecular Force Fields

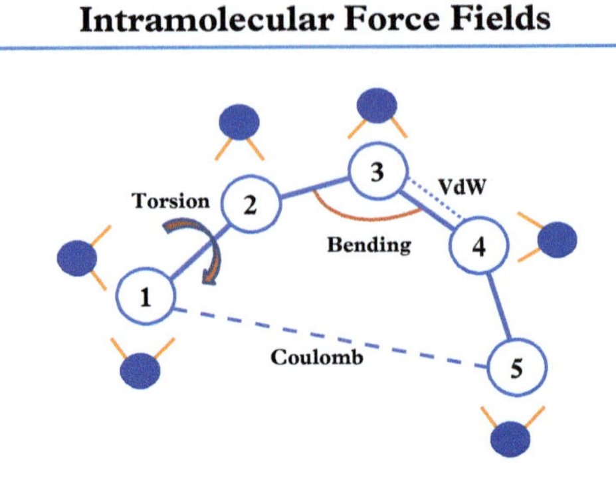

Figure 16.9 *Top: sketch of the intramolecular forces within a five amino acid cartoon protein. Bonding: bond length (thick solid lines), torsion (1–2) and bending (2–3–4). Non-bonding: VdW (3–4) and Coulomb (1–5). The non-bonding are significantly weaker, whence the dashes. The five amino acids are surrounded by water molecules, here represented as solid circles (oxygen) with two antennas (hydrogens).*

residues, while interactions between residues of the same protein give rise to the secondary structure of the protein, the so-called beta sheets and alpha helices. The schematics of the intra-molecular forces acting on the protein structure are reported in Fig. 16.9. Two basic families of interactions can be distinguished: those associated with chemical bonds (bonding) and those resulting from electrostatic forces (non-bonding). The former consists of three terms, the first describing longitudinal oscillations of the bond lengths around their equilibrium position, the second associated with rotations around the bond axis (torsions), and the third associated with the bending of the structure, i.e. changes of the angle between three 'atoms' in the chain. The non-bonding terms consist of unscreeened Coulomb interactions plus dispersion VdW forces, generally much weaker than bonding forces. It may be useful to condense the story into a 'Newton-like' equation, one for each 'atom' in the protein:

$$ma = \sum_{bonds} F_{bonds} + \sum_{angles} F_{angle} + \sum_{torsions} F_{torsion} + \sum_{atoms} F_{VdW} + \sum_{atoms} F_{coulomb} + \sum_{watermolecules} F_{water} \tag{16.1}$$

where m is the mass of the 'atom' and a its acceleration.

It should be appreciated that the previously used force fields represent a major approximation of the real story, for the actual chemical interactions within the protein, and most notably the chemical bonds, are most definitely quantum mechanical in nature. The trouble is that a fully quantum treatment of such a complex structure, consisting of several thousands of atoms, remains a distant chimaera for any foreseeable computational scheme. Eventually, the effect of the electrons is taken into account, through a sophisticated coupling between classical and quantum mechanical methods, known as QM/MM for quantum mechanics/molecular mechanics. All of the these calculations, of course, must be summed up with inter-molecular interactions, specifically with water molecules which form the surrounding environment of the protein. This gives rise to a very complex chicken-egg electro-chemical-mechanical coupling, whereby the forces that dictate the protein structure and dynamics are in turn dictated by such structure and dynamics in return. This provides the mechanistic background of protein folding.

This molecular-mechanics' view of the protein may convey the idea of the protein as a sort of rigid mechanical toy and indeed at low temperatures, this picture would not be far from the truth. As the temperature is increased however, the various parts of the 'mechanical toy' acquire their own flexibility, because each component starts 'vibrating' around its own equilibrium position, each with its own tempo. Like in orchestra, however, each instrument, although vibrating at its own rhythm, must cleverly coordinate with every other one, in order to produce a symphony instead of a cacophony or plain sonic dullness. Likewise, the various parts of the protein must vibrate in sync so as to produce purposeful motion. One could literally say that the protein moves from the state of a 'hard solid' to a state of 'soft matter'. And as the temperature is raised further, the protein is disrupted by thermal heat, a process called 'denaturation'. Once again, the delicate balance between Order and Disorder in full action!

I wish to call the reader's attention on the fact that molecular mechanics is nothing (explicitly) beyond Newton and electrostatics: good old physics. Yet, the harmony is a subtle one and remains computationally too complex to be unveiled in full. A very unfundamental, yet paramount problem [90].

16.9 Funnel explained?

We promised a molecular underpinning of the free-energy landscape and in particular of the existence of the deep corrugations called funnels which act as portentous accelerators towards the happy valleys of the landscape. Did we fill our mandate?

In the reader's shoes, my answer would be yes and no (amphiphilic answer ...), but probably more no than yes. Sure enough, we do have a sensible grasp of the basic forces which control protein motion, but the specific mechanisms by which the collective action of these forces manages to drive the protein from its primary (linear) structure to the native shape, *in matter of milliseconds and under heavy molecular storms*, are still open. In particular, we still do not know what degree of specificity needs to be injected into the model forces before quantitative predictions can be made to the desired level accuracy. As mentioned previously, just a few missing hydrogen bonds may turn a correct fold into a wrong one, which is all but a negligible mistake.

Leaving aside emotions, one could say that it is 'just' a problem of molecular mechanics, but one that involves too many deftly orchestrated actors to be solved exactly, or even to a conclusive degree of approximation. This is why, despite spectacular progress in the field, we still do not have a comprehensive theory of protein folding. Besides sheer intellectual fulfillment, such a theory is paramount for practical applications, most notably the development of new drugs and pharmaceuticals. And part of the problem is because we still do not have a full grasp of the interaction of the protein with the surrounding water molecules. As my friend and colleague Gene Stanley used to put it: if you don't understand water, you don't understand biology!

16.9.1 Caveat

Before leaving protein funnels, a word of caution is in order. In this book we have glorified the beauty and power of the funnel in providing a very effective route to the proverbial hay in a truly astronomical haystack. The reader should be informed that, not withstanding its beauty and power, the funnel idea is not exempt from criticism among professionals, and not secondary calibers .... In particular, Martin Karplus, a founding father of computational biochemistry and a Chemistry Nobel 2013, (see Fig. 16.10), notes that the notion of a guiding funnel is a misconception. In his own words [55], 'The essence of the misconception is that the decrease in configurational entropy, which gives the

Figure 16.10 *From the left: Arieh Warschel, Michael Levitt and Martin Karplus, being awarded the 2013 Nobel Prize in Chemistry 'for the development of multiscale methods for complex chemical systems'.*
Source: credit Nobel Media AB.
Source: photo: A. Mahmoud.

diagram it funnel-like shape and aids the polypeptide chain in finding the native state. In fact, the opposite is true; that is, the decrease in the number of available configurations, as the native state is approached, actually tends to slow the folding'.

In other words, if I interpret correctly, Karplus points out that the funnel representation hides the fact that entropy always opposes the attainment of the native state, and particularly so in its vicinity, as the number of available configurations becomes increasingly small. While he concedes that the funnel picture has stimulated useful work in the field, he concludes that 'there are hidden complexities not evident in the experiment nor present in the folding funnel diagram'.

The bottom line of our story stays unchanged: the key point, to quote again from Karplus, is that 'the peptide folds to the native state and solves the Levinthal paradox by having to visit only an infinitesimal fraction of the denatured configurations'. This 'infinitesimal' fraction is precisely the Ozland we referred to in Chapter 13.

16.10 The Theory of Something

Time to wrap up this long ride across molecular interactions and make a concession to some 'philosophical' remarks. As discussed in the opening of this book, it is hard to escape the deep fascination of exploring the ultimate frontiers of the physical world: the ultra-small, elementary particles, and the ultra-large, the overall Universe. This is what we call Fundamental Science (FS), for Fundamental (capital F) it really is. Possibly, the most popular spearhead of FS is the so-called Theory of Everything (ToE), the ultimate instance of reductionism: a single equation which would 'explain away' the entire Universe. Simple to the point of fitting on your T-shirt (European Council for Nuclear Research (CERN) and International Centre for Theoretical Physics (ICTP) Trieste are my favoured shops ...). This single, all-embracing equation would contain it all: from elementary particles to atoms and molecules all the way up to stars, galaxies and ..., yes, also the reason why you fell in love with X instead of Y. It is a fascinating dream, and even though an increasing number of scientists are inclined to believe that this is more of an inspiring chimaera than a scientific reality, the dream still lives on, with undiminished appeal for the general public. And again, understandably so.

For all this beauty and fascination, though, there are other beauties in town. Perhaps less bold and glamorous, but no less impactful, as they bear directly upon the stuff we are made of, as we have endeavoured to explain in the last four chapters. These beauties are (apparently) humbler in scope and instead of aiming at the ToE, they rather content themselves with a ... Theory of Something (ToS)!

ToE asks big questions: how did it all begin? How do we shape the future? Will artificial intelligence outsmart us? Just to pick some from Hawking's book *Brief Answers to the Big Questions*.

The key questions of ToS are typically very specific instead: how does DNA translocate across the cell? How do proteins fold? To this regard, I stand by the side of the Russian chemical-physicist (my guess would it was Boris Derjaguin (1902–1994), but I am not sure) who reportedly asserted that, 'it is much harder to have a Theory of Something than a Theory of Everything'.

In fact, the history of science shows that, more often than not, grand questions often end up in the Theory of Nothing (ToN), while grand answers typically result from 'small' and specific questions. There is a very peculiar process called serendipity, namely unplanned fortunate discoveries which eventually disclose entirely new scenarios. Einstein didn't develop his relativity theory with the intent of revolutionising our notion of space and time: he was 'just' puzzled by the fact that the speed of light was the same for observers at rest and those in motion.

Paul Dirac did not ask whether our world may contain anything like antimatter, he was rather preoccupied with the problem of making quantum mechanics formally compliant with Einstein's relativity. And when he succeeded, lo and behold, antimatter popped out as an inevitable consequence of the marriage between the two theories.

A more recent case in point is the theory of spin glasses, which was originally devised to study magnetic impurities and ended up providing fruitful tools for the study of a broad class of optimization problems, including neural networks and models of the brain!

16.10.1 And ... The Theory of Everything Else

I should add that complex system research is not immune to the fascination of ToEs either. For instance, a celebrated pioneer of the science of complexity, Stuart Kauffman is fond of the TAP equations as the basis for what he calls ToEE, the Theory of Everything Else! (see Fig. 16.11).

16.11 Dispelling prejudices: Can you imitate birds?

Soft matter is a major playground for ToSs, reflecting the penchant of biological matter, and complex systems in general, for a subtle mix of *Individuality* and *Universality*. Elementary interactions are universal, but once embedded in complex and self-consistently changing geometries, they generate specificity and diversity. This is a beautiful, deep, and powerful organizing principle which lies

Figure 16.11 *Stuart Kauffman, proudly wearing his own ToEE T-shirt on the scenic backdrop of Crane Island, Puget Sound, Washington!*
(*Source*: courtesy of Kathy Peil and Stuart Kauffman.)

at the physico-chemical basis of the structure-function paradigm which pervades biological science.

By the time they get 'contaminated' by geometry and statistics (coarse-graining), forces lose status of fundamental and become emergent instead. For a long time, this has been treated as a ticket for second-row science: this is the first prejudice we are talking about.

The point is that the more forces become emergent, the more they acquire relevance to biological functions, hence to us. This is by no means original thinking by this author, as the seeds of the idea were laid down some four decades ago in the epoch-making essay 'More is Different', by P. W. Anderson, one of the most influential physicists of modern times. I will not spoil the pleasure of the original reading, so I will content myself with the upshot: at any level, nature presents different categories of organization and so-called emergent phenomena, which *cannot* be inferred from the laws of the microscopic world they are based upon. By 'cannot be inferred', we imply precisely what we discussed for protein folding. We believe we have a good grasp at the basic forces acting between single atoms of a given protein, but when we put them all together, to analyze the collective motion taking the protein from the primary to its native form, we are still unable to calculate the process to the desired level of accuracy at the space-time scales of biological interest; which means that we are unable to pin down how it happens for real.

This takes me to the second die-hard prejudice. 'Calculating', a poor way of indicating the art of solving complex equations, is often dismissed as a second-tier task, a useful service to the cause of more noble scientific disciplines. Finding the right equations is obviously a paramount task, but the fact remains that knowledge does not spring from the equations but from their solutions instead! This is why, calculating, i.e. *solving* equations, is by no means less important than finding them out in the first place. The proof is that most complex phenomena remain opaque, if not impenetrable at all to this day, 'just because' we are unable to calculate them. The 'just because' is precisely the prejudice I am referring to. But, since we have already touched on this point before, I will stop here and close with a story I first heard many years ago in Boston from my colleague and friend Victor Yakhot.

Here goes the story.

A circus has an opening, and a young man walks in for the job. 'What can you do?' asks the entrepreneur. 'I can imitate birds', replies the young man. 'Come on', returns dismissively the boss, 'anyone can imitate birds'! 'Ok', says the boy, and without any further ado, he ... flies out the window! This is the 'just' I am talking about: everyone can calculate, in words, i.e. until the time comes to do it for real. And if you cannot, the front of knowledge is stuck, period. This is the essence of the Theory of Something, a Cinderella which works, every day, across virtually all walks of science. Silently and far from the limelight enjoyed by her shiny and far more celebrated ToE step sister.

16.12 Summary

The world is a book, and those who don't travel read only the first page.

(Saint Augustine).

In the last four chapters we have taken a view of the Complexity which inhabits the frontier between physics and biology through a small but representative window. I hope that, through this window, the reader may now appreciate *two* distinct and magic gifts which sit on this frontier.

The first is a very practical one: the forces acting on the protein, guide its dance 'against the entropic wind', to use the language inspired by free-energy landscapes. As discussed in this chapter, water has a decisive role in this play. This highly orchestrated dance keeps our body going *no matter whether we understand it or not.* To me, this inspires reverence, praise, and gratitude.

But there is a second astounding gift, and the gift is that, with all due limitations, we can make good sense of why and how this orchestrated dance works! Nature solves the problem the hard way, she has been given enough power to 'compute-away' the protein ab-initio, i.e. through the action of the quantum mechanical forces acting on every single atom in the protein. We just cannot go down this path all the way, even our most powerful computers pale away in front of such an ordeal. And yet, we can make sense of what's going through the abstract but very effective language of the free-energy landscape, backed up by

Figure 16.12 *The world is a book, and those who do not travel read only one page (Saint Augustine of Hippo). Two beautiful pages of my worldbook: sunset on the beach of Talamone (Tuscany) and the Boston waterfront on the Charles river.*

a variety of coarse-grained models of the molecular interactions which subtend these landscapes. These models are our best proxy to nature's way of action.

The first magic is of course what matters in practice, for otherwise we wouldn't be here to discuss the second at all! But in my view, the second commands almost as much praise and gratitude, because of the joy and sense of fulfillment that it breathes in our life. Here, my feelings align with Saint Augustine of Hippo (354–430), as he writes: 'The world is a book, and those who don't travel read only the first page', (see 16.12). I don't know whether Saint Augustine referred to physical travel alone, but I like to think that his words apply in greater generality, namely to any form of travel: physical, intellectual and spiritual alike. Under this light, Sailing the Oceans of Complexity between physics and biology is a bit like reading the book of the world within us.

This, at least, is the spirit that inspired this book.

And with that, we are ready for the final part of this book: Complexity and the Human Condition.

Part IV

Complexity and the Human Condition

17

Time, Complexity and the Human Condition

Quid est ergo tempus? Si nemo me quaerat, scio; si quarenti explicare, nescio. What is time? (If nobody asks me, I know, if somebody asks to me explain, I would not know.)

(Saint Augustine, Confessions, XI, 14, 400)

17.1 What is time?

Time isn't the main thing. It is the only thing.

(Miles Davis)

Time is perhaps the most elusive concept of all in science and philosophy alike, one that has captured mankind's attention since its dawn, as it is well echoed by Saint Augustine's words above. It also is a most distinctive trait of the human condition, as the awareness of the irreversible passage of time, and its finiteness, aka mortality, is a major burden that we humans are probably the only creatures on this world that are called to share. But, why should we be talking about time in this book and why at this late stage?

The point is that time has profound connections with Complexity, for a number of reasons, but here we shall focus mostly on a single one, and namely the fact that, arguably, *time can only be experienced by systems beyond a minimal threshold of complexity*. A threshold that human beings are guaranteed to pass, as witnessed by the Complexity of the biological processes we have been describing in the previous chapters, not to mention the neurological and psychological sphere, which we leave to more competent authors.

Briefly, time is profoundly ingrained within the human condition .

17.1.1 The transformations of time

Before we expand on the connections between time and complexity, let us remind the reader that the concept of time has gone through an amazing series of radical transformations under the drive of modern science. From Newton's absolute time, the one we still perceive as most familiar to us, to Einstein's revolution of a relative time which depends on the state of motion of the observer, to the quantum time of modern days allegedly ticking in unthinkably tiny discrete units as short as 10^{-44}

Sailing the Ocean of Complexity. Sauro Succi, Oxford University Press.
 DOI: 10.1093/oso/9780192897893.003.0017

seconds, known as Planck time. And we are not finished with time: according to most speculative fringes of modern physics, such a series of transformations might even end up as an ironic nil: time has no physical meaning, hence it simply does not exist: no time! If this is true, one can't help but being baffled at how much has been thought, said, and written throughout the history of humankind about something that just ... does not exist! Much ado about nothing ... .

But let's get back to our own story, namely the way we perceive time. There's hardly any question that, to us, the most defining feature of time and the most vexing as well in many respects, is its *one-sided* nature. As we discussed in Chapter 8, time goes one-way only, in the direction we use to call future, with no chance to reverse it back to what we use to call past.

At this stage, it is convenient to distinguish between forward time $t_f = t$ and backward time $t_b = -t$, the former pointing to the future and the latter to the past, (see Fig. 17.1). Whenever forward time, from present to future, and backward time, from present to past, can be swapped without violating any physical law, we say that the system is *reversible*. In formal terms, reversible systems obey the so-called T-symmetry:

$$t_f \leftrightarrow t_b \tag{17.1}$$

T-symmetric systems can be rolled indifferently back and forth, like in a movie, without violating any physical law.

In many complex systems, and most certainly us, the T-symmetry also known as time-reversal, is patently broken: these systems are *irreversible*.

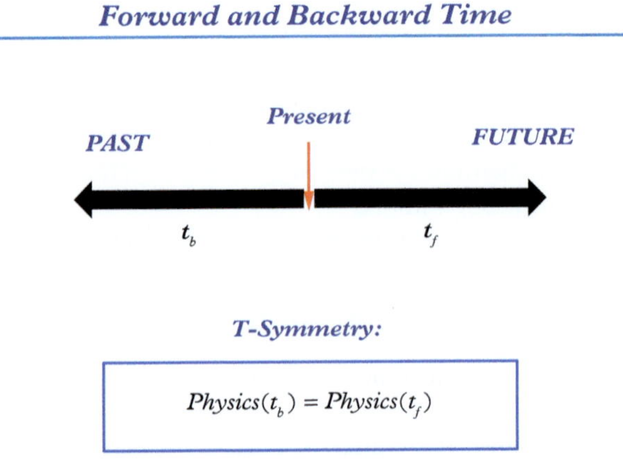

Figure 17.1 *The T-symmetry between forward and backward time. Upon swapping them, the basic equations of microscopic physics remain unchanged.*

This is no brainer: once we break eggs to prepare a tasty omelette, we can't unbreak them anymore. Better yet, we can in principle, but then we are like monkeys trying to type the *Divina Commedia*: it just takes too long. We can post-dict the past (when we can) but, discounting fortune tellers, we cannot pre-dict the future, at least not without a great deal of uncertainty. Likewise, we age but not rejuvenate, not spontaneously, at least, a good business case for the cosmetic industry after all ... . And it is precisely through this broken symmetry that times acquires a meaning to us, i.e. we start perceiving it. Hence, the broken symmetry between forward and backward time is the root of irreversibility.

It is often heard that philosophy is not the art of living but of graciously coping with our mortality instead. Indeed, for as much we may play savvy philosophers about it, the fact remains that time carries the main burden we all humans are called to share. It is finite, it has a beginning and it has an end, it comes with a deadline and it cannot be regained back.[55] And, let's face it, although there might be rewards in ageing, they are not easily appreciated. 'Time is an Ocean, but its ends on the shore', sings Dylan, the shore called mortality. So, to stay with rock stars (U2), the question is 'are we really helpless against time?'

I am not a philosopher, less of a psychologist, but I sense that pretty much of what and who we are, depends on the way that we respond to the above burden. For many, limited time means seizing every possible opportunity to bite the apple, some perhaps with more decency than others, but the idea remains that life is essentially a sequence of *Carpe Diem* (seize the day, one by one), as vividly expressed by the the roman poet Horace (65–8 BC) over two thousand years ago. For others, time is a unique opportunity to leave a wake behind our mortal body. Personally, I find the latter category more inspiring, even though I can't help but sympathizing with Woody Allen's 'I don't like to achieve immortality through my work. I want to achieve through not dying'. So, Woody is among those who want to live forever, an item we shall return to at the end of this chapter. After such somewhat philosophical digression, let's go back to science and try to discuss how time emerges from the broken symmetry between forward and backward time.

17.2 Emergent time

Time is an illusion, lunch-time doubly so.

(Douglas Adams)

In his condolence letter to the sister of his lifelong friend Michele Besso, Einstein wrote something around the lines of 'Michele preceded me by a little in leaving this strange world. It does not really matter, for us convinced physicist, the distinction between past, present and future is only an illusion, albeit a deep-rooted one'. Just an illusion writes Einstein: despite the revolutionary outcome of

[55] In fact it can, but only by mental travel, as described in the famous Madeleine episode of Marcel Proust '*A la recherche du temps perdu*, volume 7: *Le temps retrouvé*.

his theories, Einstein stands right next to Newton in looking at time as to a necessary but sort of dull event sequencer. True, his time is 'elastic' and deforms in response to the motion of matter, but these deformations just make the labelling system (much) more complicated, yet still a label. And a similar statement can be made for Schroedinger's and Dirac's quantum time (see Appendix on Quantum Mechanics) which can be rolled back and forth with no loss of information. This flies in the face of our everyday experience: time goes one-way only; inexorably, present follows past and precedes future. And it also stands in stark contrast with space: we can go left and right, up and down in space with no restrictions (not in the traffic!), but we can't travel back in time to erase our mistakes (or possibly make even worse ones ...). In a word, time is *irreversible*. The giants, Newton, Einstein, and the whole string of quantum heroes, all keep the same embarrassing silence towards this basic clash between their fundamental equations and our daily experience.

We mentioned in a previous chapter that thermodynamics, and more generally, statistical physics, deal with macroscopic quantities which make sense only as applied to groups of atoms or molecules. Temperature, pressure and entropy itself, all belong to this class of so-called collective, emergent quantities. What about time, could it be that it is also part of the same club? What happens to time as it walks across the bridge linking the microworld of atoms to our macroscopic world? In an act of hommage to Ludwig Boltzmann, we have called this the *Boltzmann bridge*.[56] Could it be that it is precisely in the act of crossing the Boltzmann bridge that backward and forward time split for good giving rise to 'our' time? This is the question we are going to explore next.

17.2.1 Crossing the Boltzmann bridge

We have discussed before in this book the notion of coarse-graining as an effective procedure that lumps many microscopic variables into collective quantitis which we called order parameters. We also mentioned that some information is inevitably lost in the process, the idea being relinquishing as much inessentail information as possible, without compromising the essential one. It is readily shown that irreversibility can be traced back precisely to this loss of information. Consider the following Gedanken experiment (thought experiment) as applied to a particle trajectory, (see Fig. 17.2):

1. *Run the trajectory from time $t = 0$ to a future time $t > 0$,*
2. *At time t, instantaneously reverse the velocity,*
3. *Run the trajectory in reverse from time t to time $2t$,*
4. *Reverse the velocity again.*

[56] I am not ware of any bridge dedicated to Ludwig Boltzmann in Vienna, but I think there should definitely be one. In case Vienna's mayor reads ....

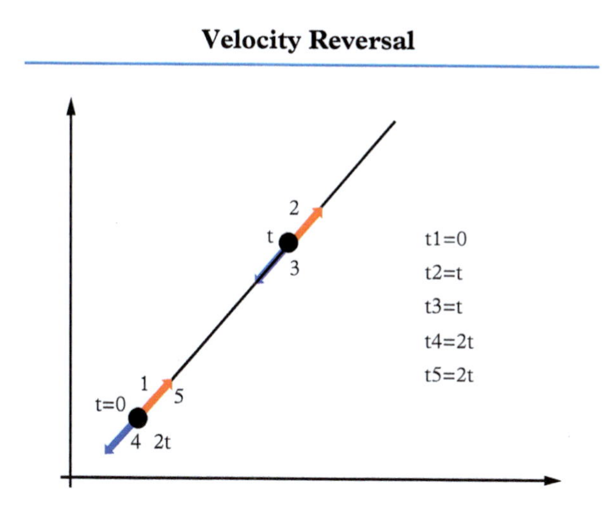

Figure 17.2 *Upon fine-tuned reversal of the molecular velocity, the configuration at time 2t is exactly the same as the one at time* t = 0. *In terms of forward and backward times, the sequence (1,2,3,4) can be interpreted as going from* 0 *to t in forward time first and then again from* 0 *to t, but in backward time. If the two times are equivalent, the system must return to its initial state and, conversely, if it does return to the initial state, the two must be equivalent.*

If the sequence $(1, 2, 3, 4)$ leaves the system completely unchanged, Future (time $2t$) is a perfectly indistinguishable copy of the past (time $t = 0$). *Time is a plain label* and no measurement would ever been able to tell the two configurations at time 0 and $2t$, apart. The critical reader might plausibly argue that this is trivial because we followed just one single particle. The key point, though, is that the same remains true if instead of a single particle we would repeat this Gedanken experiment with ten, hundred, an Avogadro of them, or as many *interacting* particles as we wish. This is the power of the T-symmetry: as long as the equations of motion of the particles comply with the relation (17.1), reversing the velocities at time t and running them up to time $2t$ returns *exactly* the starting state at time 0, (see Fig. 17.3). The past *can* be literally regained from the future, and time is only a mere event-sequencing label.

Next, suppose the many-particle procedure is repeated by a macroscopic observer that can track only one particle, say the average of ten (coarse graining again). The trouble with such a setting is that, in order to track this single average trajectory exactly the way it would result by averaging over the motion of the underlying ten, he would need to know exactly the motion all of them, which by definition he cannot. The best he can do is to proceed by guess work (a model), hoping he can 'divinate' the correct position of the ten 'phantom' particles beneath all along their trajectories. Since there are many micro-configurations which correspond to the same average velocity, the chance that he just picks up the correct microscopic configuration are really slim, and the larger the number of particles, the slimmer this chance. Hence, an error, strictly due to this ignorance of the

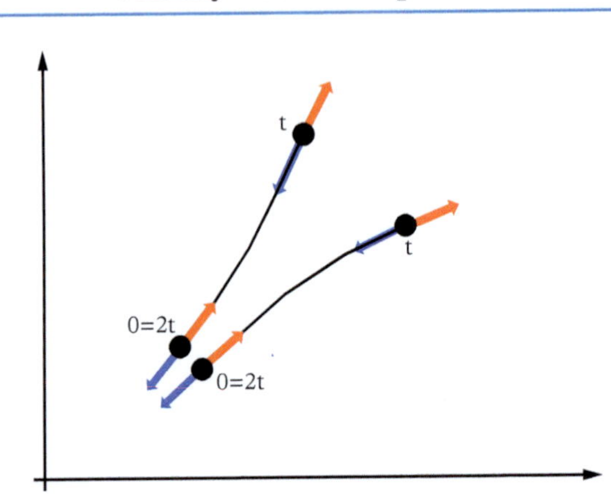

Figure 17.3 *Reversal of the molecular velocity for two interacting particles. No matter how complex their trajectories, if the equations of motion are compliant with the tome symmetry, upon reversing their velocity at time t, at time 2t the two particles regain exactly the same position and velocity they had at time 0.*

microscopic information, is introduced. Due to this error, the single particle he can track will not end up at time t in the same location where the correct average over the ten would be. He can still reverse the average velocity at time t, and incur additional errors on its way back, from time t to $2t$. At this point, reversing the average velocity at time $2t$ will *not* replicate the initial value; not unless the errors incurred in the 'backward' branch of the procedure would cancel *exactly* the ones accumulated in the 'forward' branch. A chance which gets exponentially dim with the number of particles involved, as well as the duration $2t$ of the procedure

The reader may contend that this is all very artificial, but the fact remains that, as noted many times in the course of this book, macroscale systems, including our organs, are tireless coarse-graining machines. We have no option but responding to averages over large groups of molecules. And, as discussed earlier on in this book, whenever we respond to molecular details rather than to collective behaviour, as it is often the case in medicine, Complexity rockets to the sky. By construction, coarse graining erases information, increases entropy and that's how time becomes one-sided. In this respect, irreversibility is a quintessential emergent property, *time changes its nature as it crosses the Boltzmann bridge*: a 'burnt bridge' that shuts down the very moment you leave it behind, with no way back. This smells a lot like real life, doesn't it?[57] Based on that, instead of calling entropy the arrow of time, it would perhaps be more appropriate to say that time is the

[57] For the sake of completess, I wish to add that the idea of emergent time is not universally accepted. For instance, Iliya Prigogine spent the last part of his career in an effort of showing that irreversibility is already present in the microscopic law of motions. However, his ideas have not met with any major consensus in the scientific community.

arrow of entropy, because the real driver of irreversiblity is the loss of information in crossing of the Boltzmann bridge. And, by the same argument, we can also say that the emergence of time is a measure of our ignorance of the microscopic world.

The fact of being much bigger than the atoms we are made of buys us several privileges, including robustness against molecular vagaries. In addition, the concerted action of zillions of molecules keeps us alive through highly organized bio-physical complexity (metabolic functions). But the very principles which fuel such organized complexity also imply the one-sided nature of emergent time. And by the time we realize that the one-sidedness results from the tendency to evolve towards configurations of increasing probability, it becomes natural to accept that, once the maximum probability configuration is attained, there is no room left for change, which is an operational definition of death. This is how one-sidedness implies finitess: not a theorem, of course, but a pretty plausible argument.

Now we see why time and Complexity are sort of brother and sister. And we probably also see why Complexity relates to aging and finiteness in time, aka mortality, to which we shall return in the final part of this chapter.

But let us first spend a few words on what we may call psychological time versus the physical time we have been discussing so far.

17.3 Psycho time

Paul Dirac (1902–1984), the man who earned perpetual fame (for some, a form of effective immortality!) for anticipating the existence of antimatter through his eponymous equation, was less than renowed for wordiness and social skills in general. Yet, in one circumstance, he was reported to initiate discussion with Abbé Georges Lemaître (1894–1966), a key figure of modern cosmology and a priest of the Catholic Church for a time. Dirac admired Lemaître, so he intended to pay a compliment to him by stating (unsolicited!) that cosmology is probably the most fundamental science of all. Much to his surprise, Lemaître demurred, by replying that the most fundamental science of all is psychology, as it deals with the mysteries of the human mind. So, psychologists might be elated (maybe) that their discipline could be named the *cosmology of the human mind*. And a mystery the human mind still is indeed, as the ongoing difficulty of treating mental illnesses clearly proves.

17.3.1 Brain and tears: The human condition

In my young days, there was a popular song by Demis Roussos (1946–2015) named *Rain and Tears*. Place a B in front of rain, brain and tears, and you encapsulate a basic trait of the human condition, namely that the brain holds the key to both happiness and suffering, the latter having a much higher entropy than the former (remember *Anna Karenina*? ...). Depression stands out as evidence of the emergent Complexity we still have such a hard time to understand and master. I was amazed and kind of scared to learn that the transport of neurotransmitters across our body can be inhibited by molecular malfunctioning at the level of a fraction of nanometre, a far cry from the situation discussed before, namely that we do not respond to single molecules, but large groups of them. A tiny fraction of a nanometre can change our perception of the world, and hence who we are. This said, discussing time, as it is perceived by our brain, is obviously another book, and not one which belongs to this author's range of competence.

Yet, a few words before closing are in order.

17.4 Biological time

The age of a car is its mileage.

Biological systems are based on a number of periodic clocks, which respond to driving periodicities outside it, the sunrise and sunset in the first place (circadian cycles). Modern life, particularly electricity, separated us from our biological clocks, but ultimately we still respond to them to a large extent. However, some other fundamental clocks, such as cardiac and respiratory rythms are independent of external periodicities. In the case of cyclic phenomena, the fundamental notion is not time itself, but how many clocks a given system can support before it fails, best known as *time-to-failure* in the manufacturing industry. I mean, about three billions heartbeats delimit the mortality shore for humans: since one heartbeat lasts about 1 second, this makes our life about a hundred years long, a generous ballpark. There are of course compelling biological and physiological reasons why the heartbeat should last about one second, but that's another story.[58] Should basic physics dictate one heartbeat per minute, we would live six thousand years, and time would be left the same, just a matter of prefactors (of course this true only if we restrict our attention to the cardiovascular system). This 'dilatation' property, scale-invariance in physics parlance, maybe less trivial once we enter the territory of neuroscience, i.e. the science which occupies itself with the way our allegedely most valued organ, the brain, works.

Neuroscientists justly take their work very seriously. A neuroscientist friend of mine once told me, in a very serious tone, that if they replace your heart, you are still the same person, but if they (could) replace your brain, you would be another person. I would have liked to object, but, honestly, I couldn't see how. So, the brain dictates the way we perceive the world around us, hence time might be just a brain-made 'illusion', as surmised by Einstein. It would be outrageously reckless

[58] Amazingly, the number of cardiac and respiratory clocks does not change much across species: rats and elephants, whose lifespans are about 2 and 80 years respectively, feature moreless the same number of heartbeats, about one billion, which means that rats hearts beats 40 times faster than elephant ones .... For an interesting discussion on this amazing point, see [68]

for this author to enter such vast topic without the due minimum of acquaintance with the subject. So, I will simply note that the way we perceive time may indeed be liable to mechanisms which differ from the objective measurements (Einstein's clocks and mirrors), those that we must take in order for physical time to make concrete sense for the formulation of the physical laws. In this respect, time could be indeed an exquisite construct of our brain.

17.4.1 Le temps de philosophes

And indeed, our brain/mind system merrily transcends physical boundaries: we can move back into the past, or jump ahead to places unthinkably far away in a blink of an eye, (see Figs. 17.6 and 17.8). We can even think of events which never happened or never will: *A factory of illusions*, to borrow from Buonomanno's book [18]. Of course, thoughts are immaterial, and even though they are definitely supported by physical processes, they enjoy virtually unlimited freedom. To be more precise, mental processes are subject to physical constraints, but their output is not subject to the same constraints of the physical world. Remember Picasso and the proteins, as we discussed in Chapter 14 .... Hence, since whatever we perceive is filtered by our brain, it is hard to discard the idea that time might be a mere mental derivative. At the same time, it is also hard to accept this idea, until the basic mechanisms by which time takes form in our brain are pinned down. This is a most interesting research avenue of modern neuroscience, with obvious implications for psychology and psychiatry, but for the time being (no pun intended), I think we are far from a microscopic theory of the brain.

Among others, this line of thinking is meeting with increasing interest from those who feel like the weirdest aspects of quantum mechanics, quantum entanglement in the first place, might play a major role in the way we perceive time.[59] For instance, the world-renowned cosmologist and Physics Nobel 2020, Roger Penrose (1931–), together with neuroscientist Hameroff, speculated that our brain might work like a quantum computer. A computer which (once in place) takes advantage of quantum entanglement to compute at hyper-speeds, far superior to the ones afforded by present-day electronic computers. Such hyperspeeds might then account for the most sophisticated brain functions, such as self-awareness and consciousness. To date, this is only a wild speculation, but the subject of quantum biology, i.e. the question whether biology makes use of the weirdest aspects of quantum mechanics, is the object of a very vibrant sector of modern science [75].

[59] For the record, quantum entanglement is the property by which quantum objects affect each other over distances which exceed the travel capabilities of any physical signal, including light. Einstein called this 'spooky action at distance' and always refused to endorse it. Yet, modern quantum mechanics has provided compelling evidence in its favour.

17.4.2 Time travel and the fear of uncertainty

Time Is What Keeps Everything From Happening At Once.

(Ray Cummings)

Einstein crowned Bergson's defeat with his famous *Il n'y a pas le temps de philosophes*: there is no such thing as the philosopher's time. Yet, it might be that Bergson's views will in a way be revived in a more solid form, the day physics shakes hands with neurosciences. In fact, this is already happening. For instance, renown neuroscientist Friston's free-energy principle (briefly discussed in Chapter 14), surmises that the brain is basically a model generator for the world around us, and it evolves in such a way to make the prediction of the model as close as possible to reality. In other words, the brain confers the ability to project our thoughts into the future, so as to anticipate it first and actually modify it based on these projections. Once we realize that the sky does not provide enough water for the whole year, we design dams. That's how humans progress and make the world a more hospitable place for themselves. This is a wonderful gift which places us in a unique position among other creatures (to a point, after all birds build nests, but certainly they don't design dams).

This gift does not come for free, though: the very ability to project into the future and build dams in draught regions, is the same ability that generates fear and anxiety in front of the chance, and we *know* that such chance exists, that things might just go wrong. In a word: *the fear of uncertainty*. Not coincidentally, anxiety and depression are typical conditions of advanced societies: as long as you need to care for tomorrow's meal, projecting into the far-future is a luxury you can't afford.

The idea of the brain as a model generator of the world around us sits well by the modern trends of scientific research known as 'machine learning', i.e. the discipline of instructing computers to run algorithms that improve their ability to perform 'intelligent' functions, a strong revival of artificial intelligence. In this respect, time may well fit the picture of a brain-made device aimed at minimizing the uncertainity that invariably accompanies the Complexity of the real world [80]. Under this light, it appears very plausible to speculate that sequencing events in a one-dimensional causal order is an enormous asset to make sense of things yet to happen, i.e. *before* they do happen. Briefly, to tell the future based on the past.[60] It is worth noting that the emergence of such neurological time commands a level of Complexity and organization that goes way beyond the mere act of coarse-graining described in the initial part of this chapter. Coarse-graining may explain the one-sided nature of time and its finiteness, but not our ability to send our mind on travel in forward time and back. Differently restated, physical time advances sequentially bit by bit under the constraint of the physical laws, while psycho time

60 Incidentally, this may also offer an answer to the question of why there are three dimensions in space and only one in time.

can break free from such constraints and take jumps into the future and back to the past. This ability, which seems to be unique to our species (is it?), can literally change the course of our future.

This wondrous future-shaping ability is a source of pride for the human species, sometimes verging on hubris, as expressed by catch phrases such as 'making the future'. But the future 'we make' is still wrapped with uncertainty, for the time-travel exercise described is far from being fault-free.

This is well-known in the scientific practice as *extrapolation*: i.e. predicting data *outside* the domain where they have been collected. This stands in contrast with *interpolation*, in which data collection and prediction take place within the same domain (see Figs. 17.4 and 17.5). There is no need of a PhD in math to realize that extrapolation is way more error-prone than interpolation, a statement which is particularly true in the case of the time evolution of complex systems in which, as we discussed in the early part of this book, the past does not necessarily teaches any useful lesson to divinate the future. The point is that interpolation takes place within time fences, because the data at the lower and upper end of the interpolation window serve as protection constraints to the prediction (remember the guard rails of Chapter 14). Extrapolation, on the other hand, enjoys no such protection: it is a jump in the blue. The bottom line is that we should justly be proud of the time-extrapolation ability of our brain, without ever losing sight of the fact that in front of Complexity, even our best science may well remain wrapped within a great deal of uncertainty. It is a sure bet that there'll be important discoveries along this road in the years to come.

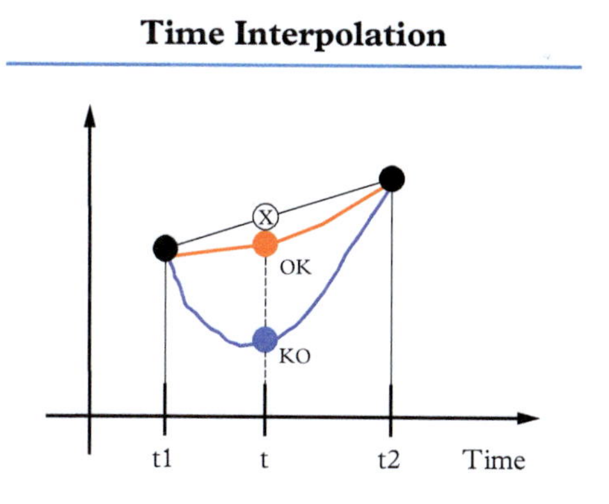

Figure 17.4 *Interpolation: the values at subsequent instants t1 and t2>t1 are known and a linear interpolation is used to predict the value at an intermediate time t1<t < t2 between the two (crossed circle). If the quantity under study does not change significantly within the time interval [t1, t2] (red upper curve), the interpolation error is small. If on the other hand, the quantity goes through significant changes (blue lower curve) the error can be large. Interpolation is comparatively safe because it can count on both forward and backward projections in time, i.e. from t1 to t and from t2 back to t. This 'fence' in time provides significant protection against inaccuracies.*

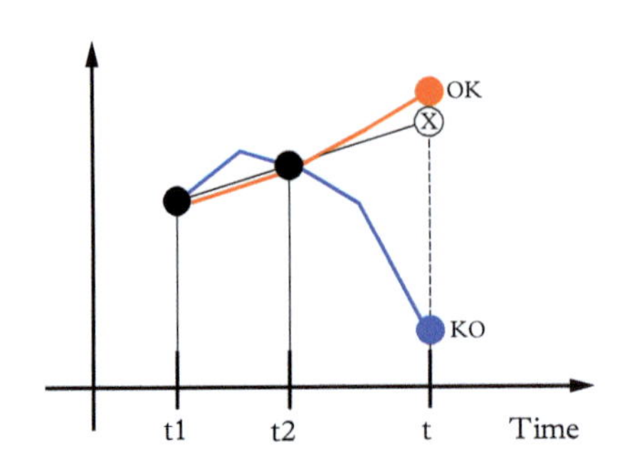

Figure 17.5 *Extrapolation: the values at subsequent instants t1 and t2 > t1 are known and a linear interpolation is used to predict the value at a future time t > t2 > t1 beyond the two (crossed circle). The extrapolation procedure is much more error-prone than interpolation because both data must be projected forward, hence at variance with interpolation, there are no protection 'fences'.*

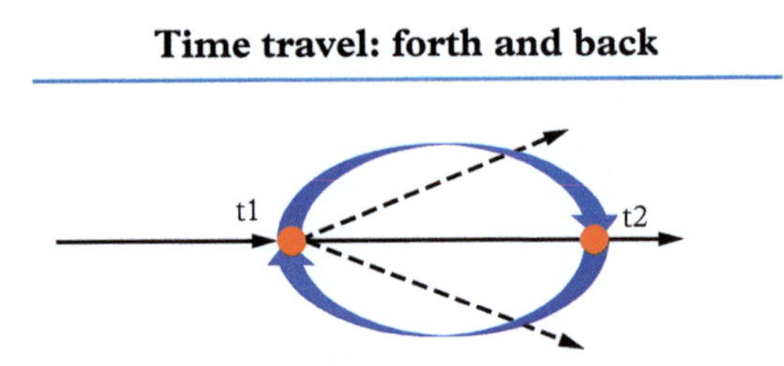

Figure 17.6 *The two times: physical and psychological in action. Physical time proceeds sequentially from past to future, possibly ticking in units of Planck time (a continuous flow for macroscopic observers like us). Psycho time, being free from the constraints of physical time, can take jumps into the future, 'see' what happens there and bring the information back to the present, so as to take a different course (dashed lines) in case this is deemed appropriate. By this time-travel procedure, we can literally change the course of the future (dashed arrows). Note that backward-time travel is as important as forward-time travel.*

17.5 Putting the times together: Time is us

Time is possibly not needed at the microlevel and more and more scientists, from both classical and quantum shores of the river, seem to agree that time, like temperature, is probably an emergent quantity: it only makes sense as applied to a large collection of atoms. In other words, as we have commented before, time would emerge only beyond a given level of Complexity, certainly the one associated with living creatures like us, and much more so to the history of mankind.

This is appealing in many respects. But we humans are endowed with a brain which is capable of both forward and backward time travel, hence we have somehow to put the 'two times' together. I can't help but borrow from the inspiring book by Carlo Rovelli [97]: *'Time is dull at microlevel, it simply is not needed there: it emerges as a perception of macroscopic bodies, once they get complex enough to perceive it as a ruler of their emotions'.* I do subscribe. Think a moment, how could we possibly experience memory, hence nostalgia, or hope for the future, without a built-in arrow of time? Time brings sequential Order, if we could say and unsay, do and undo all the time, without restrictions or penalties, it is hard to imagine how we could possibly orchestrate and plan our actions in a sensible way altogether. Too much freedom, in the sense evoked by Stravinski (Chapter 14), would impair purposeful planning.

But time is also the enabling mechanism for what we regard as most deeply human: emotions, including the dreams of the previous chapters. It really is inextricably related to our capacity of experiencing feelings and emotions: happiness, sadness, fear, love, nostalgia .... It is only through such emotional states that time takes shape and meaning to us. This is Rovelli speaking, not me, but, as I just said, I largely subscribe.

17.5.1 Tears in the rain

So, *time is basically us*, it is the very reason why pain and joy are two faces of the very same, beautiful and scary coin which goes by the name of life. Our life, yours and mine. As noted before, time also carries the major burden of humans, unlikely to be shared by our pets: the awareness of mortality. The mind turns quickly to the Ecclesiasts: *'There is a time for everything, a time for sowing and a time to reap, a time to live and a time to die'.* So go the sages, in a wise and peaceful vision of the finite-time human journey. Less serene is the heartbreaking (see Fig 17.7) 'tears in the rain' monologue by the self-aware replicant Roy Batty of Blade Runner, moments before he dies. In the famously improvised poignant words of the iconic Rutger Hauer (1944–2019): *'I've seen things you people wouldn't believe. Attack ships on fire off the shoulder of Orion. I watched C-beams glitter in the dark near the Tannhauser Gate. All those moments will be lost in time, like tears in rain. Time to die'.* Rovelli calls our innate fear of death a mistake of evolution. It's a mistake, he argues, because mortality, besides being all too natural, brings a lot of precious fruits to life. It is arguably true that immortals would never bother leaving a wake which would last longer than them, simply because *nothing* could last longer than them! That's after all how cathedrals, paintings, scientific theories, and the most beautiful crafts of mankind have come to existence. And, probably, even if he allegedly does not care, that's probably also how Woody Allens's masterpieces came into existence. This is probably all true, but, in my view, Roy runs even deeper than that. More than afraid of death, it seems to me, Roy is full of sorrow

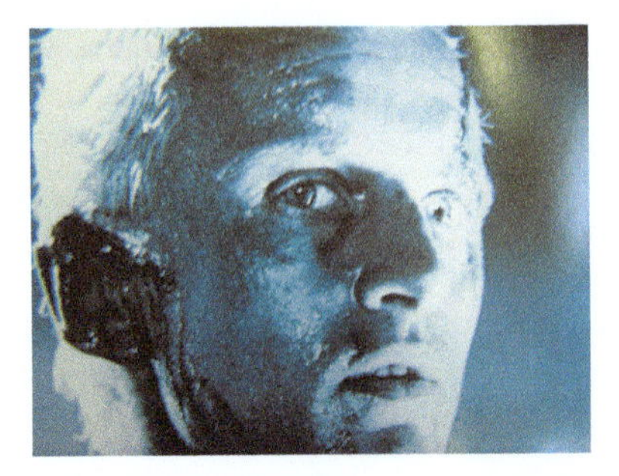

Figure 17.7 *The iconic Roy Batty (Rutger Hauer), 'more human than humans...'. Reprinted from commons.wikimedia.org.*

at watching the marvels he has been exposed to during his brief artificially given life, go down the bridge with the rain. We all share deeply in this feeling and if it is true that Ridley Scott apparently advised Hauer to make Roy 'more human than humans', I must say that this monologue is a truly well done job. Life and death are often portrayed as dual faces of the same medal, and they are indeed, but with a deeply broken symmetry too. Death is the closest proxy to certainty we have, while life is miraculously close to just the opposite. In this duality, which is certainly a very complex one, I personally see a sort of awe-commanding inseparability, and I doubt I am alone.

Coming back to Roy, perhaps we can attempt an answer to his sorrow via a revisited version of Alice's question to the white rabbit: how long is forever, How many attack ships on fire would Roy have enjoyed before they too would spell just boredom to him? How many C-beams glitter in the dark near the Tannahauser gates, before he would feel 'full of days', like the lamenting Job? When we are happy, forever is not long enough, when we are in pain, a single moment is one moment too many. This, I guess, is what the white rabbit meant for an answer to Alice's question. And the answer is that there is no single answer, but possibly infinitely many, which means that we simply don't know. So, my first guess is that Roy the replicant, would not know, and the second is that we, the humans, don't know either. Maybe it just isn't for us to know, as simple as that: as the per the famous song 'Whatever will be, will be', interpreted by Doris Day in Alfred Hitchcock's 1956 movie *The man who knew too much*: 'The future is not ours to see, que sera, sera'.

But not everybody shares: just think of the lords of Silicon Valley, the super-elite of the so-called GAFA (Google, Amazon, Facebook, Apple) billionaires [43]. This takes us to our next question: who wants to live forever?

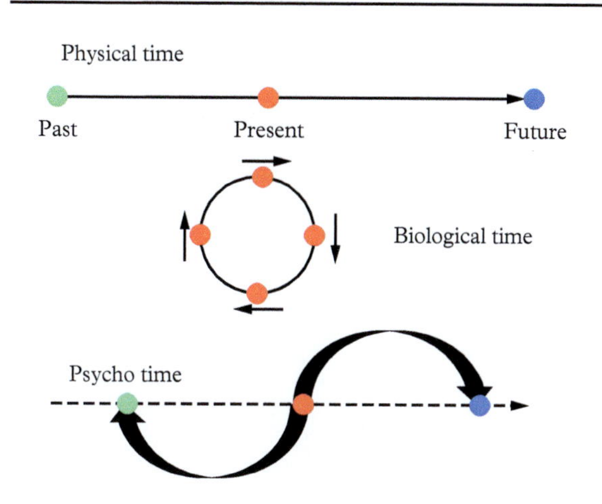

Figure 17.8 *The three times: physical, biological, and psychological. Physical time flows forward only, biological time turns around in repeated cycles like a clock, and psychological time is capable of jumps both forward and backward in time. For the sake of simplcity, physical time is taken here in its classical form, the absolute time of Newton. Whichever of the three we may chose as the most relevant one, none of them can last forever.*

17.6 Who wants to live forever?

> *I am not interested in immortality in terms of legacy, I am interested in immortality in terms of living forever.*
>
> (Woody Allen)

Humans have been interested in living forever for as long as they have existed. One of the oldest records is the famous Epic of Gilgamesh, a Mesopotamian king, one part mortal and two divine [45] (see Fig. 17.9). Gilgamesh's quest for everlasting life begins with the unexpected death of his dear friend Enkidu, which marks the start of a long journey to beat the inevitable. After a long and dangerous journey, Gilgamesh finally meets Utnapishtim, allegedly the only mortal who can confer immortality. Failing to convince Gilgamesh of the futility of his pursuit, Utnapishtim grants him a chance: Gilgamesh will get immortality provided he can stay awake for 6 days and 7 nights. But, guess what? For all his ardent lifelong wish of immortality, he can't help but failing the test. Our body needs sleep, no way out. So goes Gilgamesh story, pretty much as does ours.

Figure 17.9 *A statue of Gilgamesh subduing a lion. Reprinted from en.wikipedia. commons.*

17.6.1 Are atoms immortal?

To be immortal is a commonplace, except for man, all creatures are immortal, for they are ignorant of death.

(J. L. Borges)

Atoms can do better than that: to begin with, for all we know, they don't need sleep .... It is generally accepted that inanimated matter and living creatures, including ourselves, are made by the same very same basic constituents, the elementary particles and most notably the fundamental trio: proton, electron, and neutron. Taken in full isolation, the first two are virtually 'immortal'. The place of pride is for the proton, which is reported to live no less than 10^{32} years, vastly in excess of the age of the Universe (roughly 10^{10} years). The tiny electron does not score badly either, with a lifetime of roughly 10^{28} years. The neutron does not follow suit, and by a huge distance: if left in isolation, it takes just a mere 15 minutes to decay into a proton, an electron, and an exotic particle known as anti-neutrino. This is puzzling, since the proton is basically the uncharged version of the proton, with just a bit more mass (about one part in thousand). This is enough to spoil 'immortality'![61]. The neutron stands tall to witness that 'elementary' does

[61] Quarks say it even better. The proton is made of two 'up' and one 'down' quarks, in symbols $p = uud$, while the neutron is $n = udd$, one 'up' and two 'down'. The 'up' quark has charge $+2/3$ and

not mean 'immortal', the state of affairs is way more complex than this. The stability of matter is a serious and difficult scientific subject, but the evidence shows that the three basic bricks, neutrons, pronts, and electrons, can combine to form states of matter of exceptional longevity. For better or for worst, living creatures do not share in this: they require a level of organization and Complexity which does not allow immortality.

Stones don't need sleep, Gilgamesh did, and so do we. Such is state of affairs with the Complexity of the (soft) state of matter that supports life, as we know it.

17.6.2 The quest for immortality, in Silicon Valley

There's no time for us. There's no place for us. What is this thing that builds our dreams, yet slips away from us?

(Bryan May)

So run the lyrics of *Who Wants to Live Forever*? by Queen (Brian May, to be sure, who, besides playing occasional music as a non-lucrative hobby, holds a PhD in Astrophysics). So, who wants to live forever? We have already commented on the dream of immortality, but lately some people, typically and not surprisingly, the lords of Silicon Valley, famed billionaire PayPal founder Peter Thiel in the first place, seem to take it more seriously than ever before. In a jesty mode, don't quote me, I tell my friends that people who manage to accumulate monster fortunes, the ultra-billionaires, must be genuinely exposed to the subliminal sense of being invincible to the extreme, namely immortal. And when they occasionally come back to their senses and realize that (most likely) they are not, falling from a multi-hundred-billion-dollar high tower must surely hurt harder than it does for the rest of us. And there's arguably something running deeper than mere, if extraordinary, affluence. People who, by factual evidence, are led to think of themselves as invincible, must have a truly hard time accepting *any* sort of defeat, including in the ultimate battle with death.

So, in a way, I am sympathetic. Besides, if their frustration with mortality could help by improving healthcare, as per their philantropist and humanitarians claims, it might even do real good to humankind. I might be wrong, but, alas, I don't think so. I really don't, for many reasons, but mostly out of a 'basic instinct', namely that I simply don't trust enterprises whose prime motors are hybris and greed. If they really cared for the good of mankind, the lords of Silicon Valley would probably be OK with a bit less financial reward for themselves and their companies, at the cost of letting other, arguably less talented people (shall we say the 'mortals'?) keep their outdated jobs going, possibly in a bettered version. So, while I am certainly fond of people who try to reach for the stars and go for the impossible, the impossible I am fond of is a very different one. It is rather

the 'down' quark has $-1/3$, which means that the proton has charge 1 and the neutron has charge 0. The amazing fact is that just changing a single 'up' into a 'down' destroys 'quasi-immortality'!

the one portrayed by Saint Francis of Assisi, which is totally devoid of hybris, greed, and self-centering. Incidentally, the day he could reach near-immortality, say thousand years disease-free lifespan, (outdoing Methuselah by a mere 31), would Thiel still dare to walk on the street, driving a car or fly a plane again? My mind goes to C. S. Lewis's (1898–1963) 'The abolition of man', where he prophetically writes 'what we call man's power over nature turns out to be a power exercised by some men over other men with nature as an instrument' [65]. The GAFA billionaires want to separate life from death and possibly get rid of the latter altogether: good luck with it, I wish in the process they manage to relieve as much pain as they can from this suffering world. For my part, I sincerely hope that immortality will stay at the top of the fragile and precious list of things that money can't buy.

17.7 Summary

The notion of emergent time is still under debate in modern physics, but one thing is clear beyond doubt: the notion of emergence is fairly general as it applies whenever a micro-macro connection is involved. In biology this is just another way of saying always. Given that our own Universe connects no less than 62 decades in space, from the Planck scale to the current size of the Universe, Emergence is hardly a footnote in the book of science. Physical time is arguably discrete below the Planck scale and one-sided and finite beyond a certain threshold of Complexity. Psychological time, on the other hand can jump forward to the future and back again to the present. A wondrous ability which buys our species a lot of privileges, but also the burden of consciousness and the awareness of our mortality.

This is possibly the most distinctive trait of the human condition. The way we react to this burden defines to a large extent who we are and what we do. Nothing to be cheaply dismissed, as it takes sanctity to handle it in full grace, possibly with a smile in front of the great chasm. We are made of the same quarks and electrons as stones, but stones cannot look ahead into the future and bring the information back to take new course as they see fit. Stones, though, need no sleep. Like Gilgamesh, we do.

18

Harness the Hybris: Hallelujah!

This world is full of conflicts and full of things that cannot be reconciled. But there are moments when we can ... reconcile and embrace the whole mess, and that's what I mean by 'Hallelujah'.

(L. Cohen)

18.1 Escape from Helsinki: Eyjafjallajökull!

On Thursday 15 April, 2010, I was attending a meeting of the European Physical Society in Helsinki, eager to catch my flight back to Rome, as I had (family, as far as my recollection goes) affairs to take care of the day after. Around lunchtime, rumours started to spread around that Helsinki Airport was going to be shut down because of a major eruption from an Icelandic volcano, with a tongue-twisting name, Eyjafjallajökull! As the minutes went by, it became painfully clear that this was no ordinary affair, the wind-driven plume was obscuring most north-European skies, with no clue as to how long for. Obviously, panic and a major stampede resulted.

After frantic consultations, to me, this meant the following sequence (see Fig. 18.1):

1. Helsinki–Turku (still Finland) by bus;
2. Turku–Stockholm, night boat (comparatively fun);
3. Stockholm–Copenhagen, overnight bus (less fun);
4. Copenhagen–Hamburg, taxi drive, and I take full credit for negotiating a remarkable discount from 2,000 to 600 Euro for the whole group of five, which I regard as one of my most glorious achievements ever;
5. Hamburg–Basel, car rental and one-day, top-down drive through Germany
6. Basel–Zurich, the Southern Europe group (two Italians, two Spaniards, and a Portuguese) split up the Italians from Basel to Zurich by train, where much to our despair, we caught the wrong train!)
7. Zurich–Milan, train, no mistake;
8. Milan–Rome, train, still no mistake.

Sailing the Ocean of Complexity. Sauro Succi, Oxford University Press.
© Sauro Succi (2022). DOI: 10.1093/oso/9780192897893.003.0018

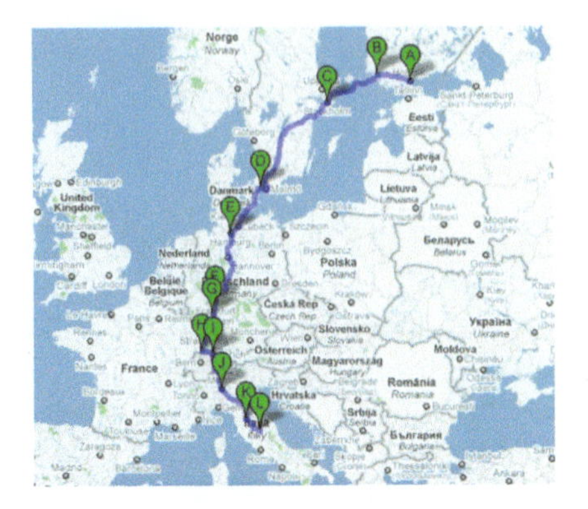

Figure 18.1 *The multi-step journey which took me from Helsinki on Thursday, April 15, 2010 to Rome on Sunday evening, 18 April. Every and each move was taken 'on-the-fly'. Antifragility showed its best side when I managed to negotiate a taxi from Copenhagen to Hamburg for 600 Euros instead of the initial request of 2,000!*

All of the previous journeys took 3 very long and eventful days after our departure. My Italian mate still had an additional stretch by train from Rome to Perugia, which makes nine steps to his credit against my eight.

This is a rather remarkable sequence in itself, but the true pathos, a kind of James Bond feeling I must confess (in hindsight, though), was that we never knew the next move ahead of time: we were really 'making the future' on the fly! All of this because all of a sudden, a totally outlandish volcano in Iceland, with a name I cannot pronounce, decided to air out his hot vapours without asking permission to the busy humans ... .

I have always been very conscious of our fragility, not much need to be reminded. But that episode really hit me with how little we are. Harness the hybris, this, besides Hallelujah, is the leit motif of this closing chapter.

18.2 Harness the hybris

It is still a rather widely held opinion that scientists should mostly be atheist, and sometimes even with a vitriolic vengeance, biologist Richard Dawkins being probably the most outspoken example in point. I have experienced this first hand with Lawrence Krauss, a 'celebrity' physics alter-ego of Dawkins.

One evening in Fall 2016, he lectured at Harvard University on 'A Universe from nothing', echoing the title of his successful book. The basic claim was that the Universe (something) may have arisen from a quantum fluctuation of the vacuum (nothing), thus ruling out the idea of or need for a Creator: something from nothing, against the time-honoured *Nihil ex nihilo*, (nothing from nothing).

To begin with, I see an elementary flaw in the very premise: even if the vacuum were empty (otherwise, it would be 'something' by itself) the fluctuation around it, by definition, *is* something! So, although it may well make a catchy title for best-selling books and red-carpet events, 'Something from nothing', is hard to make any sense of, as applied to quantum fluctuations of the vacuum. So much for plain common sense. This said, the first part of the seminar, cosmology, was pretty brilliant and informative, the man was clearly on top of his craft, and so was the notion of the spiritual enchantment which the beauty of our Cosmos commands on all of us. Agreed. Less enchanting, though, was the glorification of 'stupor' as the 'true and only' form of spirituality, as opposed to religion (the lowercase is no accident), punctuated by a relentless series of sarcastic comments throughout the entire talk.

At some point I felt this was enough.

18.2.1 Should I be doing science?

So, at the end of the seminar, I couldn't help but airing my point, namely that, in my modest view, nothing I heard in his talk would rule out the existence of a Superior Being. Reaction: 'sure, I cannot rule it out, but if you can provide a rational explanation for what you see and you still invoke a Superior Being, then you should not be doing science!' I get the point, not without noting that the closing sentence reminds me the typical intolerance of those who preach tolerance, until it is their turn to practice it. I returned that his aggressive claims showed very little knowledge and understanding of what Religion (capital R now) really is. If ever he would take the time of reading the encyclic 'Fides et Ratio' by Pope John Paul II, he would probably know a little better. Just to quote the opening 'Faith and reason are like two wings on which the human spirit rises to the contemplation of truth' [32].

And, for that matter, it is a good thing for cosmology, long before Religion, that Abbè Georges Lemaître (1894–1966), the founder of Big Bang Theory, did not abide by Krauss's veto ... As mentioned in Chapter 11, Georges Lemaitre famously dissuaded Pope Pius XII from making claims on Big Bang Theory being a confirmation of the 'Fiat Lux' in the initial page of the Book of Genesis . His point was that, like any other scientific theory, Big Bang Theory might have proved wrong in the future, hence the tie with the Bible would have put the Latter on a very slippery slope. Coming back to Krauss, seems to me that he fell in the typical trap that Francis Collins (head of the group that decoded the human genome on public funds and a believer) in his book 'The language of God' [27] calls 'The God of the gaps'. The trap is to assume that rational explanations rule out a Creator. In other words, The Creator lives in the gaps of our knowledge, so that, once the gaps are bridged, there is no place left for Him. This strikes me as a very narrow view: to me, the Creator is even more felt and seen in the beauty, charm and elegance of natural explanations than in their gaps! The few things I do understand speak to me more than the far many more I don't. As a scientist, you are committed to Saint

Thomas, don't believe until you see. Religion is a standing invitation to believe without seeing. At first sight (no pun!) this seems a head-on clash, but personally, I've never seen it this way [1]. Again, this is very personal, but to me, believing without seeing is like taking a flight to extra dimensions which are not accessible to the brain. And, whenever you discover something new about the world around us, the emotion of the scientist is nothing short of a prayer of gratitude. For those who believe, not only did the Creator provide us with a wonderful architecture, He also granted us with the immense gift of being able to read off some of its basic ingredients! I see no place for hybris here, but only for humbleness and gratitude.

18.3 These are the two wings

In my view, these are the two wings advocated by 'Fides and Ratio'. I have scarce interest and less patience for heated debates on the conflict between Religion and Science and even less for those who seem to thrive on such alleged conflict to achieve a (square-root) star status. Personally, I don't think science will ever be able to either prove or disprove the existence of a Creator, and I'm certainly making no original statement here. Yet, if I have to express in full honesty the bias I've got from my career in science, I would say that the task of disproving seems to me much steeper than that of proving. This said, the fact remains that Religion and Science do indeed portray a very different approach to reality. Religion is committed to revealed truth, whereas Science is based on factual verification; as we said previously, like Thomas, scientists would not believe until they see. And they make of this a defining point, which makes perfect sense in this context. A subtlety not to be missed: 'see' must be taken here in its broadest sense, you can see with your mind and not your eyes, that's precisely what we did in this book when we discussed the funnel. To the contrary, the New Testament proclaims beatitude for those who believe without seeing. The scientist is rational, Religion, capital R again, invites to go beyond rational thinking, not against it, as far as I can tell. Surely no invitation to believe that Earth is flat.

Fides and ratio is again adamant on this point: 'Faith unmoored from reason runs the grave risk of withering into myth and superstition'. However, revealed truth is absolute and unchangeable, whereas Science is by definition open-ended and welcoming (in principle ...) of changes: the truth of today is the mistake of tomorrow. A very productive mistake, though ...Yet, as we have discussed time and again in this book, both change and unchange carry essential and complementary roles in science. Hence, why should they be mutually exclusive in the Religion versus Science framework?

18.3.1 Typecast stereotypes

Both scientists and believers suffer some significant degree of stereotyping. Scientists are often portrayed as cold and arrogant 'nose-ups', who only believe in the rational power of their brain-mind super duo, often aiming at nothing shorter than reading the mind of a Superior Being. Believers, on the other hand, are sometimes still depicted as over-naive, rather uneducated, and narrow-minded poor fellows, ready to surrender their (narrow) mind right at the outset. Both clichés are germane to me. I have never seen it this way, on either side of the river. To me, Science has never been about any wish or desire to read a Superior's mind, much less so about biting the infamous apple, but rather as the innate pleasure of finding things out, to go with Feynman (not a believer). A growing tree of wonder for the breathtaking beauty, subtlety, and elegance of the natural world. Wonder, hence gratitude and, above all, a great sense of humility. Saint Ignatius of Loyola (1491–1556) says it best:

For it is not the abundance of knowledge, but the interior feeling and taste of things, which is accustomed to satisfy the desire of the soul.

Granted, you don't have to be a believer to experience and share these feelings, but they certainly do not prevent you from being one either. Sure enough, we (scientists) are bold, in that we feel that we can get a grip on most things we see around and within us. But this should not prevent us from being very humble and grateful in recognizing that this is just a never-ending wondrous trip. Maybe, to go with Freeman Dyson (1923–2020), this trip really is 'infinite in all directions' [31]. In our ephemeral life, we are granted the wondrous chance of doing things that can last longer than our mortal journey: take Mozart or Einstein, to name perhaps the two most iconic geniuses of all times: their work will be with us for as long as mankind is around.

More importantly, there is no need to fly as high as Einstein or Mozart to taste the wondrous beauty of discovery and professional fulfillment. I will never forget the first time I met with the Schroedinger equation, the one which governs the behaviour of atoms and molecules, hence the molecular basis of life. This equation struck my previous me, then aged twenty-two,[62] as having enchanting beauty, nothing short, for those who can read it (I'm sorry, some math can't be escaped), of Mozart's music. Forty-plus years down the line, it still carries the same enchantment, possibly even more .... So, no question to me that the brain-mind asset is indeed a gorgeous gift that must be practiced and enjoyed as much as possible. This brings me straight back to the Matthew's parable of talents that we discussed before earlier on in this book.

18.4 A jump into the blue

Here we are facing a bifurcation, though; you may either believe that reason is all we have to figure things out or take a step of ultimate humility and admit that

[62] As a matter of curiosity, Heisenberg wrote down his uncertainty principle aged 23 ... that tells you a depressing lot on the difference between him and me!

there is (way) more to this world than our marvellous mind-brain system can grasp. 'There are more things in heaven and Earth, Horatio, than are dreamt of in your philosophy', recites Shakespeare's Hamlet: this is a poet, not a scientist, do we really care at this point?

It seems to me that being humble in the precise sense of exposing willingness to surrender reason (at some point) does not make a goofy simpleton of you. In fact, I believe the opposite is true. It may not be rational, not even natural, but I find it more challenging and charming, even from the sheer intellectual viewpoint to believe things you just can't see, let alone prove. At some point, surrendering your mind might well be the most fulfilling intellectual act, something close, if I understand correctly, to the sentiment that prompted Leonard Cohen's most famous piece, Hallelujah! There comes a point where you no longer understand, no matter how hard you try, so ...let it be, and eventually light will shine in on its own.

Some see this as a recipe for failure. I don't. Not that I can prove it, but what I sense here is a path to handle complexity beyond the means of the spectacular brain-mind jurisdiction. I rather sympathise with C. S. Lewis, who notes that reason is in fact powerful to the point of informing us when she is no longer capable of shedding further light on our way: job done, from now on, you rely upon something else. In Lewis's own words [66] 'When it becomes clear that you cannot find by reasoning whether the cat is in the linen-cupboard, it is Reason itself who whispers: go and look. This is not my job, it is a matter for the senses'.

The Italian journalist-writer-monk Tiziano Terzani (1938–2004) walks on a similar trail as he writes: 'Science is a great route to knowledge, the only mistake is to think it is the only one'. For all my passion for science, I do agree. '*Le coeur a ses raisons que la raison ne connait guè re*', (the heart has reasons that reason does not know)', famously reasoned Blaise Pascal (1623–1662) three and a half centuries earlier. I doubt he meant sentimental melodramas; besides pumping blood across our body, the heart is a *practical* route to knowledge, although a different one. If I had to borrow from modern physics, I would say that Religion is a sort of extra dimension, which partly projects on what we can grasp through science and reason, but leaves most beyond their reach. You may call it mystery, or miracle, if you wish. Recognizing this extra-dimension is the jump in the blue called 'Faith' [66]. This is the Hallelujah which eventually comes, when it comes, after one or many failed Eurekas.

18.5 Of sparrows and lilies

Of course I do see the slippery slope: the very moment you are willing to surrender reason, you open up to irrational and potentially very dangerous attitudes, of which we see all too many aggressive forms in our modern times.

Agreed, but, please, listen a bit longer. A Russian-Jewish colleague of mine once commented to me on the piece of the Gospel, inviting us not to over worry, since the Father takes care of sparrows in the sky and lilies in the fields, all the more so He would take care of us.

To him, this sounded like an invitation to a party of laziness and anti endurance, surrender to weakness and sloppiness, another recipe for failure. Again, I really don't see it this way, one must dig a bit deeper: the invitation is not to laziness but to a state of inner lightness of mind/spirit. Strive to be the very best version of yourself, but always keep your spirit light, never let it be held hostage by your ambition. Play the game, don't be played by it: when you find yourself systematically eating and sleeping in front of your computer screen, it is probably time to take a step back.[63] I always made a point of following this habit: I am always out for my best and my job is done when it's done, not by the hours I spent on it. It is a matter of respect for myself, before it is for my peers, for I am here to succeed and not to fail. But I still need to keep surveillance of the fact that some jobs might be beyond me (harness the hybris) and when I know I have done all I possibly could, and perhaps a bit more, that is it: *que sera sera*. With the sparrow and lily parable on my mind (the international version quotes 'flowers' but the one I learned in Italian explicitly mentions lilies, which I like better. I am indebted to Kathy Fearn for pointing out that the New King James Version, quoted in Appendix, also speaks of lilies).

The impossible is my no-fly-zone, when/if it comes, it comes by itself, *but not unless you prepare for it*, i.e. by working hard, precisely as per Saint Francis's statement in the heading of this book: first do what is Necessary, then do what is Possible and soon you will be suprised to be doing the Impossible. Now I hope you see that 'Hallelujah' is by no means a commitment to uncommitment or an endorsement of flat-Earth views, but just the opposite!

Incidentally, this fully resonates with human greatness: minor characters (!) such as a certain Michelangelo or Wolfgang Amadeus Mozart, made it crystal clear that they were not in control of the wondrous processes that gave birth and shape to their masterpieces. They rather felt as if they were instruments vibrating to some stronger breeze. And if you think that this is art and not science, please think twice: even on my little scale, I know for a fact that there is no good science without emotions, where emotion means that you 'known' it's gonna work even though you really cannot explain why.

There is a narrow divide, if any, between Mozart's music and the Schroedinger equation, they are both part of the same 'stronger breeze'. And if the Schroedinger equation pre-existed its discoverer Erwin, so did Mozart's music!

Beware, this is Mozart speaking, not me.

Here comes the timesheet of Mozart's Sonata FaM K332, side-by-side with the Schroedinger's equation (see Fig. 18.2):

[63] For the record, in its ending, the same parable adds that once you have endorsed this lightness of spirit, the material things (clothes) shall eventually be given to you on top of the rest, for the 'Father knows that you need them'.

Mozart and Schroedinger

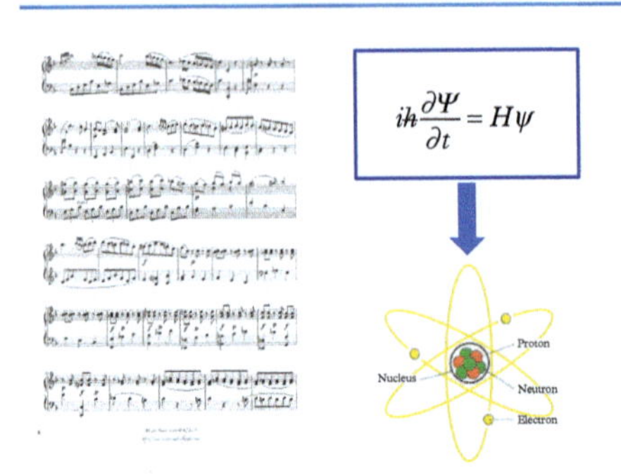

Figure 18.2 *The music sheet of Mozart's Sonata FaM K332/1 (left), the Schroedinger equation, describing the structure of atoms of molecules.*
Source: commons.wikimedia.org.

The language is not the same, the Schroedinger's equation is more compact as it lays down the rule which generates the 'music of Physics', while Mozart's timesheet gives the full set of notes, one by one. But they both breath in the same 'stronger wind'.

Dear reader, I do see your question coming: fine, but where should the bar be set, when do I know that the time has come for me to let it go and listen to the 'stronger breeze'? In the first place is that you have to listen, which is really hard if you systematically eat and sleep in front of your screen.

This said, the real answer is that I really don't know, but ... you do! It is like asking where the bar is to be placed for falling in love: you don't know, but when you pass it, you definitely do. This is all highly personal, I have no proofs and fewer theorems to 'impose', and even if I had any, I would not impose anyway: you may or may not be willing to surrender reason and take a jump in the blue at some point. I do, with due respect to those who don't, but with no apologies either: Hallelujah!

18.6 It's April again, COVID-19 time

It's April again, 2020 this time, and for the five past weeks I'm locked down the hard way. This time is not in Helsinki, not my hometown, Rome, either, but my native city, Forlì, 60 km south of Bologna, where we happened to be for family reasons. Like every second citizen in the world, they tell me, I am locked because of a nasty little piece of 'crystal' (such are viruses), called COVID-19, which is good at nothing but colonizing human lungs for his best survival and prosperity, eventually killing its host in the process. It all broke out in Wuhan, China, early this year, unleashing the worst global crisis since World War II. We didn't see this coming, or perhaps we did, the New England Complex Systems Institute (NECSI) informed us that our modern hyper-connected global society is hyper-exposed to this kind of catastrophic events in dangerous proportion to the three hyper-pillars of pre-COVID society: hyper connection, hyper-speed and hyper-efficiency. Hence, a pandemic was expected, and Bill Gates is certainly the most famous entry in the list of people who allegedly foresaw it.

Well, we might need NECSI science to put it into numbers, but surely we don't need any deep brain-twister to grasp the basics, namely that if an infected citizen in China flies to Lombardy, and Lombardy is, for some whatever genetic-atmospheric-climatic reason, particularly prone to contagion, Lombardy may well suffer more devastating consequences than Wuhan. Speed deflates the world; it brings Wuhan in close touch with Milan notwithstanding the thousands of miles which separate them geographically. In some sense, speed erases space and time. This is precisely the distinction between topological and metric networks we discussed in Chapter 7. With hyperconnectivity on our desk, topological networks call the shots: speed brings immense benefits and devastating risks as well. This is the terrific/tremendous complexity of our modern society.

I thought my Helsinki story was a nice and personalised vehicle to convey the idea of human fragility, and I think it still is, even though it patently pales in front of COVID-19. Fragility stands out as never before in my generation lifetime. Wise people say that instead of aiming at recovering the pre-COVID state of affairs, we should be targeting a better return point, a better society, more conscious and respectful of the overall environment fragility, humans included, of course, but not humans alone.

I see no way to possibly disagree.

But I do have additional memories and feelings by the side. Recently, (Summer 2020, lockdown released) I took my daughter to a spectrally deserted Fiumicino airport. My mind couldn't help but flashing back to the 'hustle and bustle' place from which I took innumerable flights over the last three decades. Flights that landed me to new countries, new people, different habits and different cultures. How many lessons and emotions I absorbed thanks to these flights! My heart was full of sorrow, nearing pain, and I really think that this sorrow and pain are also part of the new equation, i.e. part of the respect we owe to this world's fragility and our own as well. Airplanes are polluting the skies, we know this, and we should do the best of our best to curb the impact, but they also help us opening our minds and our hearts. As I am writing, I can see Saint Augustine's words written on Sydney's Opera's walls (see Fig. 18.3): 'The world is a book and if you don't travel, you stop reading at the first page'.

I am immensely grateful for the privilege I had of *not* stopping my reading at the first page of the marvellous book of our world.

Figure 18.3 *Picture of a magnificent Sydney Opera under multicoloured artistic beams. The picture was taken by this author on a magic Saturday night of June 2019.*

18.7 Do not worry

For those who wish, here comes Matthew's 6:25–34, New King James Version (NKJV)

25 'Therefore I say to you, do not worry about your life, what you will eat or what you will drink; nor about your body, what you will put on. Is not life more than food and the body more than clothing? 26 Look at the birds of the air, for they neither sow nor reap nor gather into barns; yet your heavenly Father feeds them. Are you not of more value than they? 27 Which of you by worrying can add one [a]cubit to his [b]stature?

28 So why do you worry about clothing? Consider the lilies of the field, how they grow: they neither toil nor spin; 29 and yet I say to you that even Solomon in all his glory was not [c]arrayed like one of these. 30 Now if God so clothes the grass of the field, which today is, and tomorrow is thrown into the oven, will He not much more clothe you, O you of little faith?

31 Therefore do not worry, saying, "What shall we eat?" or "What shall we drink?" or "What shall we wear?" 32 For after all these things the Gentiles seek. For your heavenly Father knows that you need all these things. 33 But seek first the kingdom of God and His righteousness, and all these things shall be added to you. 34 Therefore do not worry about tomorrow, for tomorrow will worry about its own things. Sufficient for the day is its own trouble.'

Do not worry!

Epilogue

This completes the roundtrip of what I had to say. I could have said it better, I could have said it worse, but this is as it is at wrap-up time.

I sincerely hope I managed to offer a somehow informative and possibly entertaining walk through the Complexity that inhabits the frontier between physics and biology, just through a small but representative window. We have covered some of the basic principles which rule a broad class of complex systems, and we have then moved to a discussion of the fundamentals of the science of change, thermodynamics, in its historical and then its modern biological and cosmological vests. Thermodynamics is often regarded as a 'good old science', but I hope my reader now sees how departed this view is from the actual state of affairs. Both biology and cosmology present plenty of new thermodynamic challenges which are still awaiting for a thorough understanding, especially in connection with kinetic phenomena far from equilibrium. That is, the phenomena most crucially connected with living organisms. We have also detailed some of these challenges, with specific examples of direct biological relevance, such as deoxyribonucleic acid (DNA) translocation, and protein folding. We have provided a rationale for the mechanisms by which some biological processes manage to accomplish tasks that would appear doomed at the outset, guided by powerful beacons which steer their way across oceans of combinatorial complexity. We have also provided a description of the "unfundamental" forces responsible for such an enterprise and their pivotal place in soft matter, the state of matter most relevant to biological systems. Finally, we concluded with a few reflections on the profound relation between time and complexity, highlighting how deeply enmeshed such a relation is within our very human condition.

This book was prompted by the many emotions I experienced in my multi-decade walk through science, always a passion long before it became a profession. For that, all I can say is the single six-letters magic word which never hurts: thanks.

Acknowledgements

To my family, Claudia and Caterina in the first place and to my friends and colleagues that shone light along my way, too many to be mentioned, without risking embarassing omissions.

Specific thanks to Pablo Tello and Stuart Kauffman for taking the pain of reading through the entire manuscript and providing a lot of insightful remarks.

I am also indebted to Sonke Adlung of OUP for his friendly and impeccable assistance throught this work and to Kathy Fearns for her witty remarks in the process of making my English look better than it is.

19

Appendices

19.1 Numbers

Numbers might be dry to many, they may or may not lie, but always matter, a lot. The Science of Complexity confronts us with truly huge numbers, huge to the point of surpassing and defying not only our intuition but even imagination. In this Appendix, we present some examples, but let us first familiarise with the representation of very large (and very small) numbers, the so-called scientific notation.

19.1.1 Scientific notation

We say a billion and we write $1,000,000,000$, 1 followed by 9 zeros (commas for readability). A billion billion is a billion multiplied by itself, which makes 1 followed by 18 zeros, $1,000,000,000,000,000,000$. Manifestly, this notation gets rapidly impractical for large numbers. An expedient way out is to use just two numbers, the so called *base*, usually 10, and the exponent, 9 for the billion and 18 for a billion billion. In this notation, we write:

$$1,000,000,000 = 10^9$$

pronounced 'ten to the nine'.

Likewise, a billion multiplied by itself, gives

$$1,000,000,000,000,000,000 = 10^{18}$$

pronounced 'ten to the eighteen'.

We saved significant space, didn't we?

Let us then take a look at multiply and divide operations. The rule is really easy: multiply means adding the exponents and divide means subtracting them. Hence $1,000$ times 100 means $10^3 \times 10^2 = 10^{3+2} = 10^5 = 100,000$, while $1,000$ divided 100 means $10^{3-2} = 10$.

How about negative exponents? Actually, the same rule, but let's go by example: take 10 and divide by 1,000, $10^1/10^3 = 10^{1-3} = 10^{-2} = 0.01$. Zero, dot,

Sailing the Ocean of Complexity. Sauro Succi, Oxford University Press.
 DOI: 10.1093/oso/9780192897893.003.0019

another zero for a total of two, and then 1. Pronounced ten to the minus two. Likewise ten to minus five is 0.00001, zero, dot, another four zeros, and 1 in the fifth position after the dot. Note: ten to the power zero is one: $10^0 = 1$, and same goes for any other number as a basis, a very popular one being 2, the one used in digital computers.

19.1.2 Exponential and logarithms

Numbers transform into other numbers under the effect of so called *functions*. In mathematical notation:

$$y = f(x)$$

means that the number x is transformed into the number y under the effect of the function f. The simplest function of all is the constant, $y = c$, which returns the same number c for any value of x. The next simplest function is the linear one $y = cx$, which returns the number multiplied by a constant c. Note that when $c = 1$, y and x are the same, in which case the function is called identity. Much more complex functions can be generated by adding further terms, quadratic, cubic, and so on, eventually up to infinity

$$y = c_0 + c_1 \times x + c_2 \times x^2 + c_3 \times x^3 + \ldots$$

This series is known as Taylor expansion of the function $f(x)$.

Two remarkable functions are the exponential and the logarithm. The exponential transforms small numbers into big ones, while the logarithm does just the opposite. In fact, they are precisely the inverse of each other, meaning by this that the exponential of the logarithm of a given number is the number itself, and vice versa (see Figs. 19.1 and 19.2).

The scientific notation for the exponential is:

$$y = e^x$$

where $e = 2.71828 \ldots$ is known as the Euler constant after the famous Swiss mathematician Leonhard Euler (1707–1783). At $x = 0$, $e^0 = 1$. For positive x, e^x grows much faster than x unbounded to infinity, and for negative x, it decreases to zero.

The mathematical notation of the logarithmic function is:

$$y = log(x)$$

At $x = 1$, $log(1) = 0$, while for x larger than 1 the logarithm grows, but very slowly and remains well below x. For x between zero (excluded) and 1, the logarithm is negative and tends to minus infinity as x goes to zero, just to give an idea,

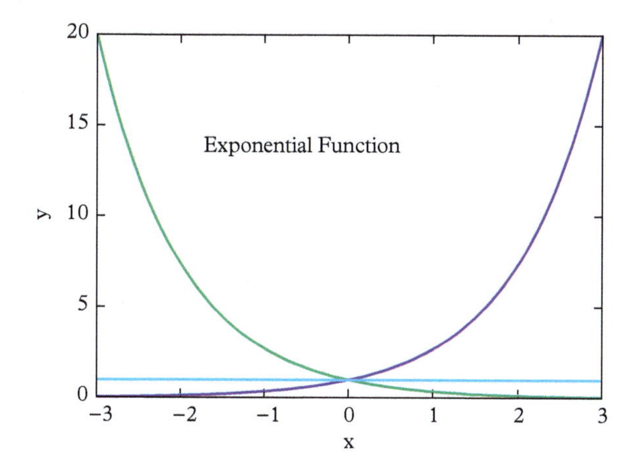

Figure 19.1 *The exponential function $y = e^x$ and its reciprocal $y = e^{-x}$. The product of the two is a constant equal to 1 (horizontal line). The range of x is limited between −3 and +3 for graphic purposes.*

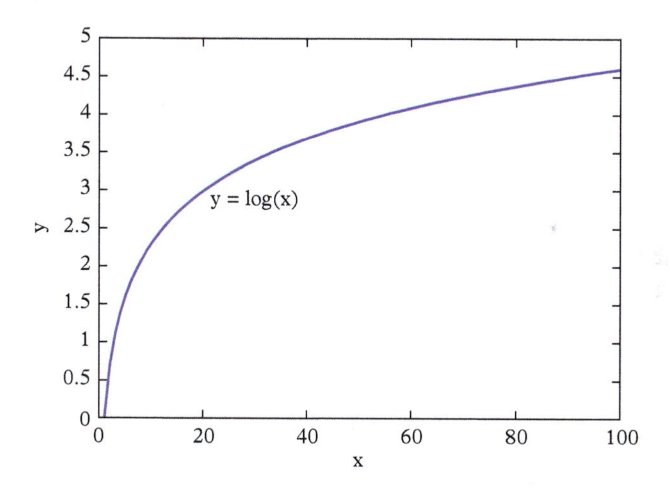

Figure 19.2 *The logarithmic function $y = log(x)$. Note that y remains almost constant at increasing the value of x.*

$e^5 \sim 148$ and $e^{10} \sim 22026$, showing that the exponential turns positive numbers into much larger ones. Conversely, $log(22026) \sim 10$, hence the logarithm converts large numbers in small ones. For negative x, the function does not return any real number (it requires an extension to the domain of complex numbers, not covered here).

19.1.3 Factorial

Take an integer number n, multiply by its lower next $n - 1$, then again by its next lower next $n - 2$ and so on down the line, until you reach the end of the ladder, namely 1. The result is called n-factorial, usually denoted as

$$n! = n \times (n-1) \times (n-2) \cdots \times 1$$

This is a huge number as n is increased, for instance $n = 10$ already gives 3628800, more than three million. In fact, it grows slightly faster than exponentially, an excellent approximation being the so-called Stirling's formula

$$n! \sim \sqrt{2\pi n}(n/e)^n$$

which is often used in combinatorics and statistical physics.

19.1.4 Large numbers: Gates, Bezos, and Avogadro

In ordinary life, a billion sounds like a pretty large number: a billion dollars is a huge sum, seven billion humans on planet Earth are a lot of people, and so on. Bill Gates is credited with about a hundred billion dollars, that is 1 followed by eleven zeroes. So, let us define the 'Gates number' as:

$$Ga = 10^{11} = 100,000,000,000 \qquad (19.1)$$

Hence, in scientific notation, his exponent is 11. A more updated account would replace Bill Gates with Jeff Bezos, credited about two hundred billion dollars. It changes nothing to the exponent, still eleven. A lot, but natural and biological sciences exceed this hands down, as we are going to discuss next.

The natural unit for the number of molecules in a macroscopic piece of matter, say a glass of water, is the so-called Avogadro number, after the Italian chemist, Count Amedeo Avogadro (1776–1856). It's a huge number, 6 followed by 23 zeros. In its full splendour:

$$Av \sim 6.022 \times 10^{23} = 6,022,000,000,000,000,000,000,00 \qquad (19.2)$$

A more telling way to convey the sense of the Avogadro number, is the famous story of the kid trying to empty the sea with his sea bucket, except than instead of sea bucket the kid is given just a cup of coffee [104]. The story begins with the kid dropping a cup of coffee in the ocean. Then the waters of all oceans on Earth get mixed up, and finally the kid fills up the cup again. What are the chances that he regains back one of the molecules he poured into the Ocean? Our intuition responds with a laugh, until we stop laughing once we realize that chances are that the newly filled cup of water would still contain at least one molecule of the original cup of coffee! Why? Because it takes approximately an Avogadro number of cups to fill up all the oceans on Earth, meaning that, after mixing, each cup contains, on average, one of the molecules of the original cup!

Let's move from Earth to the whole cosmos: there are about ten billion galaxies in the observable Universe (10^{10}) and each of them contains about hundred billion stars (10^{11}). As a result there are one billion trillion (10^{21}) stars in the observable Universe. Exponent 21, nearly a thousand times *less* than Avogadro. There are more molecules in a cup of water than there are stars in the entire Universe!

These are different large numbers, which, if perhaps with a little stretch, can still be mastered by our imagination.

Next to huge numbers.

19.1.5 Huge numbers: Googol

The large numbers discussed before have an exponent approximately between 10 and 20. Let's step on it and make it 100: the result is 10^{100} for which we need to invent no new name, since this is known as *Googol*, not exactly the same, but closely related to the famous Silicon Valley giant. Googol reads as 1 followed by a hundred zeroes and, although huge, we can still afford writing it down in just two lines:

$$\begin{aligned} Googol = 1, &0000000000, 000000000, 000000000, 000000000, 000000000 \\ &, 0000000000, 000000000, 000000000, 000000000, 000000000 \end{aligned} \tag{19.3}$$

What we have done, here, is type ten sequences of ten zeros, for a total of hundred, after 1. If you think that this is too huge to make any practical sense, you might be surprised to learn that this is nothing but the number of sentences of a mere 100 characters you can write with an alphabet of just 10 letters. A single page of this very book offers many more possibilities, since I am using an (English) alphabet of 26 letters, not to mention commas, blanks, and other associated symbols ....

19.1.6 Towards monster numbers: DNA

Leaving books and turning to biology instead, we have seen in this book that human deoxyribonucleic acid (DNA) consists of about three billion base pairs, each coming in four flavours, adenine, guanine, cytosine, and thymine, AGCT. These are rather complex molecules, but for many purposes they can be treated as simple beads of four different colours. By this analogy, the human DNA is a long necklace with three billion four-coloured beads.

The question is: how many such necklaces can we assemble? The calculation is easy. If there were, say, just two beads in the chain, it would be $4 \times 4 = 16$, with three it would be $4 \times 4 \times 4 = 64$, and so on down the line. With three billion, the number is 4 multiplied by itself three billion times, $4^{3\times10^9}$. Here we go far beyond intuition, but math still assists us. Some algebra first: since $4^3 = 64$, our number is $64^{10^9} \sim 10^{1.8\times10^9}$, which means 1 followed by about 1.8 billion zeroes! Compare this with the huge numbers of the previous section which counted at most hundred zeroes past 1. If we had to write it explicitly, this entire book would not be long enough! We are clearly nearing a class of numbers which defy not only our intuition but also our imagination.

For all its unimaginable size, the number of DNA necklaces pales in front of still much larger numbers.

19.1.7 Monster numbers and beyond: Googolpex and Penrose

So, let us scale it up even further, and define the so-called *googolpex*, namely:

$$Googolpex = 10^{Googol} \tag{19.4}$$

That is, 1 followed by a googol of zeroes! Since *Googol* is far bigger than one billion, googolpex is far larger than the number of DNA necklaces!

For all its unimaginable size, the Googolpex is still relevant to the natural world, possibly even falling short of reaching the numbers involved in modern cosmology!

The distinguished mathematical cosmologist Roger Penrose, estimates that the probability that the Universe began precisely in the conditions that are required to reach our current Universe as we know it today, is about 1 in $10^{10^{123}}$, namely 1 followed by 10^{123} zeroes. The extra factor in the exponent is curiously close to the Avogadro number, so that we can write the 'Penrose number' as googolpex raised to the Avogadro number!

$$1 Penrose \sim Googolpex^{Av}$$

What does it mean? As far as I can tell, nothing, but consider the following example. By some accident, you have been poisoned, but, happily enough, an antidote is available. You rejoyce at the good news, until you learn that the antidote is served in a glass, in which every and single molecule is picked up from a googolpex of possible different chemical species. Any of the exterminate multitude of possible glasses will let you down with your poison, and only one of them will save your day. Notwithstanding the desperate odds, you proceed to pick up your glass and you succeed to cherry-pick the right one out of the exterminate crowd of glasses in front of you. According to Penrose, that's exactly what our living Universe did at the act of the Big Bang .... The proverbial needle in the haystack, but what a haystack indeed!

19.1.8 Logarithmic survival

These are the numbers which describe the inconceivabe and breathtaking vastness of the Ocean of Complexity, as we meet it in biology and many other domains of the natural world. These numbers spell doom for our capacity to sail the Ocean of Complexity, though we are always sure to cast away. Yet, we can survive. How, and why? Because the cosmos has been equipped with a structure which makes itself intelligible, to some extent at least, without having to deal with the monster numbers, but basically only with their logarithms, and sometimes even logarithms of logarithms! The number of DNA necklaces is one followed by 1.8 billion zeroes, but the number of the actual individuals (phenotype) is in the billions. A truly thin sieve applies in going from genotypes (what is *conceivable*) from phenotype (what is *compatible* with physical constraints). This thin sieve is the set of constraints imposed by the laws of the physical world. The lesson is that the Ocean of combinatorial Complexity can be navigated because, thanks to the physical constraints, we don't depend on the monster numbers themselves but on their logarithms instead. Well, this is a drastically simplified view of the actual picture, but it captures nonetheless an essential aspect of our ability to make sense of the complex world around us. This is greatness encoded within Boltzmann's most famous equation

$$\mathrm{S} = \mathrm{k}\ \mathrm{logW}$$

If there is a representative equation for the survival in the Ocean of Complexity, this is the one.

19.2 Thermodynamics

We have already discussed thermodynamics in some length in the main text. In the following, we provide just a few additional notions that might be of some use for the reader interested in digging deeper into the subject.

19.2.1 Thermodynamic variables

The thermodynamic state of a given macroscopic body is characterized by three independent quantities: the number of molecules N, the volume V, and the temperature T, or, alternatively, the pressure P. At equilibrium, these four variables are linked via a so-called equation of state.

For the case of an ideal gas, i.e. a collection of molecules with no potential energy interactions, the equation of state read simply as:

$$PV = Nk_BT \tag{19.5}$$

where k_B is the so-called Boltzmann constant. The internal energy of an ideal gas is given by:

$$E_{int} = \frac{3}{2} N k_B T \tag{19.6}$$

indicating that each molecule contains an amount of energy $k_B T/2$ along each of the three spatial directions. Note that this is the energy contained in the *disordered motion* of the molecules, the one responsible for temperature and no macroscopic motion. Macroscopic motion contributes an extra term of the form:

$$E = \frac{1}{2} M U^2 \tag{19.7}$$

where $M = Nm$ is total mass of the system moving at a macroscopic speed U and m is the mass of the single molecules. Boltzmann noticed that if a body can be heated up or cooled down, it must necessarily be made of microscopic units. In other words, temperature is a direct proof of the existence of atoms!

Also, note that for non-ideal gases, in which atoms and molecules interact via potential energy, both pressure and the internal energy no longer depend linearly on the number of molecules. Such nonlinear dependence results in Complex thermodynamic phenomena, such as condensation and evaporation, namely phase transitions between the gas and liquid states and vice versa.

19.2.2 The First Principle of Thermodynamics

The first principle can be mathematically stated as follows:

$$dE = \delta Q - \delta W \tag{19.8}$$

where dE is the change of the internal energy of a given body, upon absorbing an amount δQ of heat and performing an amount δW of work towards the external environment.

Note that the symbol d stands for thermodynamic state functions, i.e. quantities that depend only on the initial and final states of the given thermodynamic transformation, whereas the symbol δ denotes dependence on the specific transformation.

Remarkably, the difference of two path dependent quantities gives rise to a path-independent one! The heat absorbed by the body is also expressed in the terms of the so-called heat capacity, C, through the following expression

$$\delta Q = C\delta T \tag{19.9}$$

In the above δT is the temperature change due to the absorption of the amount of heat δQ. Big thermal capacity means small temperature changes and conversely. Note that C depends on the thermodynamic transformation, taking different values for instance for the case of constant pressure or constant volume transformations. The champion of big capacity is called a 'reservoir' in the thermodynamics literature, standing for a body so big that it can take any amount of heat, in or out, without any noticeable effect on its temperature.

Now to work. The work done by the body to the exterior at constant pressure is given by the following expression:

$$\delta W = P\delta V \tag{19.10}$$

where P is the (constant) pressure and δV the change in volume. The work done on the exterior lowers the energy, hence the temperature of the body (see equation of state) unless some compensating heat is supplied.

19.2.3 The Second Principle of Thermodynamics

The second principle states that the entropy change in any spontaneous transformation taking an equilibrium state to a different one is bound to be positive or zero.

$$dS = \frac{\delta Q}{T} \geq 0 \tag{19.11}$$

The latter case defines so-called reversible transformations, a useful idealization which cannot be realized in real life. It is sometimes stated, including in this book, that entropy can only grow monotonically in time. This is not entirely correct; entropy can occasionally decrease in time due to local instabilities, for instance, in the case of formation of structures. The point is that once these instabilities are settled, the new equilibrium state must necessarily contain more entropy than the initial state. As explained in the text, this statement applies to the closed system made by any given body and its surroundings, what we have called the Universe. Single subsystems can decrease their entropy at the expense of others.

19.2.4 The Third Principle of Thermodynamics

The Third Principle of Thermodynamics, often attributed to the Prussian physicist Walter Nerst (1864–1941), states no material system can ever reach zero temperature in degrees Kelvin. This empirical observation received a theoretical explanation with the advent of quantum mechanics, which showed that the minimum energy attained by a quantum system, the so-called ground state energy, is given by:

$$E_0 = hf_0/2 \quad (19.12)$$

where f_0 is the base frequency of a particle trapped in a parabolic potential (harmonic oscillator for the experts). The minimum energy for a classical particle would be the bottom of the parabola corresponding to zero energy. But in a quantum world the 'particle' cannot stop moving. In fact, it keeps oscillating forever with frequency f_0 (dictated by the amplitude of the potential). And up in absorbing energy, higher frequencies are excited, but still on a discrete sequence. As a result the energy is never zero, and neither is the temperature.

In very compact form:

$$T > 0.$$

19.2.5 The Fourth Principle of Thermodynamics (in jesting mode ...)

As mentioned in the text, the Fourth (unofficial) Principle of Thermodynamics, asserts that anytime probability of success is not next to one, it is necessarily next to zero. No room for compromises. This a typical case of 'extreme statistics', which does not admit anything but extreme values: rich or poor, '*tertium non datur*', the middle class gets extinguished. Such behaviour is often displayed by systems subject to increasingly heavy constraints, which only fewer and fewer individuals can carry, leaving everybody else behind. It is typical of complex systems at higher levels than the molecular ones, such as finance and economics, and even though it often shows up in the absence of a regulating mechanism, this is not a necessary condition either. Although they are not part of classical thermodynamics, these matters hold a major place in the modern theory of complex systems.

19.3 Quantum physics

In this Appendix, we provide a brief survey of the laws of quantum mechanics, the physics that governs the behaviour of matter at the atomic and nuclear scale, at about a tenth of a nanometre and below, since quantum mechanics developed from classical mechanics, by reviewing the basic ideas behind this latter first.

19.3.1 Classical mechanics

Classical mechanics is a world of particles and trajectories, described by the famous Newton equation:

$$F = ma \tag{19.13}$$

In words, a material body of mass m, subject to a force F, undergoes an acceleration $a = F/m$. But what exactly is acceleration?

For a long time it was confused with the velocity, namely the rate of change of space and time. Using infinitesimal calculus:

$$v = \frac{dx}{dt} \tag{19.14}$$

where dx is the change in position over a small interval of time dt. It took the genius of Galileo Galilei to realize that a material body can also move in the absence of any force, because forces are not responsible for the velocity itself, but for its change in time, which is precisely the acceleration.

Using infinitesimal calculus again:

$$a = \frac{dv}{dt} \tag{19.15}$$

Combining the relation (19.13) with (19.14) and (19.15), we obtain:

$$\frac{dv}{dt} = F/m \tag{19.16}$$

This means that once we know the initial position and velocity of the material particle at time $t = 0$, say x_0 and v_0 respectively, equation (19.16) provides the velocity at a subsequent time $t = dt$, and with this new velocity one can proceed to compute the new position $x(t + dt)$ via the equation (19.14). This is how Newton's law permits us to compute the trajectory, i.e. the position and velocity of the particle at any subsequent time. Hence, the Newton chain is:

$$(x_0, v_0) : F_0 \to v_1 \to x_1 \to F_1 \to v_2 \to x_2 \ldots \tag{19.17}$$

where the subscripts indicate the sequence in time.

This suggests that with enough calculational skills, the 'Newton chain' can compute the position and the velocity of any number of particles at any subsequent instant of time.

This pictures a world of *deterministic certainty*, epitomized by the famous sentence of Pierre Simon de Laplace (1749–1827), often regarded as the manifesto of reductionistic determinism. We may regard the present state of the universe as the effect of its past and the cause of its future. An intellect which at a certain moment would know all forces that set nature in motion, and all positions of all items of which nature is composed, if this intellect were also vast enough to submit these data to analysis, it would embrace in a single formula the movements of the greatest bodies of the universe and those of the tiniest atom; for such an intellect nothing would be uncertain and the future just like the past would be present before its eyes. (Pierre Simon Laplace, *A Philosophical Essay on Probabilitie*[64].)

By now we know that Laplace's dream gets shattered against the cliffs of chaos, namely the fact that in order to accomplish the task envisaged by Laplace, one would need infinite precision, since trajectories which start arbitrarily close at time zero may depart from any arbitrary distance as time unfolds. But leaving chaos aside, Newtonian mechanics, remains the symbol of deterministic certainty, the dominating paradigm of the Age of Enlightenment.

19.3.2 Classical waves

In Laplace's times physics was one with mechanics, material bodies moving under the effects of applied forces, as per the chain discussed previously. Yet, about at the same time, another group of scientists, Alessandro Volta (1745–1827), Andrè Marie Ampère (1775–1836), Charles Augustine Coulomb (1736–1806), Michael Faraday (1791–1867), culminating with the prodigious synthesis of James Clerk Maxwell (1831–1879), who developed another paramount chapter of physics associated with the propagation of waves, particularly electromagnetic ones, including light.

Unlike particles, waves 'fill out' space everywhere and propagate mass, momentum, and energy via undular vibrations of a supporting media, like the chord of a violin.

Particles and waves stand as paradigms for *localized* versus *extended* objects, respectively. Despite this fundamental difference as to their spatial extent, both particles and waves were thought to possess virtually any value of energy, what the physicist call a continuum spectrum, no gaps or jumps in between. Like space and time, energy is a continuum.

19.3.3 Quantum mechanics

This view was under heavy scrutiny at the beginning of the nineteenth century, mostly on account of the efforts to understand the energy radiated by black bodies. These are empty cavities, inside which light bounces back and forth on the internal walls of the cavity, until it finds its way out through a tiny hole on the surface. Based on the law of classical thermodynamics, one would predict that the amount

[64] Laplace was not a piece of cake, he did not leave any doubt on the fact that he regarded himself as the best mathematician around, the effect on his peers being mitigated only by the fact that he was right. He also took very important political charges, apparently with much less success.

of energy radiated by the back body would be infinity, the culprit being the highest frequencies in the spectrum.

Manifestly nothing physicist would take for an answer.

19.3.3.1 Max Planck

It was for Max Planck (1858–1947) to first figure out that, in order to tame this infinity and the explain the laws of black body radiation, one should assume that energy can only by emitted in discrete packets or grains, called *quanta*. '*Natura non facit saltus*' (nature does not make jumps), recites Gottfried Leibniz (1646–1716), the famous contender to Newton on the discovery of infinitesimal calculus, and Planck felt no urge to depart from this view, in that he regarded his *quanta* as a mere calculational device.

To the contrary, it actually turned out that, with good peace of Leibniz, that *natura -facit- saltus* and, far from being a mere calculational trick, Planck's assumption unveiled a very profound truth about the way nature works. For this major discovery, Planck was awarded the 1918 Physics Nobel. However, the one who provided indisputable evidence for the reality of quanta was ... guess who, Albert Einstein, still him! In his prodigious year (1915), besides the theory of special relativity, he also published two more epoch-making papers, one on the motion of pollen in suspensions, which firmly established the existence of atoms, and one on the photoelectrical effect, which achieved the same goal for the case of light, photons being the name of the quanta of light.

And even though Einstein's fame with the greater public rests chiefly with relativity, what earned him the Nobel Prize was precisely the photoelctrical effect (relativity was still judged too mathematical for a Nobel in physics). Be that as it may, the point is that in the quantum world the energy carried by a single vibration is proportional to its frequency, the faster the vibration, the highest its energy. In equations,

$$E = hf \tag{19.18}$$

where h is a universal number, known as Planck constant (no surprise!), and f is the frequency, i.e. the number of vibrations per unit time, typically measured in hertz (one hertz corresponds to one oscillation per second). As an example, our radios, the few that still exist, work at around 1 megahertz, one million oscillations per second. It is named after the German scientist Heinrich Hertz (1857–1894). The corresponding wavelength, i.e. the distance travelled by the wave in the time it takes to perform one oscillation ($1/f$), is the inverse frequency times the speed of light, $\lambda = c/f$. For radio waves at 1 MHz this corresponds to about 30 metres. For a cell phone, the radiofrequency is about a thousand times higher, giving a wavelength of about 3 cm. For visible light (photons), the frequency is still one million times higher, from 400–800 terahertz (1 terahertz = 10^{12} vibrations/second) or red/violet respectively, corresponding to a wavelength between

400–800 nanometres, a fraction of one micron (few hundred times thinner than our hairs).

Like the notes of a musical instrument, energy is *quantized*, it comes in discrete units of h. Thus, nature takes jumps indeed.

The jumps are however tiny, to the point that they cannot be appreciated on a macroscopic scale. Like stairs with infinitesimally tall steps, which we would not even perceive upon walking up the stairs. Or, like blood in our veins, which appears to be a continuum fluid as long as we look it from a distance far above the size of red cells (about 10 microns). A single photon of visible light carries an energy of about 1.5–3.5 electron volts. This is a very tiny amount in macroscopic units: it takes about 10^{19} for such photons to make just one joule, the unit of energy relevant to the macroscopic world, named after the British scientist James Prescott Joule (1818–1889).

The attentive reader may argue that the energy of ordinary waves depends on their amplitude and not on their frequency: a giant wave on the sea certainly looks scarier than the tiny oscillation after you throw a small stone in the pond. This is indeed true; the point is that the amplitude measures *how many* quanta are participating in the process, so that the energy is the sum of all hfs carried by each quantum.

Thus, light as we perceive it, is a fluid of photons, and its granular nature is entirely concealed by the multitude of individuals. That light is made of granular units, the photons, came as an exciting discovery, which earned Einstein his Nobel Prize in 1921. However, even more surprising and in fact disconcerting, was the other side of the medal, namely that matter behaves like a wave at the atomic scale! This mind-boggling *matter-wave* duality was proven beyond doubt by crystallographic experiments in the early part of the former century, which clearly showed that electrons passing through solid crystals produce interference patterns precisely like waves in a pond: the undulatory nature of matter!

All this implied revolutionary changes in the way we look at nature and consequently radical innovations in the mathematical description of matter at the atomic scale. This prepared the ground for our next hero of quantum physics, the Austrian Erwin Schroedinger.

19.3.3.2 Erwin Schroedinger

Erwin Schroedinger (1887–1961) was the first to write down his eponymous wave equation, accounting for the bizarre effects discussed above, and particularly the dual wave-particle nature of matter at the atomic scale. In particular, Schroedinger introduced the equation that describes the time and space change of the so-called wavefunction $\Psi(x, t)$. This is a rather mysterious object, whose main property is the fact of measuring the *probability* $P(x, t)$ to find the quantum particle at position x at time t (see Fig. 19.3).

In equations:

$$P(x, t) = |\Psi(x, t)|^2 \tag{19.19}$$

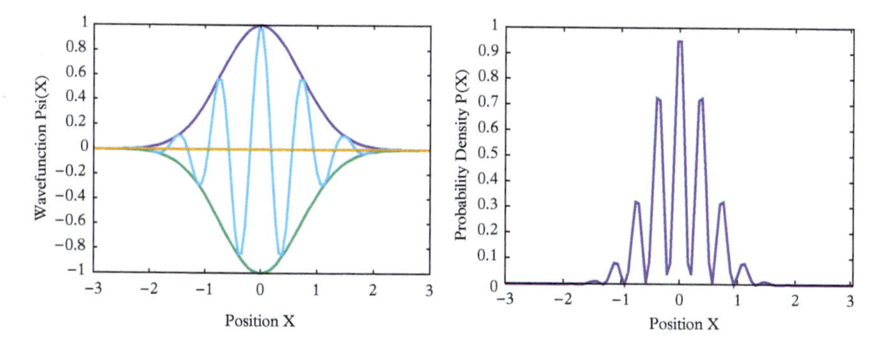

Figure 19.3 *The wavefunction $\Psi(x)$ of a quantum particle (top) and the associated probability density $P(x)$ (bottom). The probability of finding the particle at a generic position x is localized in a finite region of space, centred around the origin $x = 0$, where it attains its peak value. This is where the classical particle would be found, with probability 1, i.e. without uncertainty. The oscillations of the wavefunction are a measure of its velocity (energy), the more it oscillates, the higher its energy, according to the Equation (19.18).*

This probability is the quantum replacement for Newton's trajectories and its physical interpretation was first due to the Polish-born German physicist Max Born (1882–1970), who was awarded the 1954 Nobel Prize.

The Schroedinger equation still stands as one of the most precious cornerstones of modern science and it provides a possibly unsurpassed combination of mathematical beauty and practical usefulness (you find it in full display in Chapter 18 of this book). Since its inception, it has served as the workhorse for computing the energy of atoms and it still forms the basis for the computational methods of modern chemistry, earning two Chemistry Nobel Prizes in recent years (1998 and 2013). It is unquestionably one of the most useful equations in the whole history of science.

This is the operational side of quantum mechanics, one making less headlines than its philosophical cousin, the one typically associated with another giant of the quantum saga, the German Werner Heisenberg (1901–1976), whose fame trespassed physics, mostly on account of his famous principle of uncertainty.

19.3.3.3 Werner Heisenberg

Heisenberg's principle states that you cannot know both the postion x of a given particle and its velocity v, to an arbitrary degree of accuracy. If you are demanding a lot on the position, you must surrender on the velocity side, and vice versa.

This stands in stark contrast with the very milestone of Newtonian mechanics, namely the concept of particle trajectory, in which both position and velocity are known to any desired degree of accuracy (discounting of course experimental limitations, but this is not matter of principle).

Expectedly, this came like a death knell for classical determinism, as glorified by Laplace's dream. Beware, no need to invoke large numbers or chaos, it is an inevitable consequence of the wave-particle duality of quantum mechanics!

In mathematical terms

$$\delta x \delta v > \hbar/m \tag{19.20}$$

where δx is the uncertainty on the position, δv on the velocity, m is the mass of the particle, and $\hbar = h/2\pi$ is another way of writing the Planck constant.

We have learned that relativity remains hidden to our senses because the speed of ordinary objects is much smaller than the speed of light. Likewise, quantum mechanical uncertainty remains hidden to macroscopic creatures because the Planck constant is totally inconspicuous in macroscopic units, $\hbar \sim 10^{-34}$ (kilograms per metre squared per second). For a human being with, say $m \sim 100$ kg, the uncertainty $\delta x \sim 10^{-36}$ metres, 1 part over a billionth of a billionth of a billionth of a billionth of its size! In fact smaller than the Planck length, below which spacetime itself makes no physical sense anymore.

This holds for humans, but protons tell another story. For a proton at room temperature, $\delta x \sim 10^{-10}$ metres, one tenth of a nanometre. This looks like very small number, until one realizes that the estimated size of the proton is about 1 femtometre, i.e. 10^{-15}m, namely five orders of magnitude smaller! This means that the size of the proton as a classical particle is a hundred thousand times smaller than the cloud of quantum uncertainty around it. Under such conditions, the proton is much more a wave than a particle and the very notion of Laplacian trajectory, a sequence of positions and velocities, crumbles in pieces. Laplace dream goes shattered against the wall of quantum uncertainty.

19.3.3.4 Louis de Broglie

As discussed above, the transition from classical to quantum physics is governed by the Planck constant, whose smallness makes quantum effects usually negligible at a macroscopic scale.

A more precise measure of this classical to quantum transition is provided by the De Broglie wavelength, from the French scientist and aristocrat Louis de Broglie (1892–1987), the 7th Duc de Broglie, who first postulated the particle-wave duality of material particles in his 1924 PhD thesis.

The De Broglie length is defined as

$$\lambda_B = \frac{\hbar}{m V_T} \tag{19.21}$$

where $V_T = \sqrt{k_B T/m}$ is the thermal speed of a particle of mass m at temperature T.

Physical phenomena occurring at scales much larger than λ_B obey classical mechanics, while those that take place at scales comparable or smaller than

λ_B must be treated quantum-mechanically. In terms of Heinseberg's uncertainty principle, we can make the following identifications: $\delta x \sim \lambda_B$ and $\delta v \sim V_T$. Incidentally, the relation (19.21) informs us that the De Broglie wavelength decreases with the square root of the temperature and becomes virtually infinite in the limit $T \to 0$. This is yet another indication that quantum mechanics is incompatible with the limit of zero (Kelvin) temperature, i.e. the Third Principle of Thermodynamics, from yet another perspective.

De Broglie also introduced the so-called 'De Broglie pilot wave' picture, whereby the quantum probability wave not only measures the chance to find the particle at a given position in space and time, but actually takes an active role in 'guiding' it along its classical trajectory. His work earned him the 1929 Physics Nobel, not without raising a hot scientific and philosophical debate, since the pilot-wave picture clearly opened the door to the interpretation of the wavefunction as a sort of statistical collection of underlying classical particles, somehow in the spirit of Boltzmann's statistical physics. In other words, *classical* probability as the result of our ignorance of microscopic details (remember Boltzmann?) versus *quantum* probability as an ontological limitation to our knowledge of the microscopic world, even at the level of a single particle.

The De Broglie picture met the initial sympathy of Einstein, who never accepted the probabilistic interpretation of the wavefunction first brought up by Max Born and later vigorously propagated by the Copenhagen school, led by another major father of quantum mechanics, the Danish Niels Bohr (1882–1970), Physics Nobel 1922 for his profound contributions to the understanding of the atomic structure.

The Copenhagen interpretation won hands down over the De Broglie pilot-wave picture, often with a vengeance. Yet, modern developments in quantum physics are providing increasing evidence in favour of De Broglie's interpretation. This is still a very open field of research in modern quantum physics.

19.3.3.5 Paul Dirac

We cannot close this brief Appendix on quantum mechanics without mentioning yet another towering hero, the British Paul Adrien Maurice Dirac (1902–1984). His perennial fame is associated with the eponymous equation which reconciles quantum physics with Einstein's special relativity (no gravity). That is, quantum particles which travel near the speed of light. The most stunning outcome of Dirac's equation was the prediction of the existence of antimatter, which was experimentally confirmed in 1932, just a mere four years after Dirac published his equation. Dirac and Schroedinger were jointly awarded the 1933 Nobel Prize in Physics, just the year after Werner Heisenberg, who didn't have to share.

It is not often the case that Nobel Prizes leave such a monumental legacy as those mentioned in this Appendix. These were the truly heroic and perhaps unrepeatable heydays of theoretical physics.

References

1. A. D. Aczel (2014) *Why Science does not Disprove God.* NY: Harper Collins.
2. https://en.wikipedia.org/wiki/Anton
3. P. W. Anderson. (1972) More is different. *Science.* 177(4047): 393.
4. P. W. Anderson and D. L. Stein. (1984) Broken symmetry, emergent properties, dissipative structures, life: Are they related? In *Basic Notion of Condensed Matter Physics,* p. 263. Frontiers in Physics. NY: Addison Wesley.
5. W. B. Arthur. (2007) *Complexity and the Economy.* Oxford: Oxford University Press.
6. P. Atkins. (2007) *Four Laws that Drive the Universe.* Oxford: Oxford University Press.
7. J. Bahcall, T. Piran, and S. Weinberg (eds) (2004) *Dark Matter in the Universe.* Singapore: World Scientific.
8. A. L. Barabasi, (2014) *Linked, How Everything is Connected to Everything Else and what it Means for Business, Science and Everyday Life.* New York: Basic Books. (First published by Perseus, 2002).
9. R. Albert and A. L. Barabasi. (2009) Statistical mechanics of scale-free networks. *Review of Modern Physics.* 74(47).
10. A. L. Barabasi. (2018) *The Formula: The Universal Law of Success.* Little Brown and Company.
11. R. Benzi, A. Sutera, and A. Vulpiani. (1981) The mechanism of stochastic resonance. *Journal of Physics A: Mathematical and General.* 14(11): 453.
12. A. Ben-Naim. (2014) *Statistical Thermodynamics: With Application to the Life Sciences.* Singapore: World Scientific.
13. A. Bejan and J. Peder Zane. (2012) *Design in Nature, How the Constructal Law Governs the Evolution in Biology, Physics, Technology and Social Organization.* NY: Anchor Books.
14. G. Bianconi. (2018) *Multilayer Networks, Structure and Function.* Oxford: Oxford Univerisity Press.
15. J. D. Bryngelson, J. N. Onuchic, N. D. Socci, and P. G. (1995) Wolynes funnels, pathways, and the energy landscape of protein folding: A synthesis. *Proteins: Structure, Function, and Bioinformatics.* 21(3): 167–195.
16. Anna D. Broido and A. Clauset. (2019) Scale-free networks are rare. *Nature Communications.* 10(1017).
17. C. Bustamante, J. Liphardt and F. Ritort. (2005) Thermodynamics of small systems. *Physics Today.* 58(7): 43. https://doi.org/10.1063/1.20124
18. D. Buonomanno. (2017) *Your Brain is a Time Machine, the Neurosciences and the Physics of Time.* London: W. Norton.
19. G. Caldarelli and M. Catanzaro. (2012) *Networks, a Very Short Introduction.* Oxford: Oxford University Press.
20. P. Castiglione, M. Falcioni, A. Lesne, and A. Vulpiani. (2008) *Chaos and Coarse-Graining in Statistical Mechanics.* Cambridge: Cambridge University Press.
21. C. Cercignani. (2008) *Ludwig Boltzmann, The Man who Trusted Atoms.* Oxford: Oxford University Press.

22. S. Carroll. (2016) *From Eternity to Here, the Quest for the Ultimate Theory of Time.* Penguin Random House.
23. P. Charbonneau. (2017) *Natural Complexity, a Modeling Handbook.* Princeton NJ: Princeton University Press.
24. M. Chaplin. *Water Structure and Science. http : //www1.lsbu.ac.uk/water/hexagonal ice.html*
25. B. Chopard and M. Droz. (1998) *Cellular Automata Modeling of Physical Systems.* Cambridge: Cambridge University Press.
26. B. Clegg. (2019) *Dark Matter and Dark Energy, the Hidden 95 Percent of the Universe.* Hot Science.
27. F. Collins. (2006) *The Language of God, a Scientist Presents Evidence for Belief.* New York: Free-Press.
28. P. V. Coveney and R. Highfields. (1995) *Frontiers of Complexity, the Search for Order in a Chaotic World.* London: Ballantine.
29. P. Domingos. (2018) *The Master Algorithm, How the Quest for the Ultimate Learning Machine will Remake our World.* Basic Books.
30. F. Dyson. (2007) *Reflections on the Place of Life in the Universe, A Many Colored Glass.* Virginia: University of Virginia Press.
31. F. Dyson. (1989) *Infinity in all Directions.* New York: Harper Collins.
32. E. Reilly and C., Mc Grath. (2018) Fides et Ratio: The Pursuit of Faith and Reason. In the 21st Century Catholic University. *Jesuit Higher Education. A Journal.* 7(45): The original: *Encyclic: John Paul II, Fides et Ratio.* Encyclical Letter (Vatican City, Italy: Libreria Editrice Vaticana, 1998). accessed May 4, *https : //w2.vatican. va/content/john − paulii/en/encyclicals/documents/hf*$_{j}$*p − ii*$_{e}$*nc*$_{1}$4091998$_{f}$*ideset − ratio.html.*
33. F. Franks. (2012) Water. A comprehensive treatise. Vol. 6. *Recent Adavances, Springer Science & Business Media.*
34. H. Frauenfelder, S. G. Sligar, and P. G. Wolynes. (1991) The energy landscapes and motions of proteins. *Science.* 254(5038): 1598–1603, 3116.
35. D. Frenkel and B. Smit. (2014) *Understanding Molecular Simulation, from Algorithms to Applications.* 2nd edn. Cambridge, USA: Academic Press. (1st edn. 2001).
36. K. Friston. (2010) The free-energy principle: A unified brain theory? *Nature Reviews Neuroscience.* 11(1): 127–113.
37. K. Friston.(2012) A free energy principle for biological systems. *Entropy.* 14: 2100–2121; doi: 10.3390/e14112100 ISSN 1099–4300.
38. Georgi Yordanov Georgiev, John M. Smart, Claudio Flores, Martinez Michael, and E. Price. (eds) (2019) *Evolution, Development and Complexity, Multiscale Evolutionary Models of Complex Adaptive Systems.* Springer Proceedings in Complexity.
39. M. Bernaschi, M. Fyta, S. Melchionna, S. Succi, E. Kaxiras, (2008) Quantized current blockade and hydrodynamic correlations in biopolymer translocation through nanopores: Evidence from multiscale simulations. *Nanoletters.* 8(4): 115. Berlin.
40. *https : //en.wikipedia.org/wiki/Folding*$_{f}$*unnel*
41. J. Gleick. (2008) *Chaos. Making of a New Science.* London: Penguin Books.
42. C. Gomes and P. Faisca. (2019) *Protein folding, an introduction.* Springer Briefs in Molecular Science.
43. L. Greene. (2018) *Silicon States, the Power and Politics of Big Tech and what it means for Our Future.* Berlin: Counterpoint.

44. G. Gallavotti, W. L. Reiter, and J. Yngvason. (eds) (2008) Boltzmann's legacy. ESI Lectures in Mathematics and Physics. *European Maths Society*. Vienna.
45. George, Andrew R. (2010) The Babylonian Gilgamesh Epic. In: *Introduction, Critical Edition and Cuceiform Texts* (in English and Akkadian). Vols 1 and 2 (reprint ed.) Oxford: Oxford University Press, p. 163. ISBN 978-0198149224. OCLC 819941336.
46. M. Gell-Mann. (1994) *The Quark and the Jaguar*. NY: Penguin Books.
47. J. Gleick. (1998) *Chaos, Making a New Science*. NY: Penguin Books.
48. N. Goldenfeld and L. Kadanoff. (1999) Simple lessons from complexity. *Science*. 284(54–1): 87–89.
49. H. J. Jensen. (1998) Self-organized criticality,emergent complex behaviour in physical and biological systems. In *Cambridge Lecture Notes in Physics*. Cambridge: Cambridge University Press.
50. N. F. Johnson. (2017) *Simply Complexity, a Clear Guide to Complexity Theory A*. London: One World Group. (1st edn. 2007).
51. E. Johnson. (2018) *Anxiety and the Equation, Understanding Boltzmann's Entropy*. Cambridge, USA: The MIT Press.
52. S. A. Kauffman. (1995) *At Home in the Universe, the Search for the Laws of Self-organization and Complexity*. Oxford: Oxford University Press.
53. S. A. Kauffman. (2019) *A World Beyond Physics, The Emergence and Evolution of Life*. Oxford: Oxford University Press.
54. L. Kadanoff. (1986) On two levels, complex structures from simple systems, *Physics Today* 39(9): 7. https://doi.org/10.1063/1.2815134
55. M. Karplus.(2011) Behind the folding funnel diagram, *Nature Chemistry and Biology*. 7: 404.
56. S. Klein. (2018) *How to Love the Universe, a Scientist Odes to the Hidden Beauty behind the Visible World*. New York: Springer Verlag
57. J. Hadamard. (1922) The early scientific work of J. Hadamard, Rice. Institute Pamphlet. 9(3): 111.
58. S. Harris. (2011) *An Introduction to the Theory of the Boltzmann Equation*. Dover Books in Physics.
59. S. Hawking and R. Penrose. (1996) *The Nature of Space and Time*. Princeton: Princeton University Press, and references therein.
60. C. E. Hecht. (1991) *Statistical Thermodynamics and Kinetic Theory*. NY: Freeman and Company. (Reprinted by Dover Publications, 1998).
61. J. Holland (2014) *Complexity, a Very Short Introduction*. Oxford: Oxford University Press.
62. Y. Holovatch, R. Kenna, and S. Thurner. (2017) Complex systems: Physics beyond physics. *European Journal of Physics*. 38(2): 1–19, 023002.
63. K. Huang. (2005) *Lectures on Statistical Physics and Protein Folding*. Singapore: World Scientific.
64. L. Leibovitz. (2014) *A broken Hallelujah, Rock and Roll, Redemption and the Life of Leonard Cohen*. NY: W.W. Norton and Company.
65. C. S. Lewis. (1944) *The Abolition of Man*. NY: Harper and Collins.
66. C. S. Lewis (1947) *Miracles, A Preliminary Study*. NY: Harper and Collins.
67. S. Lloyd. (2006) *Programming the Universe, a Quantum Computer Scientist takes on the Cosmos*. NY: Vintage Books, Random House.

68. G. Longo. (2021) Confusing biological rythms and physical clocks. Today's exological relevance of Bergson-Einstein debate on time. Invited lecture at 'What is time? Einstein and Bergson 100 years later.' Campo, Ronchi (eds) Berlin: De Gryuter.
69. D. K. Lubenski and D. R. Nelson. (1999) Driven polymer translocation through a narrow nanopore. *The Biophysical Journal.* 77(1824).
70. J. B. Mandumpal. (2019) *A Journey Through Water: A Scientific Exploration of The Most Anomalous Liquid on Earth.* Bentham Science, Doi: 10.2174/9781681 084237117010011
71. B. Mandelbrot. (1977) *The Fractal Geometry of Nature.* San Francisco: W.H. Freeman and Company.
72. M. Mitchell Waldrop. (1992) *The Emerging Science at the Edge of Order and Chaos.* Touchstone books.
73. M. Mitchell. (2009) *Complexity, a Guided Tour.* Oxford: Oxford University Press.
74. J. Monod. (1970) *Le hazard et la necessite.* Paris: Gallimard.
75. J. Mc Fadden. (2015) *Life on the Edge, the Coming Age of Quantum Biology.* London: Crown hardcover.
76. G. Musser. (2015) *Spooky Action at Distance, the Phenomenon that Reimagines Space and Time and what it Means for Black-Holes, the Big-Bang and the Theories of Everything.* NY: Scientific American.
77. M. Newman. (2010) *Networks, An Introduction.* Oxford: Oxford University Press.
78. H. C. Van Ness. (1983) *Understanding Thermodynamics.* Dover books in Physics.
79. M. Nowak. (2006) *Evolutionary Dynamics, Exploring the Equations of Life.* Cambridge USA: Harvard University Press.
80. H. Nowotny. (2016) *The Cunning of Uncertainty.* Cambridge, UK: Polity Press.
81. E. Ott, C. Grebogi, and J. Yorke. (1990) Controlling chaos. *Physical Review Letters.* 64: 1196–1199.
82. G. Parisi. *Complex Systems: A Physicist's Viewpoint.* arXiv:cond-mat/020-5297v1,[cond-mat.stat-mech], 14 May 2002.
83. G. Parris. (2017) *Thermodynamics: Heat Capacity, Enthalpy, Entropy, Free Energy and Free Energy of Activation,* 2nd ed.
84. P. J. E. Peebles. (2020) *The Large Scale Structure of the Universe.* Princeton NJ: Princeton University Press.
85. P. Bak, C. Tang, and K. Wiesenfeld. (1987) Self-organized criticality: An explanation of 1/f noise. *Physical Review Letters.* 59(4): 381–384.
86. P. Bak. (1996) *How Nature Works, the Science of Self-organized Criticality.* NY: Springer Verlag.
87. J. P. Boon, P. V. Coveney, and S. Succi. (2016) *Bridging the Gaps at the Physics-Biology-Chemistry Interface.* Philosophical Translations of the Royal Society. 374, 2080 special issue of the 26th Solvay Symposium 'Multiscale modeling at the Physics-Chemistry-Biology interface', Brussels.
88. A. Pentland. (2015) *Social Physics.* Penguin London: Books.
89. *https://en.wikipedia.org/wiki/Protein_folding*
90. R. Piazza. (2015) *Soft Matter, the Stuff Dreams are Made Of.* Cambridge: Cambridge University Press.
91. I. Prigogine. (2017) *Non-equilibrium Statistical Mechanics.* Dover: Dover Publications in Physics.

92. I. Stengers and I. Prigogine. (1986) *La Nouvelle Alliance*. Gallimard
93. D. Kondepudi and I. Prigogine. (2014) *Modern Thermodynamics: From Heat Engines to Dissipative Structures*. NY: Wiley.
94. D. Ruelle (1992) *Chaos et Hazard*. Paris: Gallimard.
95. M. Rees.(2000) *Just Six Numbers*. NY: Basic Books. (First published by Weidenfeld and Nicolson, 1999).
96. J. Rivet and J. P. Boon. (2005) *Lattice Gas Hydrodynamics*. Cambridge: Cambridge University Press.
97. C. Rovelli. (2018) *The Order of Time*. New York, Penguin Random House.
98. https://www.santafe.edu As per their own highlight, a place where people 'ask big questions'.
99. E. Schroedinger. (1944) *What is Life?* Cambridge: Cambridge University Press.
100. J. Sethna. (2007) *Statistical Mechanics: Entropy, Order Parameters and Complexity*. Oxford: Oxford University Press.
101. https://www.deshawresearch.com
102. K. Sneppen and G. Zocchi. (2005) *Physics in Molecular Biology*. Cambridge: Cambridge University Press.
103. L. Smith. (2007) Chaos, a very short introduction. In *Oxford Very Short Introductions*. Oxford: Oxford University Press.
104. J. D. Stein. (2012) *Cosmic Numbers*. NY: Basic Books.
105. D. L. Stein and C. M. Newman. (2013) *Spin Glasses and Complexity*. Princeton: Princeton University Press.
106. R. Livi and P. Politi. (2017) Non Equilibrium Statistical Physics, a Modern Perspective. Cambridge: Cambridge University Press
107. P. L. Krapivsky, S. Redner and Eli Ben-Naim. (2010) *A Kinetic View of Statistical Physics*. Cambridge: Cambridge University Press.
108. S. Strogatz. (2016) *Nonlinear Dynamics and Chaos*. Avalon Publishing.
109. D. J. Watts and S. Strogatz. (1998) Collective dynamics of small-world networks. *Nature*. 393(6684): 440–442.
110. S. Succi. (2001) *The Lattice Boltzmann Equation for Fluids and Beyond*. Oxford: Oxford University Press.
111. S. Succi. (2018) *The Lattice Boltzmann Equation for Complex States of Flowing Matter*. Oxford: Oxford University Press.
112. N. Taleb. (2001) *Fooled by Randomness, the Hidden Role of Chance in Life and Finance*. Paris: Random House. (Reprinted Texere, London, 2010).
113. N. Taleb. (2010) *The Black Swan, the Impact of the Highly Improbable*. Penguin Books Ltd.
114. S. Thurner, R. Hanel, and P. Klimek. 2018. *Introduction to the Theory of Complex Systems*. Oxford: Oxford University Press.
115. S. Toppaladoddi, S. Succi, and J. Wettlaufer. (2017) Roughness as a route to the ultimate regime of thermal convection. *Physical Review Letters*. 118(7): 074503
116. A. Turing. (1952) Chemical Basis of Morphogenesis. *Philosophical Transactions of the Royal Society of London. Series B, Biological Sciences*. 237(641): 37–72. Stable URL: http://links.jstor.org/sici?sici=0080-4622.
117. G. A. Voth. (ed.) (2009) *Coarse-Graining of Condensed Phase and Biomolecular Systems*. CRC Press, Taylor & Francis Group.

118. D. J. Wales. (2003) *Energy Landscapes, with Applications to Clusters, Biomolecules and Glasses*. New York: Cambridge University Press.
119. S. Weinberg. (1992) *Dreams of a Final Theory: The Scientist's Search for the Ultimate Laws of Nature*. Vintage Books.
120. E. Wigner. (1960) The unreasonable effectiveness of mathematics in physics. *Communications in Pure and Applied Mathematics* 13(1).
121. F. Wilczek. (2016) *Physics in 100 years*. Physics Today.
122. S. Wolfram. 2016) *A New Kind of Science*. Wolfram Media Inc.
123. S. Wolfram. (2020) *A Project to find the Fundamental Theory of Physic*. Champaign, Illinois: Wolfram Media Inc.
124. P. G. Wolynes, J. N. Onuchie, and D. Thirumalai. *Navigating the Folding Routes*. AAAS-Weekly Paper Edition 267 (5204): 1619–1620, 1319, 1995.
125. Mark W. Zemansky and Richard H. Dittman. (1997) *Heat and Thermodynamics*. NY: Graw Hill Inc. 2020PG.

Index